Computers in Building:
Proceedings of the CAADfutures'99 Conference

Computers in Building:
Proceedings of the CAADfutures'99 Conference

Proceedings of the Eighth International Conference on
Computer Aided Architectural Design Futures
held at Georgia Institute of Technology,
Atlanta, Georgia, USA on June 7-8, 1999

edited by

Godfried Augenbroe
Charles Eastman
College of Architecture
Georgia Tech

SPRINGER SCIENCE+BUSINESS MEDIA, LLC

Library of Congress Cataloging-in-Publication Data

International Conference on Computer Aided Architectural Design
Futures (8th : 1999 : Atlanta, Ga.)
Computers in building : proceedings of the CAADFutures'99
Conference : proceedings of the Eighth International Conference on
Computer Aided Architectural Design Futures held at Georgia
Institute of Technology, Atlanta, Georgia, USA on June 7-8, 1999 /
edited by Godfried Augenbroe, Charles Eastman.
p. cm.
Includes bibliographical references (p.).
ISBN 978-1-4613-7294-3 ISBN 978-1-4615-5047-1 (eBook)
DOI 10.1007/978-1-4615-5047-1
1. Architectural design--Data processing Congresses. 2. Computer
-aided design Congresses. I. Augenbroe, Godfried, 1940-
II. Eastman, Charles M. III. Title.
NA2728.I59 1999
720'.285--dc21 99-24634
CIP

Originally published by Kluwer Academic Publishers in 1999
Softcover reprint of the hardcover 1st edition 1999

Printed on acid-free paper.

CONTENTS

List or Reviewers ix

Preface xi

CAD interfaces to the ARROW manufactured product server 1
Robert Amor, Leonard Newnham
Building Research Establishment, Watford, UK

The future of CAAD: from computer-aided design to computer-aided collaboration 13
Yehuda E. Kalay
University of California, Berkeley, California, USA

Exploiting the Internet to improve collaboration between users and design team: The case of the new computer laboratory building at Cambridge University 31
Paul Richens, Michael Trinder
The Martin Centre for Architectural and Urban Studies
University of Cambridge, UK

Urban texture analysis with image processing techniques 49
Carlo Ratti, Paul Richens
The Martin Centre for Architectural and Urban Studies
University of Cambridge, UK

A Web-centric CAD system for collaborative design 65
Raymond McCall
University of Colorado, Denver, USA

Immersive redlining and annotation of 3-D design models on the Web 81
Thomas Jung, Ellen Yi-Luen Do, Mark D. Gross
Sundance Lab for Computing in Design & Planning
University of Colorado, Boulder, USA
C.R.A.I., Nancy, France

How do distributed design organizations act together to create a meaningful design? Towards a process model for design coordination 99
Jeffrey Huang
Harvard University, Graduate School of Design, Cambridge, USA

A performance-inspired building representation for computational design 117
Georg Suter, Ardeshir Mahdavi, Ramesh Kirshnamurti
Carnegie Mellon University, School of Architecture, Pittsburgh, USA

About objects and approaches: a conceptual view on building models 133
Ann Hendricx, Herman Neuckermans
KU Leuven, Department of Architecture, Leuven, Belgium

The BAS-CAAD information system for design – principles, implementation, and a design scenario 149
Anders Ekholm, Sverker Fridqvist
Lund University, Computer Aided Architectural Design, Lund, Sweden

Information sharing in building design 165
James Snyder, Ulrich Flemming
Lockheed Martin Advanced Technology Laboratories, Camden, New Jersey, USA
Carnegie Mellon University, School of Architecture and Institute for Complex Engineered Systems, Pittsburgh, USA

Aspects of S2 185
Ardeshir Mahdavi, Mustafa Emre Ilal, Paul Mathew, Robert Ries, Georg Suter
Carnegie Mellon University, School of Architecture, Pittsburgh, USA

Interactive 2D constraint-based geometric construction system 197
Benachir Medjdoub
The Martin Centre for Architectural and Urban Studies
University of Cambridge, UK

Collaborative design with NetDraw 213
Dongqiu Qian, Mark D. Gross
Sundance Lab for Computing in Design & Planning
University of Colorado, Boulder, USA

The computer's role in sketch design: A transparent sketching medium **227**
Michael Trinder
The Martin Centre for Architectural and Urban Studies
University of Cambridge, UK

Making designs come alive: Using physically based modeling techniques in space layout planning **245**
Scott A. Arvin, Donald H. House
Texas A & M University, Department of Architecture,
Visualization Lab, College Station, Texas, USA

An interpolation/extrapolation process for creative designing **263**
John S. Gero, Vladimir Kazakov
University of Sydney, Department of Architectural and Design Science,
Key Centre of Design Computing and Cognition, Sydney, Australia

Feature-based high level design tools: A classification **275**
Henri H. Achten, Jos P. van Leeuwen
Eindhoven University of Technology, Eindhoven, The Netherlands

π-Resolution in design space exploration **291**
Andrew Burrow, Rob Woodbury
The University of Adelaide, Department of Computer Science
School of Architecture, Landscape Architecture and Urban Design
Adelaide, Australia

Feature-based qualitative modeling of objects **309**
John S. Gero, José C. Damski
University of Sydney, Department of Architectural and Design Science,
Key Centre of Design Computing and Cognition, Sydney, Australia

The front of the stage of Vitruvius' Roman theatre: A new approach of computer aided design that transforms geometric operators to semantic operators **321**
Giovanni De Paoli, Marius Bogdan
University of Montreal, Faculty of Environmental Design
Montreal, Canada

Experiments on animated visual forms **335**
Kostas Terzidis
UCLA, Arts and Architecture, Los Angeles, USA

Environmental life cycle assessment in an integrated CAD environment: The epilogue approach **351**
R. Ries, A. Mahdavi
Carnegie Mellon University, School of Architecture, Pittsburgh, USA

Downstream of design: Web-based facility operations documents **365**
Mark J. Clayton, Robert E. Johnson, Yunsik Song
Texas A & M University, College Station, Texas, USA

The Piranesi system for interactive rendering **381**
Paul Richens
The Martin Centre for Architectural and Urban Studies
University of Cambridge, UK

List of Reviewers

Robert Amor, BRE, England
Osman Ataman, University of Utah, USA
Olubi Babalolo, Georgia Institute of Technology, USA
Can Baykan, Middle East Technical University, Turkey
Mark Clayton, Texas A & M University, USA
Thanos Economou, Georgia Institute of Technology, USA
Anders Ekholm, Lund Institute of Technology, Sweden
John Gero, University of Sydney, Australia
Mark Gross, University of Colorado, USA
Theodore W. Hall, Chinese University of Hong Kong, China
Jeffrey Huang, Harvard University, USA
Bob Johnson, Texas A & M University, USA
Yehuda Kalay, University of California, Berkeley, USA
Uk Kim, Hongik University, Korea
Jos van Leeuwen, Eindhoven University of Technology, The Netherlands
Ardeshir Mahdavi, Carnegie- Mellon University, USA
Ali Malkawi, University of Michigan, USA
Tom Maver, University of Strathclyde, Scotland
Rivka Oxman, Technion, Israel
Paul Richens, University of Cambridge, England
Rudi Stouffs, Swiss Federal Institute of Technology, Switzerland
Bauke de Vries, Eindhoven University of Technology, The Netherlands
Robert Woodbury, University of Adelaide, Australia

Preface

Computers in Building

This is the eight *CAADfutures* Conference. Each of these bi-annual conferences identifies the state of the art in computer application in architecture. Together, the series provides a good record of the evolving state of research in this area over the last fourteen years.

Early conferences, for example, addressed project work, either for real construction or done in academic studios, that approached the teaching or use of CAD tools in innovative ways. By the early 1990s, such project-based examples of CAD use disappeared from the conferences, as this area was no longer considered a research contribution. Computer-based design has become a basic way of doing business.

This conference is marked by a similar evolutionary change. More papers were submitted about Web-based applications than about any other area. Rather than having multiple sessions on Web-based applications and communications, we instead came to the conclusion that the Web now is an integral part of digital computing, as are CAD applications. Using the conference as a sample, Web-based projects have been integrated into most research areas. This does not mean that the application of the Web is not a research area, but rather that the Web itself is an integral tool in almost all areas of CAAD research.

This year the submissions identify multiple other themes:

First, *building models* continue to be a major area of research, continuing the long-term effort to develop integrated application environments. Building model research has long emphasised the development of a single common data model around which all applications may be operated. The building model may be integral with the application, or alternatively, it may be a representation whose major role is the exchange of data between different applications, as per ISO-STEP and IAI.

A second long term research area is *design theory*. Earlier conferences have emphasised such issues as semantic languages, shape grammars, constraint-based systems, and search mechanisms. This conference has similar emphases, though including new approaches. In particular, feature-based design systems is receiving increased attention, and also logic-based, subsumption relations, as a means to classify design information.

A third long-lasting CAAD research area is *performance-based design*. Here, the goals are to support design informed by various functional and performance issues. The work this year emphasises a building model approach, integrating all applications around a shared building representation.

A fourth area of long-lasting research is in the development of new applications. However, the assumptions upon which applications are developed have changed. At the beginning of CAAD research, all applications were standalone and were developed pretty much from scratch. Today, some applications are developed this way, but a growing number are built on top of an existing CAD platform. The tool base of existing application platforms, provided by such companies as Autodesk and Bentley, provide a base for advanced work that is almost impossible to replicate, even if one does not agree with the structure and approach these platforms may have taken.

Another area of continued work is *sketching tools*. Sketching tools have from their beginning been motivated by the desire for simple, natural tools that work as easy as a pencil on a napkin, allowing a designer to record and explore ideas, but augmented by a computer. There have not been many convincing examples of what that augmentation should be. The papers this year include Web-based sketching and constrained sketching systems.

A new area of increasing research attention is collaboration, especially *Web-based collaboration*. Collaboration most commonly means collaboration among a design team. Less often it means collaboration between designer and client. However, this year, it also means collaboration between architect and builder or between builder and facility manager.

A somewhat surprising new area of research in this CAADfutures Conference is the development of *graphical and presentation methods* and algorithms. The papers by Richens and Terzides provide clear examples that there is still room to develop new graphical tools that provide valuable new presentation capabilities. This area has not seen much work in recent years, and many thought this was a fully mature research area, in which all the important capabilities had been addressed.

The conference received 105 abstracts, from which 55 abstracts were selected for complete paper submission. From the submissions, twenty-five papers were finally selected. These papers represent the best in the field. We hope you enjoy, and in some cases learn from the work presented here.

CAD Interfaces to the ARROW Manufactured Product Server

Robert Amor and Leonard Newnham
Building Research Establishment, Bucknalls Lane, Garston, Watford WD2 7JR, UK

Key words: computer-aided design, product retrieval

Abstract: The UK national project ARROW (Advanced Reusable Reliable Objects Warehouse) (Newnham et al. 1997, 1998a and 1998b) provides an Internet based framework through which it is possible to identify any of a range of manufactured products meeting specific design criteria. This open framework (based upon the IAI's IFCs) provides a mechanism for users to search for products from any participating manufacturer or supplier based both on specific attributes of a product or on any of the textual descriptions of the product. The service returns the closest matching products and allows the user to navigate to related information including manufacturer, suppliers, CAD details, VR displays, installation instructions, certificates, health and safety information, promotional information, costings, etc. ARROW also provides a toolkit to enable manufacturers and suppliers to more easily map and publish their information in the format utilised by the ARROW system.

As part of the ARROW project we have examined the ability to interface from a design tool through to ARROW to automatically retrieve information required by the tool. This paper describes the API developed to allow CAD and simulation tools to communicate directly with ARROW and identify appropriate manufactured information. The demonstration system enables CAD systems to identify the closest matching manufactured product to a designed product and replacing the designed product with the details supplied by the manufacturer for the manufactured product as well as pulling through product attributes utilised by the design application.

This paper provides a description of the ARROW framework and issues faced in providing information based upon standards as well as containing information not currently modelled in public standards. The paper looks at issues of enabling manufacturers and suppliers to move from their current world-view of product information to a more data-rich and user accessible information repository (even though this enables a uniform comparison across a range of manufacturer's products). Finally the paper comments on the likely way forward for ARROW like systems in providing quality information to end users.

1. THE ARROW PROJECT

ARROW is a UK initiative that can provide access to any construction manufacturer's product information through virtual warehouses. This enables designers and specifiers to correctly identify products that match the parameters of their particular design. Benefits of this approach will be reduced abortive design work and reworking, a more reliable design specification, designs that can be constructed quicker, greater use of off-the-shelf components, savings in cost, and improved response time for customers. To make ARROW feasible a central search engine able to handle structured data and also free-text information has been developed. This is based on data collected from all manufacturers and suppliers willing to publish electronic information. This allows fast and accurate retrieval of specified product information, as well as delivering information in a form useable by CAD systems and other design tools (e.g., thermal simulation programs).

The main parts of the system comprise the following. First, the user, who has an interface with the system directly through a web browser or indirectly through a design tool. Second, the main system or Building Object Warehouse (BOW) server. Third, the distributed product databases at numerous sites across the Internet containing product data in a variety of formats. Finally, there is provision for interaction with existing KBS systems for product advice, design considerations, etc. Again, these can be distributed across the Internet. The main points are described briefly below and the system illustrated in Figure 1.

User Environment The user can interact with the BOW system in one of two ways: either, directly through a web browser or alternatively, through a design tool. In the former case, the user interacts directly with the query handler. In the latter case the user may point to a wall and specify "window" with the design tool supplying the size parameters directly. An application built using the ARROW API residing on the user's machine accepts parameters from the design tool and then interacts with the Query Handler on the user's behalf. This is described in more detail later in this paper.

Query Handler The function of this is to help the user formulate exactly what he or she wants in terms of parameters used in the data model. The query handler contains knowledge of the product data model and so can prompt for parameters appropriate for a particular product type. There is provision for unspecified parameters, parameters within a range and exact values only, return sort order, etc. In addition, there is provision for free-text searches of the manufacturer's general product information, which is directly tied in to one query process.

Indexes, Indexers and the Search Engine The indexer has an authoritative list of all product repository addresses. Periodically these are searched, all product parameters retrieved and indexed. There are two indexes, one a structured database (an OO database) containing the minimum product information specified in the product model, the second for free text containing additional manufacturer information within html pages on a web server. The free text index is kept up to date using the Harvest indexer (Hardy et al, 1996).

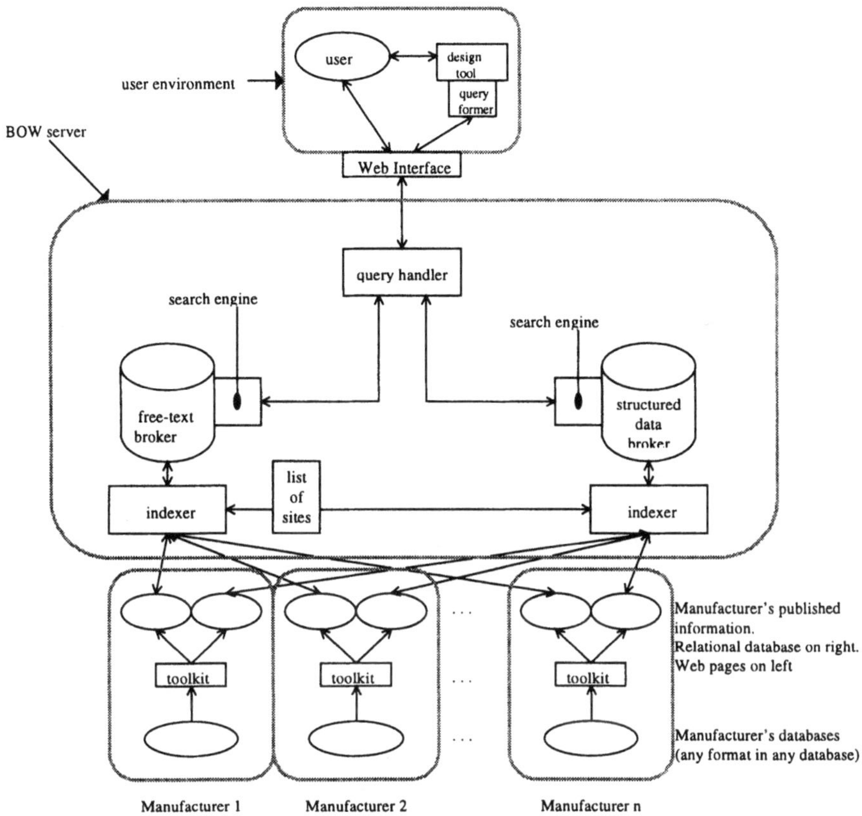

Figure 1. BOW Architecture

The results of a search when displayed on a browser consists of core information for the product together with buttons for further information, allowing the user to navigate around the full data structure. Within the data structure there is provision for electronic documents for which a mime type is stored along with the URL of the original electronic document residing on the manufacturer's machine. CAD files, VR displays etc can then easily be downloaded through the browser.

A demonstrator of the browser interface to the ARROW system is accessible at: http://www.bre.co.uk/arrow.

2. PRODUCT REPRESENTATION USING THE ARROW DATA MODEL

Analysis of the requirements of the ARROW system, led to a specification of the areas that needed model development. Subsequent analysis of the existing models showed a very low level of coverage of these areas in most developments, the exception being the IAI IFCs, which gave the building blocks for almost all areas required. The final ARROW data model is therefore built entirely upon the IFCs, but with extensions to its structure to represent more specific areas. The main areas covered within the data model are:

- Product and sub-assembly structures allow for the definition of all technical data about products. The IFCs give very basic information about building elements (e.g., door, window, distribution elements), which were specialised to contain all the parameters required for product selection. This part of the specification also defines the forms of graphical representation of products and the possible connections between related products, e.g., to ducts or walls, and through which mechanism. The product section also provides the links to all associated documentation about products, in terms of BBA certificates, manufacturer's specification sheets, CAD files of product detailing, VRML file, etc.
- Organisation structures allow for the specification of manufacturers and suppliers and the connections that exist between them for specific products. There is a triangular structure which joins a product to a manufacturer, and the manufacturer to a set of suppliers, and the supplier to a set of products (and vice-versa). See Figure 2. The organisation structures provide the means for contacting manufacturers and suppliers as well as all information about their range of products and services.
- Catalogue information ties closely to products, manufacturers, and suppliers providing a mechanism to duplicate the type of catalogue services currently offered in the industry.
- Core information structures provide the majority of the basic information required in all other parts of the ARROW system. This includes: mechanisms to allow for multiple classifications for a single product; definition of materials and the layering of materials; specification of the cost for a product, and cost models which allow special relationships between suppliers and customers to be recognised; specification of standard units, and conversion factors for non-standard units; specification of measured values (e.g., luminosity, mass, volume); and documentation types to define whether a file is a CAD detail, VRML, or a Word document and also whether it is a contract, detailed drawing, etc. This was the area which was best covered by the IFCs and required the fewest additions (the main exception being the document model specification).

No matter how complex the data model, finding appropriate fields to search upon is still a problem. To address this problem ARROW allows free-text searching of manufacturers' web pages as an integral part of the searching process. For each product, the system contains information in two forms: a) structured database product information within the above described data model, and b) a free-text description of the product in an associated web page. In addition to allowing more flexible querying, it also allows the manufacturer to:

- present a sales pitch for the product,
- differentiate the product from rival products, and
- describe aspects of the product not adequately covered by the formal data model.

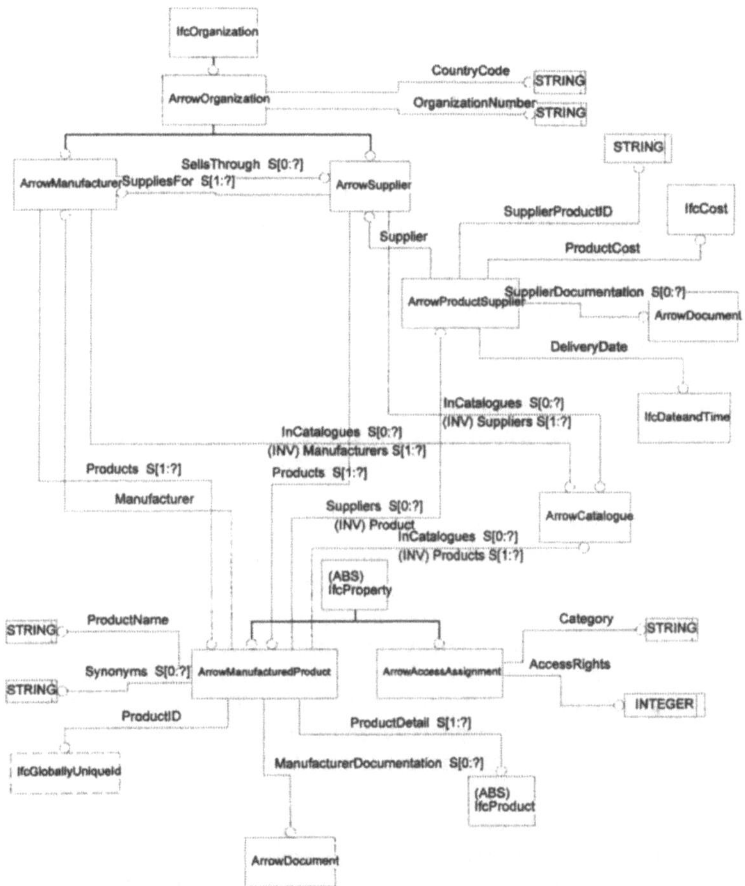

Figure 2. EXPRESS-G diagram of a portion of the ARROW Data Model

The full specification of the ARROW data model can be found in Amor *et al.* (1999). This document describes all structures in the data model as well as routes to implementation of the data model. Mappings to common implementation forms are also provided, e.g., SQL, C++ and Java.

3. ISSUES FACED IN PROVIDING INFORMATION BASED UPON STANDARDS

The development of the ARROW system, which is concurrent with the development of international standards for the construction industry, has raised many interesting issues. The current approach to the development of data models, which are focused on specific business processes and servicing specific processes in the industry, does not provide the broad definition of information on the whole range of manufactured products which are needed to construct a building. Though there is a libraries group in the IAI, there has been little visible progress in this area over the three year timeframe of this project, which vindicates a decision not to wait for international standards to provide the required model for the ARROW system.

The ARROW system is based upon the most recent published standards (IFC version 1.5.1), however, the data model extensions required to implement the ARROW system puts ARROW in the position of creating non-standard information in its dealings with manufacturers, suppliers and user applications. Though ARROW participants can provide input to likely standards in this area there is no guarantee, and indeed a high likelihood, that the final standards are different from those in the current system. Though, as there is a limited number of measured attributes for most manufactured products, the different should not be large. The hope is that through having a semantically clear data model it will be possible to map the ARROW data structures through to those which are finally produced as standards. The current IAI standards provide a mechanism called property sets to allow user specified attributes to be attached to any object in a model. Though ARROW can provide all its information in a standards compliant file through this mechanism we have resisted following this route for all of our extensions. The main reasoning is that this seems to negate the benefit of standards, as then the consensus on semantics of property set attributes needs to be agreed with every new participant using (some subset of) the user defined attributes (this impacts all information providers and all user applications). It is also not clear how homonym conflicts between different information providers can be caught and resolved. The authors look to evolving standards in the XML world, especially RDF (Resource Description Framework) (URL1) to provide possible solutions to this problem.

In attempting to gain structured information from manufacturers and suppliers the problems with the complexity of current standards caused many problems. In short, the overhead in a manufacturer or supplier understanding enough of the current standards in order to provide their information in a form suitable for ARROW is daunting and not currently justifiable financially. Hence, the development of a simplified model for the information provides alongside toolkits to map through to the full data model is seen as the only way forward in the short to medium term for ARROW-like systems. The disparity between relational database structures, which is the most that can be expected from manufacturers and suppliers, and the object-oriented data structures of current standards also poses problems. Though it is possible to map between the two technologies this is a complex task for the majority of IT service groups in an organisation to undertake, and implementing an object-oriented model in a relational system provides an inefficient system (e.g., handling polymorphic object references).

A similar problem of model complexity is seen at the user end. Teaching all users the meaning of all the attributes for objects in which they may wish to search is also a daunting task. However, if the semantic specification of attributes are continued to be made available electronically, along with the standards, this provides a means to take users to a clear definition of what they are dealing with. It is also likely that interfaces specialised for specific types of users will be developed (e.g., architects, HVAC engineers, etc.) providing a view which matches their understanding. This is similar to the approach already taken in providing the data from the central database to user applications. Views of the data are created which can represent, for example, IFC 1.5.1 data structures as well as the full data model.

4. ARROW API

The ARROW API consists of a collection of Java classes that duplicate the functionality available through the browser interface. These allow a developer to extend the functionality of a design tool so it can interact directly with the ARROW system. The functionalities of the API are:

- To return a simple YES/NO answer on whether a product with the specified parameters exists or not.
- To return a list of the products available in the database. This would usually be the first information a design tool would want.
- To return a list of the searchable fields associated with a given product. The application developer will need to know the fieldnames for a product. These include; provision for either precise or range values for numeric parameters; free-text contained within manufacturers web pages associated with products and also results sorting requirements. He can then decide whether to bind certain design parameters stored within the drawing to specific fields or whether to present some or all of the fields to the user to narrow the search.
- To return the table of results that the given query generates. The search will return details of all objects that satisfy a query. It is then up to the application developer to choose how to display these results or what to do with them. The results are returned in chunks, each being broadly associated with a class in the underlying IFC datastructure, allowing navigation through this datastructure.
- To download the specified file from the BOW Server and save it to a specified local location.
- To download the specified file from the BOW Server and returns its contents.

These last two are required for inserting any drawing file that may be stored in the database for a particular product, into a user's drawing.

5. AUTOCAD DEMONSTRATOR

A demonstrator has been built to identify the closest matching manufactured product to a designed product and then show the direct importation of a DWG of this product into an AutoCAD drawing. The demonstrator consists of several parts:

1. AutoCAD with BS Link (URL2) installed to allow the storage of object information within the drawing.
2. An AutoLisp script.
3. The ARROW API as java classes.

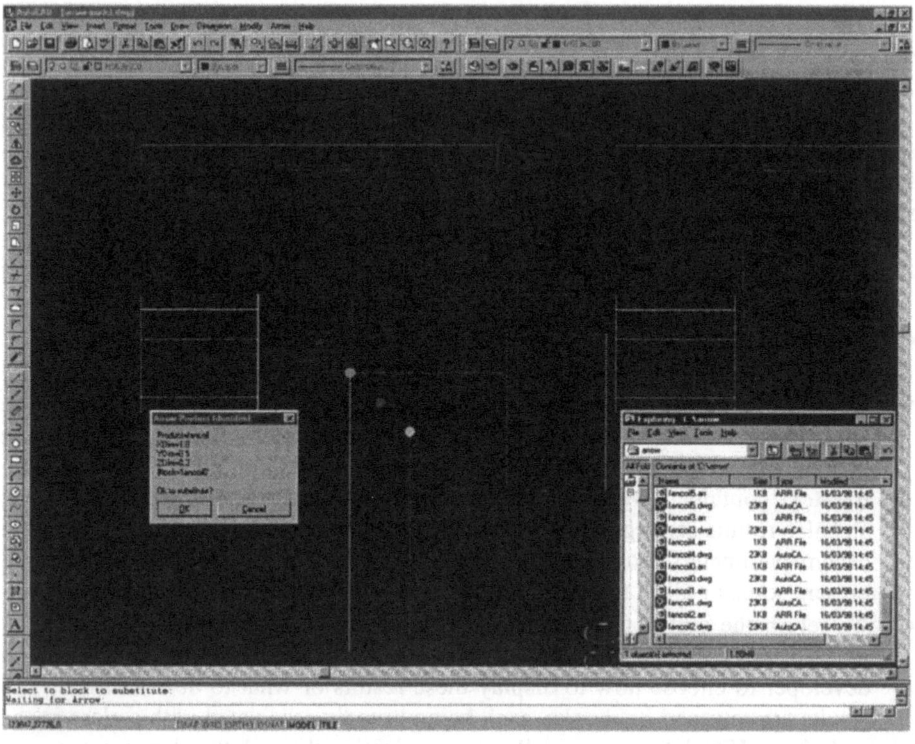

Figure 3: ARROW AutoCAD demonstrator. The user is presented with the closest match to the required fancoil unit dimensions and is being asked if he wishes to substitute the original with the manufacturer's product drawing.

The AutoLisp script provides the user interface to the ARROW system. It also reads the selected product type and dimensions from the drawing, changes the actual dimensions into a user defined acceptable range. It assembles this information into an ARROW query and passes it to the ARROW API. When the search results are returned the script selects the product with dimensions closest to those specified in the drawing. (Here the script simply looks for the product with the smallest absolute difference between sum of product dimensions and the sum of the required dimensions.) The script then submits a second query requesting the download of the DWG file for this product (if it exists) stored on the manufacturer's machine. Once downloaded the component drawing is automatically substituted into the main drawing.

6. LINKS WITH UK MANUFACTURED PRODUCT RELATED SYSTEMS

The development of the ARROW system, at the same time as other developments in manufactured product related systems, has highlighted the great overlap between the requirements of these systems. The UK National Standard Details Library (NSDL) project (URL3) has developed an infrastructure to allow access to standard details for common industry products which are usable in a number of CAD systems (e.g., DWF format to allow drag-and-drop to AutoCAD, see Figure 4). This is a process which precedes the initial selection of a manufactured product, but the attributes specified for the standard detail are the same as those then needed to select a compliant manufactured product. The same is true for the certification process. Often a manufactured product is searched for with a requirement that it has been certified as fit for use. The ARROW system already captures the link to certificates where they exist for a manufactured product, however, the certification process measures many of the attributes that a user would wish to select a product on (e.g., U-value, or fire resistance).

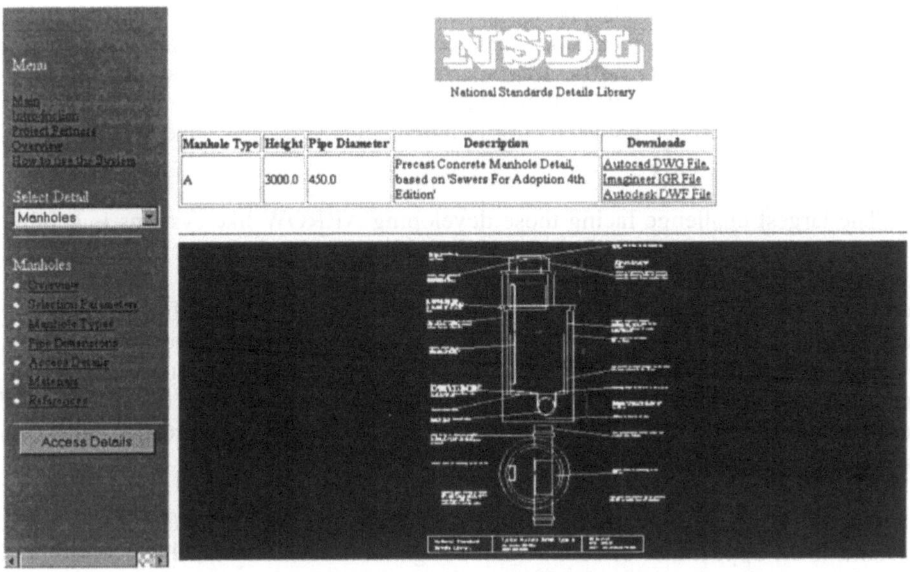

Figure 4. A DWF view of a standard detail manhole

To show the benefit of linking to related systems ARROW and the NSDL will be extended to allow the parameters used to select a standard detail to be re-used when finding a manufactured product. ARROW will also be extended to enable measured attributes from the certification process to be captured alongside a manufactured product description, as well as the attributes supplied by the manufacturer. The approach to these extensions is to harmonise the data representation utilised in all three systems so that attribute data can be re-used across applications.

7. THE WAY FORWARD

The ARROW system presented provides a view of what the future may hold. This is not an isolated view, as similar systems are being developed at both a research and commercial level in many countries. In the UK the Barbour Index (Barbour 1997) provides some measured attributes for its product catalogue as a step up from purely classification based searching for manufactured products. The model of the future that we hold is one where multiple ARROW-like systems exist in all countries in a similar way to the existence of current product catalogues, mostly aimed at different, but overlapping, segments of the market. The major difference from today's systems is that the new systems promoters will see benefit in establishing links with other systems, especially in other countries. The selling point will be access to a much greater range of manufactured products, which reflects the growing globalisation of the manufacturing market. To enable this to happen there must be broad agreement on the data models used to represent manufactured products, and there must also be agreement on a set of interface functions (API) which will make interoperability between different ARROW-like systems a possibility. In the short term these agreed models are likely to be defined ahead of the standardisation efforts. But, with an effort to build upon current and developing structures, the difference between current agreements and final standards should be small.

The largest challenge facing those developing ARROW-like systems is gaining structured data from all manufacturers and suppliers. This requires a large change in approach for the majority of manufacturers and suppliers. However, while the long-term benefits of current developments are clear, the short-term benefits are harder to identify and quantify. There will need to be a large effort expended to ensure that enough information is provided in the new form to make it a requirement for all manufacturers to publish in the new style to stay competitive. The development of toolkits for manufacturers and suppliers, to ease the cost of this process, will be a major facilitator in the development of ARROW-like systems.

The other development which will benefit the creation of manufactured product systems is the appearance of commercial design tools which utilise the standard data models now under development. These tools are starting to appear (e.g., Autodesk's Architectural Desktop with IFC interface) and with this shift to object-based systems it will be easier to interface these design tool's requirements to the information which can be offered from ARROW-like systems.

8. ACKNOWLEDGEMENTS

The authors acknowledge the support of the UK government in the development of the ARROW system through their Partners in Technology programme. The manufacturers who have provided their product information are gratefully acknowledged, as well as Advanced Visual Technology who provided the industry view in the project. The ARROW system is being further developed through EU funding in the CONNET project, in co-operation with BII and VTT in Finland.

9. REFERENCES

Amor, R., Newnham, L., Parand, F. and Nisbet, N., 1999, The ARROW Data Model Specification, BRE publication, Watford, UK, March 99.

Barbour Index plc, The Barbour Report 1997, Electronic Delivery of Product Information, 1997.

Hardy, D.R., Schwartz, M.F. and Wessels, D., 1996, Harvest User's Manual, Technical Report CU-CS-743-94, Department of Computer Science, University of Colorado, USA.

ISO (1994) ISO/DIS 10303-1:1994. Industrial Automation Systems and Integration – Product Data Representation and Exchange Part 1: Overview and Fundamental Principles.

Newnham, L., Parand, F., Amor, R. and Nisbet, N., 1997, The ARROW Framework for Object Warehouses, CIB W78 Workshop on Information Technology Support for Construction Process Re-Engineering, IT-CPR-97, Cairns, Australia, 9-11 July, pp. 319-328.

Newnham, L., Amor, R. and Parand, F., 1998, Gaining Quality Manufactured Product Information through ARROW, PDT Days, BRE, UK, 25-26 March, pp. 39-46.

Newnham, L. and Amor, R., 1998, Translation of Manufacturer's Product Data for the ARROW Product Search System, The Second European Conference on Product and Process Modelling, London, UK, 19-21 October, pp. 405-412.

URL1, http://www.w3.org/RDF/, The W3C page on the Resource Description Framework.

URL2, http://bs-link.visual-tech.co.uk/ Visual Technology's BS Link page.

URL3, http://cig.bre.co.uk/nsdl/docs/nsdl.htm The National Standards detail Library Home Page.

The future of CAAD: From computer-aided design to Computer-aided collaboration

Yehuda E. Kalay
University of California, Berkeley CA. USA

Key words: Collaborative design, distributed design environment, product modeling, performance modeling, process modeling, negotiation, integration.

Abstract: The primary uses of computers in the construction industry have been shifting, over the past four decades, from the evaluation of proposed design solutions, to their graphical (and other) representation, and more recently to facilitating collaboration among the various professionals who are involved in the design process. The paper argues that what may appear to be shifts in emphasis actually represents convergence on a single, original goal: the use of computers to help designers assess the quality, desirability, and the implications of their creations. The paper shows how the formerly independent components can be joined into an integrated collaborative design environment, where they build upon and strengthen each other. Moreover, the paper argues that this convergence represents the future of CAAD research and development, providing the appropriate answer to the upcoming needs of the construction industry, whose products have become too complex and must abide by too many requirements for any one professional to handle all by himself. The paper argues that further improvements in the overall quality of the products, and the process of their design, will only accrue when the heretofore separate solutions are considered together, as integral parts of an overall solution. The paper describes the efforts that have been made by the CAD Research Group in Berkeley over the past six years in developing an integrated collaborative design environment that can facilitate multidisciplinary, a-synchronous design of buildings. The environment includes several semantically-rich, shared product representations, a network of distributed evaluators, and graphically-enhanced collaboration and negotiation tools.

1. THE CHANGING ROLE(S) OF CAAD

Horst Rittel defined design as *"an activity aimed at achieving certain desired goals without undesired side- and after-effects"* [Rittel & Webber 1984]. He also called architectural design problems "wicked," because one cannot achieve all the desired goals (they might be in conflict with each other), nor can one avoid causing undesired side- and after-effects (either by action or no action). A 'good' design, accordingly, is one that achieves as many of the desired goals as possible, while causing the least number of undesirable side- and after effects.

But how can one achieve such a 'good' design? How can one tell if a certain design proposal will achieve a reasonable set of goals? How can one tell what undesirable side- and after-effects will it have? Such assessment requires foresight, judgement, and—most of all—considerable knowledge about the many diverse affects a design proposal might have. The more complex the design, the harder it is to predict and evaluate its 'goodness.'

CAAD was introduced in the 1950s to assist designers in assessing the 'goodness' of their creations. Initially, computers were used to assist in engineering analyses. The process of designing buildings continued to use habitual, manual methods, but at certain points along the design process quantities were taken off manually and fed into computer programs that could analyze them. The results were then applied manually to the evolving design [Mitchell 1977].

This process soon bogged down, when the difficulties associated with the input/output bottleneck outstripped the benefits that could be derived from the evaluations. The emphasis in CAAD research shifted from developing better engineering analysis programs to finding more efficient modes for bringing the emerging design solution to the computer: computational representation of buildings took center stage.

Although computer graphics were first introduced in the late 1940s, through the US Navy's Whirlwind project (a general purpose flight simulator), their application to (mechanical) design only occurred at MIT in the 1960s, through the efforts of Steven Coons and Ivan Sutherland, who developed Sketchpad—the first interactive 2D and 3D design tool. This approach simplified the input of design properties into analyses programs, which opened the floodgate for the development of CAD programs. A bumper-crop of such programs was introduced in the 1970s, and became a popular design tool in the 1980s. Research into representing information other than graphics soon followed.

Along the way, some attempts were made to use computers as design generators [Armour & Buffa 1968, Cross 1977]. These attempts have, by and large, failed to provide meaningful assistance to designers. Not only were the solutions generated by these program rather trivial, they also intruded on what designers considered to be their most 'sacred turf.' Such objections were not raised with regard to computer-aided evaluation, representation, and communication, where in most cases computer assistance was welcomed.

But drafting and modeling systems could meet only a few of the original needs, namely—visual appraisal of the emerging design solutions, and certain geometry-based evaluations (e.g., habitability) [Kalay & Shaviv 1979]. Hence, researchers resumed their quest for more powerful computer-aided evaluation programs, and better means for representing non-geometric building information [Steinfeld & Kalay 1990]. In the 1980s this search was strongly influenced by the general euphoria associated with Expert Systems [Flemming 1994]. A large number of expert and other knowledge-based systems were developed, purporting to package design expertise and to bring it to bear on the design process without the experts who generated the knowledge in the first place [Carrara et al 1994, Coyne et al 1990]. Few of these systems lived-up to their expectations, and it soon became evident that the real meaning of 'Expert Systems' was "systems for the use by the experts themselves" [Shaviv et al 1996].

The globalization of the building industry in the 1990s, coupled with the increasing capabilities of computers as telecommunication devices due largely to the rise of the Internet, brought about the birth of computer-aided collaboration. The first use of computers as facilitators of collaboration was purely technical: it was easier and faster to send digital design information through the Internet than physical drawings through the mail. But this ability, along with accelerated design/build schedules, raised serious problems of concurrency, authority, and version control. Some systems that can manage the multifarious data formats used in a typical building project have emerged, such as ProjectNet by BlueLine/OnLine and Speedicon (a German company). Still, these systems are, at best, project management tools, not collaboration tools: they facilitate the sharing of *information*, not the sharing of *understanding*, which is a necessary condition for joint efforts in design.

What do we mean by 'shared-understanding'? Why is it necessary? How can it be facilitated by CAAD? These are the questions this paper tries to answer.

2. WHAT IS COLLABORATION?

In his seminal book *The Sciences of the Artificial*, Herbert Simon argued that any body of knowledge that requires more that 10 years to master tends to fragment into specializations [Simon 1969]. Such fragmentation has occurred in medicine, law, engineering, as well as in the design of buildings. But while in other disciplines the specialist can typically deliver the service independently from other specialists, a specialist in the building industry represents only one part of an integral whole: he can rarely complete the task on his own. At the same time, the task itself cannot be completed without the contributions of all the specialists. Thus, the fragmentation of knowledge in the building industry has created a *symmetry of ignorance*, where no single professional has all the knowledge needed to design a complex facility, and where it is no longer possible to design a building without consulting many specialists (architects, engineers, construction managers, lighting consultants, mechanical engineers, acoustical experts, financial advisors, and legal experts, etc.) [Cuff 1991].

Collaboration can thus be defined as *"the agreement among specialists to share their abilities in a particular process, to achieve the larger objectives of the project as a whole, as defined by a client, a community, or society at large"* [Hobbs 1996]. By combining their abilities, a collaborative arrangement can help individuals undertake larger and more complex tasks, gain perspective on the shared enterprise they would not have been able to perceive on their own, learn from others, and be motivated by them. Collaboration, as such, is an enabling force. At the same time, it can also be a restrictive force, in the sense that the action best suited for the goals and needs of one individual may not also be best suited for the goals of another collaborator, thereby raising potential conflicts and the need to compromise or even to yield to the needs of others.

Collaboration, therefore, is a highly complex and challenging task. It has been the subject of study in virtually every field: sociology, psychology, politics, science, technology, and professional practices in law, medicine, and engineering. Yet, collaboration in the building industry is different from collaboration in other fields. First, it involves individuals representing often fundamentally different professions, who hold different goals, objectives, and even beliefs. Unlike collaborators in medicine or in jurisprudence, who share a common educational basis, architects, structural engineers, electrical engineers, clients, property managers, and others who comprise a design team, rarely share a common educational foundation, and often have very different views of what is important and what is not. Second, it involves what has been termed 'temporary multi-organizations': a team of independent organizations who join forces to accomplish a specific, relatively short-term project. While they work together to achieve the common goals of the project, each organization also has it own, long-term goals, which might be in conflict with some of the goals of the particular project, thereby introducing issues that are extraneous to the domain of collaboration [Mohsini 1992]. Third, collaboration in the building industry tends to stretch out over a prolonged time, even when the original participants are no longer involved, but their decisions and action still impact the project [Jockusch 1992].

2.1 Communication vs. shared-understanding

Communication is a prerequisite to (intentional) collaboration: it is the means by which the intents, goals, and actions of each one of the participants are made known to the other participants in the collaborative effort, thereby forming the basis for their own actions. Communication, however, is not enough: the heterogeneous backgrounds of the participating professionals in the building industry are often a source for misunderstandings and misinterpretations of the communicated information, leading to errors and conflicts. Although better means of communication have been developed over the past 500 years, in the form of drawings, three-dimensional models, and written specification, they have not overcome the problem of shared-understanding.

2.2 Misunderstanding in A/E/C

In their book *The Social Construction of Reality*, Berger and Luckmann [1967] discuss the processes by which any body of knowledge comes to be accepted and recognized as reality. They argue that 'reality' is not an objective, value-less, fixed phenomenon, shared by everyone. Rather, it is a product of *social systems* through which human knowledge is developed, transmitted and maintained. It is, in many ways, a matter of *belief.*

As such, there is no shared, objective basis for the design and evaluation of buildings (or any other product, for that matter). The importance, meaning, and value of objects, concepts, and situations can only be understood within the socially constructed reality within which they are perceived. This reality, or 'worldview,' is different for each one of the participants in the process of designing, constructing, and using buildings. It is developed through professional education and practice—the process of socializing into a specific way of thinking and acting [Cuff 1991]. Professional education teaches a 'right' way of seeing the world, and instils faith in that way, which over time becomes no longer open to challenge. Each worldview may have at its core a different set of values, or objectives, which might not be central to other worldviews, or may be completely absent from their view, as depicted schematically in Figure 1. Architects, engineers, construction managers, facilities managers, building owners, and end-users all have different worldviews, making it difficult for each to understand and to value the joint product in the same way as other participants do. One of the virtues of the collaborative process is to insure that no legitimate view is overlooked. It is also one of most critical impediments it must overcome.

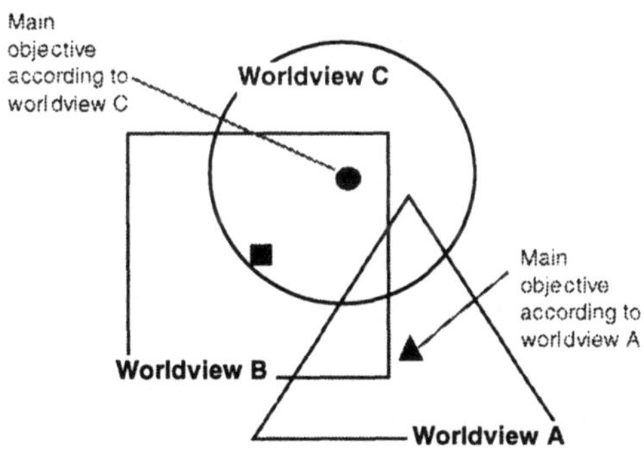

Figure 1. Different worldviews promote different objectives

2.3 Conditions for collaboration

Given that all the participants in a building enterprise have been educated into their own ways of seeing, understanding, and valuing the world, it is inevitable that there will be conflicts between their different socially constructed realities. The first step in resolving such conflicts and facilitating joint action is to recognize that different worldviews exist. The second step is to develop means that can help each participant at least to understand, if not to agree with the worldviews of the other participants. Only then can they arrive at the third step: the development of a consensual worldview that will recognize the legitimate concerns and goals of each participant and maximize the overall utility of the project (what Kuhn [1962] called the 'super paradigm').

That different worldviews exist is self-evident, though not always recognized by the participants in the collaborative process. *Understanding* these different worldviews, as the second step in developing joint action, is not so obvious. Because collaboration in design is not a matter of mere *coordination* of the inputs of the specialists who join together for the purposes of the project. Rather, it is a matter of *joint decision making*, which will deliver a better overall product. Such collaboration is much more difficult to accomplish than mere coordination, especially in the building industry, because it requires an initial *suspension of judgement* by the participating professionals [Shibley & Schneekloth 1988]. But design professionals are trained to evaluate and to judge, and immediately to seek the action that follows from their observations, while discarding information that appears to be irrelevant. They are trained to search for congruence between what they observe and the theoretical constructs of their respective professional worldviews, which they have come to accept as truth. Professionally, neither architects, nor engineers nor construction managers are rewarded for suspending judgement or for allowing other worldviews to alter their own. The suspension of judgement increases the professional's vulnerability and the risk of failure. It can only be justified if the risks taken lead to the attainment of desirable objectives.

3. HOW TO COLLABORATE?

Much effort has been put into facilitating the *communication* among the participants in a construction project and, paradoxically, into resolving conflicts among the professionals once they arise [Axelrod 1984, 1997, Protzen & Dehlinger 1972, Pruitt 1981, Raiffa 1982, Sycara 1988, etc.]. Much less effort has been put into promoting *shared-understanding*, which is the source of the conflicts.

Professional practices in the building industry have recognized the need for collaboration, and have generally adopted one of two methods to facilitate it: *hierarchically-partitioned* decision making or *temporally-partitioned* responsibilities. Hierarchically-partitioned decision making among the professionals comprising a design team takes the form of contractual arrangements, where one of the participants (often the architect) is appointed team-leader, and the rest are considered sub-contractors or consultants [Mohsini 1992]. While this arrangement may be efficient in terms of process (i.e., getting the job done), it introduces the risk of diminishing the overall performance of the product by reducing the commitment of the sub-contractors to the project because of their diminished ownership or influence on the product. The overall result, therefore, is often less than optimal.

Temporally-partitioned responsibilities represent the typical 'over the wall' practice of transferring responsibility for the project from one design professional to the next, as it moves through the design/build/use process. Thus, the responsibilities of the architect end when construction begins, and the construction manager's responsibilities end when the facilities manager takes over. While this method too is efficient in terms of process (and in terms of legal exposure), it is detrimental in terms of quality: the carefully considered knowledge which has been applied in earlier design phases is lost when the responsibilities are transferred, along with all the assumptions underlying them [Hitchcock 1996].

Thus, accepting either one of these two collaboration strategies, while superficially efficient, may result in overall failure of the project or, at best, less than optimal overall performance: the piecemeal nature of hierarchically- or temporally-partitioned decision making in habitual collaboration methods makes it almost impossible to recognize the existence of higher-level objectives, let alone develop a combined, overall view of the project. Hence, these approaches can be seen as methods of *coordination*, rather than *collaboration*.

Several computational methods, aimed at facilitating network-like (rather than piecemeal) collaboration in the building industry have been proposed by researchers in CAAD. Their objectives have generally been to assist human designers to *communicate* and to *evaluate* the evolving product in an effective, and if possible, concurrent manner. The methods and tools that were developed for these purposes can be classified as *product-sharing* methods, *performance-evaluation* methods, and *process-oriented* methods.

3.1 Product-sharing methods

These methods use some common data exchange format to facilitate the communication among the participating professionals. Efforts such as ID'EST, EDM, and COMBINE [Kim et al 1997, Eastman & Siabiris 1995, Augenbroe 1995, Amor et al 1995] have focused on developing sharable product models and databases of increasing sophistication that include factual information about the objects they describe, with particular emphasis on solving issues of concurrency, data-integrity, and data-sharing [Eastman 1994, Galle 1995, Jacobsen et al 1997, Sun & Lockley 1997, Yaski 1981]. Most of these methods assume that the data will be accessed by suitable computer programs. Some, like FCDA [Khedro et al 1993], assume that the data will be processed by human experts. An underlying assumption in all these systems is, nonetheless, that the *readers* of the data (whether human or computer programs) will interpret it correctly, using their own professional knowledge. However, Valkenburg [1998] proved this assumption to be false, because of the social and professional reasons discussed earlier. Therefore, while these efforts have made *communication* easier and more efficient, they have not, in and of themselves, improved *shared-understanding*, which is fundamental for making joint decisions and for negotiating tradeoffs among competing objectives.

3.2 Performance-evaluation methods

These methods combine separate, discipline-specific performance evaluations into a composite or an overall performance assessment of the evolving design solution, such as demonstrated by Wiezel & Becker [1992], and by Hacfoort & Veldhuisen [1992]. To arrive at an 'overall' assessment, some of these methods require setting up a weighted goal tree, stating the relative importance (according to someone's worldview) of individual design criteria [Manning & Mattar 1992]. Typically, such weighting systems must be set up prior to engaging in the design process, and cannot respond to changes in preferences arising from the dynamically unfolding design process. An alternative to relative weighting is the use of *benchmarks*, developed from case studies, as demonstrated by the BDA system [Papamichael et al 1998]. Case-based methods are very sensitive to the context of the project they come from. Systems that can account for the context of the case study are emerging, though they are yet to be integrated with collaboration methods [Oxman 1990]. Generally, performance-based systems tend to emphasize the *technological* aspects of the evolving design solution (energy, lighting, cost, etc.), while largely ignoring the *human* aspects of design collaboration. They suffer, therefore, from the same limitations as the habitual collaboration methods, namely—compartmentalization of worldviews and a tendency to communicate the *results* of the evaluations without the *objectives* they strive to accomplish, or the *assumptions* they rely upon.

3.3 Process-oriented methods

These methods emphasize the deliberative aspects of design decision making processes, in terms of design *intents, assumptions* and *arguments* in favor of or against proposed design actions. This mode of collaboration, first suggested by Musso & Rittel [1967], was implemented in case- and knowledge-based networked hypermedia systems, such as MIKROPLIS, PHIDIAS I and II, and Janus [McCall 1986, McCall et al 1990,1994]. Agent-based systems that support argumentation were developed by Pohl and Myers [1994], and implemented in their ICADS system. These systems have helped us understand the deliberative nature of the design process, but suffer from the inherent difficulty of encoding design knowledge in computational constructs, such as expert systems and agents. Therefore, they tend to work well in restricted domains, such as military ship load planning in the case of ICADS, or NASA's lunar habitat module, as demonstrated by PHIDIAS II.

4. P3: AN INTEGRATED COLLABORATIVE DESIGN ENVIRONMENT

Each of the approaches listed above has certain advantages, as well as drawbacks, with respect to its ability to facilitate shared-understanding across disciplinary boundaries, thereby enhancing collaborative design. The approach we propose is, essentially, a judicious *collection* and *composition* of several previously proposed approaches, with many adaptations and enhancements. It is their collection and specific composition that endows our approach with what we believe is needed to enhance multidisciplinary shared-understanding.

The integrated collaborative design environment we are developing, which we call P3 [Kalay 1998], consists of three complementary computational constructs:

1. Semantically-rich *representational* tools, which provide explicit reference and frame-of-reference representations for the objects comprising the evolving design solution.

2. *Communication* and *evaluation* tools, through which the values, issues, and assumptions of each one of the participants are made known to the other participants.
3. *Negotiation* tools, which help the participants adjust their respective solutions and objectives for the purpose of improving the overall performance of the project.

We have begun to implement such an integrated, distributed design environment, to support the design phase of buildings. In the following we will describe each of these constructs, and how they come together into one whole.

4.1 Representation tools

The purpose of these tools is to enable the sharing of project information (about the facility and its context) among the participating professionals. In contrast to most existing CAD tools, we have embedded a considerable amount of *semantic knowledge* in the shared information, including high-level intents, beliefs, and as much as possible, disciplinary conventions regarding the represented objects. This added semantic information augments the factual information carried by the shared product model, without which the subsequent interpretation, evaluation and negotiation processes cannot work. By providing shared semantic information we can reduce the amount of interpretation that needs to be made by each one of the collaborating experts, or at least make their individual interpretations more consistent with each other. For example, when communicating DOOR information, we include architectonic notions such as its affordance of *passage* and *privacy*, in addition to typical product-related properties such as *material*, *cost* and *fire rating* attributes.

Our approach to embedding semantics in the product model separates assembly-related information (mainly, the topology of the product) from object-related information (its attributes), and from context-related information. This separation, which is reflected in the three different databases discussed below, allows each to be optimized to meet its own objectives. Together, they provide complete product and context information.

4.1.1 Object database

The Object Database (ODB) represents *referent* information [Cohen 1944] : information that defines the nature and the character of the object, and the conditions under which this character will be altered. This information is object-specific but project-independent; that is, it is specific to an object, regardless of its frame-of-reference. Thus, the ODB can be considered a semantically-rich *digital library* and *dictionary* of objects (e.g., walls, doors, windows, etc.). An inheritance hierarchy is used to structure the information in a non-redundant form, as depicted in Figure 2.

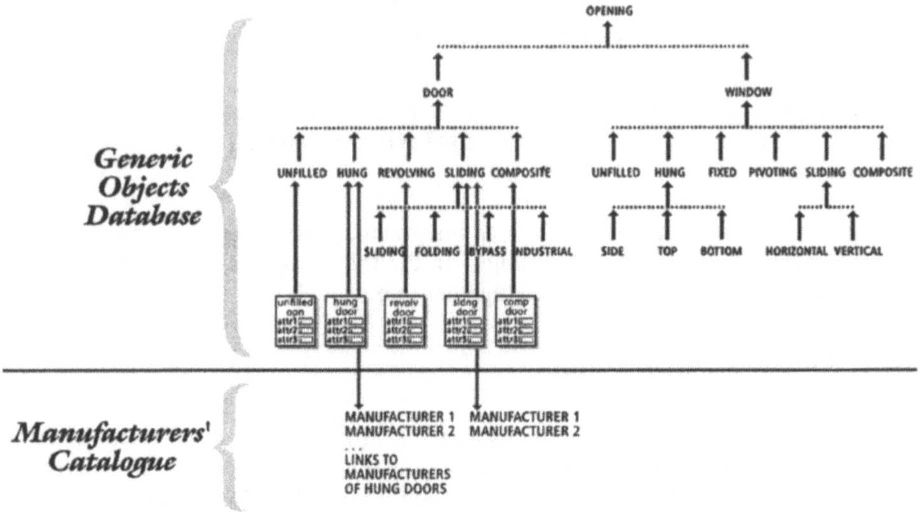

Figure 2. The inheritance hierarchy of DOORs (source: Timerman 1998)

By definition, each level of the inheritance hierarchy includes more specific information than its parent level. This information comprises both *descriptive* data (e.g., the object's name, shape, materials, and other attributes), and *functional* data (e.g., that a DOOR affords access, egress, etc.), as depicted in Figure 3. General integrity constraints, which define the conditions under which the nature of the object will change, are also included. Episodic information, in the form of design cases or manufacturers' catalogs (in multimedia form), provide anecdotal information about the object. The intent of the ODB is to be comprehensive enough so professionals who may not be intimately familiar with the type of objects it represents will be able to learn about it without having to make implicit assumptions.

We have developed, so far, ODBs for DOORS and for WALLS, each of which is a collection of databases accessible through the World Wide Web [Timerman 1998, Kalay et al 1998]. These are based on the SfB classification system, which was developed in Sweden and is broadly used in most European countries, especially in Great Britain, to classify building-related objects. Its adoption provided us with a standard classification system, and introduced uniformity into the database (e.g., the number of levels from general to specific objects). As such, it provided a good starting point for composing and extending the Object Database, and saved us the considerable effort needed to develop a classification system of our own.

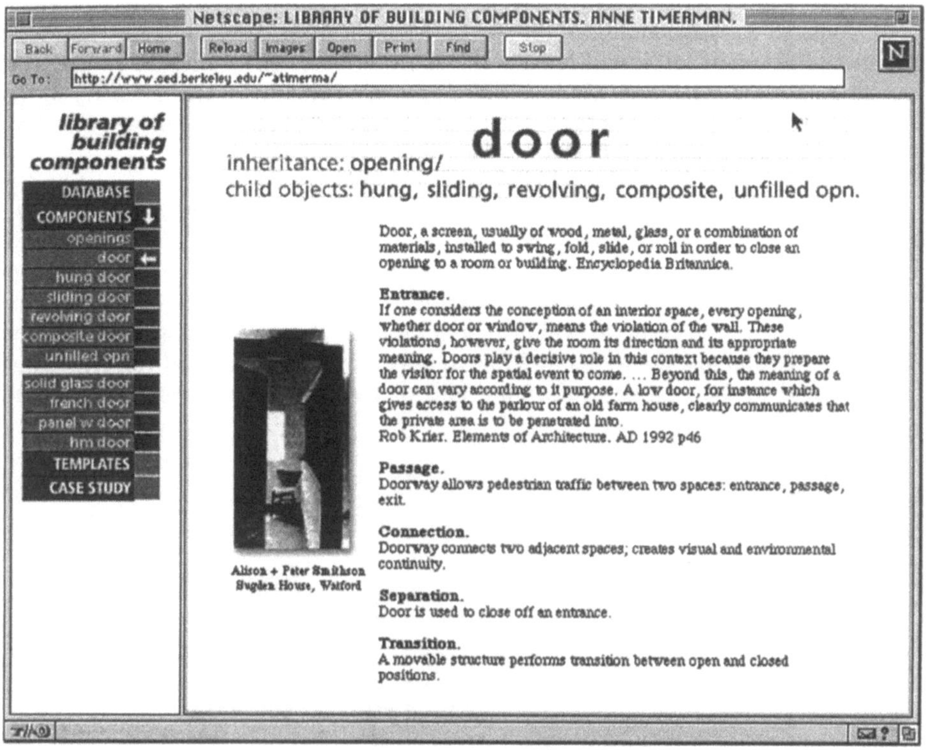

Figure 3. DOOR as a Generic object (source: Timerman 1998)

4.1.2 Project database

The Project Database (PDB) represents part of the *frame-of-reference* in which particular design objects are embedded (the other part of the frame-of-reference information is represented by the Context Database). More specifically, the PDB includes *assembly* information (information about the relationships between an object, e.g. a DOOR, and other objects, e.g. WALLS), as well as information that objects acquire when they are embedded in a particular project (e.g., dimensions, orientations, etc.). The PDB objects are linked to the ODB objects through an *instantiation* relationship, whereby they acquire the pertinent object-specific semantic information.

The specific representational model we have adopted for the PDB accommodates both structural and spatial views of the building, while facilitating a highly compact, yet well-formed, general, and complete representation of all the elements (Figure 4). The PDB can thus be viewed as an *architectural* assembly of spaces and the partitions between them, or a as a *structural* assembly of walls, columns, and beams, with the spaces they bound. Given that the underlying components are shared between the two views, it is possible to identify elements in one view that may pose conflicts with the other view. For example, if a specific WALL has been designated as a shear wall in the structural view, then this information persists when the same wall is visited from the architectural view. If the shear-resisting capability has been designated as dominant, it will prevent the substitution of this wall with a non-shear resisting partition, such as a window-wall or a row of columns. This property of the representation thus accommodates changing referents and frames-of-reference, which is important when multi-disciplinary design is concerned.

Moreover, the links between the PDB objects are themselves semantically rich: they carry logical information regarding the *nature* of the link between objects (e.g., embeddedness, support, abutment, etc.). Therefore, they provide a conduit through which changes can be propagated: for example, if a WALL has been relocated, the doors and windows it contains will also be relocated. If the WALL has been replaced by a row of columns, the doors and the windows will be eliminated (or a message will be send to the designer informing him of the consequences of the proposed change).

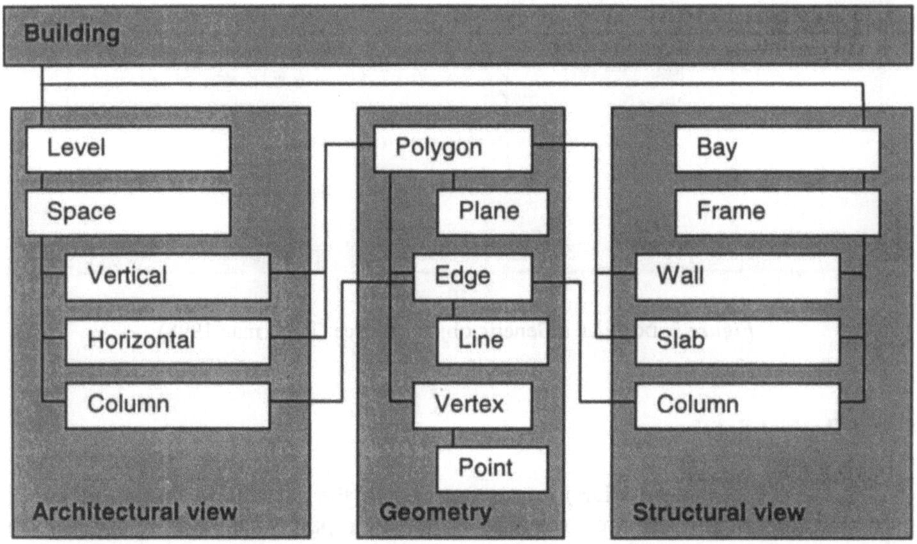

Figure 4. The architectural and structural views afforded by the PDB

We are currently working on the third version of a non-redundant representation for the PDB, using a data structure modeled after the well-known *winged-edge* model [Khemlani et al 1998].

4.1.3 Context database

The Context Database (CDB) represents information about the physical and temporal context of the project. Context is understood in its broadest sense: information the design team must respond to, and over which it has little or no control (e.g., topography, climate, views, cultural environment, economic and political environment, zoning codes, etc.). The context also comprises the predominant *activities* that the building must support, as implied by the nature of the project (e.g., medical procedures for treating patients in a hospital, the method of teaching in a school, and traditional habits of a family within its own house). As such, the CDB represents another facet of the frame-of-reference information.

We have only begun to develop the conceptual framework for the CDB, relying on principles borrowed from Geographical Information Systems (GIS), therefore it is not discussed here.

4.2 Communication and evaluation tools

Performance evaluation is very much discipline-specific. Over the past 30 years, many discipline-specific knowledge repositories, and their attendant tools (algorithms, expert systems, case-bases, etc.) have been developed. Many of these systems require their own form of input, and produce output that is often meaningful only to experts in their respective fields. It is impractical to re-design these systems, which represent considerable accumulation of knowledge and hard work (what is affectionately known as 'legacy' systems). Instead, we advocate *networking* existing evaluation systems to the collaboration environment through custom-made communication modules, which we call IDeAs (Intelligent Design Assistants). For example, an IDeA that connects an *energy expert* to the network is able to search the ODB, PDB, and CDB for the information needed to perform a thermal evaluation. It can also *translate* this shared data into the format used by a particular energy evaluation program. The (human) expert who uses the energy evaluation program can further augment the data with the necessary disciplinary information, and select appropriate defaults for the control parameters. Once the analysis has been completed, the (human) energy expert interprets and sums up the results in the form of a report, which she communicates (through the IDeA) to the other participants in the design process. The IDeA posts the report on a shared blackboard, and alerts interested individuals that it has been posted.

Such an energy-expert IDeA has been recently completed [Benne 1998]. It connects the collaboration environment to ENERGY—a passive solar energy analysis program developed by Shaviv & Shaviv [1977]. In addition to the Energy analysis IDeA, we have completed the development of a Habitability analysis IDeA, and an IDeA that can evaluate the suitability of a dwelling to the Korean traditional lifestyle [Khemlani & Kalay 1997]. We have begun to develop IDeAs to evaluate the performance of windows [Llavaneras 1996], and a Structural analysis IDeA.

4.3 Negotiation tools

Typically, the same design solution will be valued differently by the different professionals, due to their differing worldviews and objectives. To achieve the desired shared-understanding among the professionals it is necessary to communicate the results of the disciplinary valuations among the experts, in a manner that will be readily understood by all of them. Furthermore, the communication must convey not only the *parametric value* of the performance (e.g., cost, energy use, etc.), but also the *degree of satisfaction* in which each professional views the results, and the *degree of flexibility* with which the performance may vary before that degree of satisfaction is greatly affected.

The method we have chosen to use for this purpose is based on the concept of *Satisfaction Functions*, first introduced by Musso & Rittel [1967], and more recently applied by Mahdavi et al as part of their SEMPER system [Mahdavi et al 1997, Mahdavi & Suter 1998]. These are mappings that express, in functional form, the perceived relationship between some *parameter value*, indicating the performance of a system (as predicted by some performance evaluation tool, or personally by the expert) and the subjective measure of its *desirability* under specific circumstances. Figure 5 depicts some typical satisfaction curves: the horizontal axis indicates the parameter value representing the performance of some aspects of the designed system (e.g., energy consumption, cost, etc.). The vertical axis measures the degree of satisfaction each performance value elicits.

Each satisfaction curve must, of course, be set by the respective expert, because each one reflects disciplinary knowledge, such as satisfaction thresholds and sensitivity to change. The sources of such knowledge might be prevailing practices, case studies, codes or standards, results of post-occupancy studies, individual priorities of the experts or their clients, etc. Using these curves it is possible to identify needs that are not being satisfied, and those that are being over-satisfied. A design solution can thus be sought that better achieves *all* the needs. The satisfaction curves comprise the core of the *Project Management* (PM) negotiation module of the P3 system. When fully implemented, they will consist of a blackboard displaying these satisfaction curves, means to identify possible tradeoffs among them (and alert the respective experts), and a threaded discussion tool to record the negotiations process.

Figure 5. Typical satisfaction curves: the dotted lines represent how the current design solution performs, according to each point of view

4.4 The overall P3 environment

Figure 6 depicts, schematically, the overall distributed collaborative design environment. As the lighter shaded area illustrates, the various databases and the Project Manager(s) reside on, or are accessible through the World Wide Web. The disciplinary knowledge repositories and their respective IDeAs reside with the participating experts.

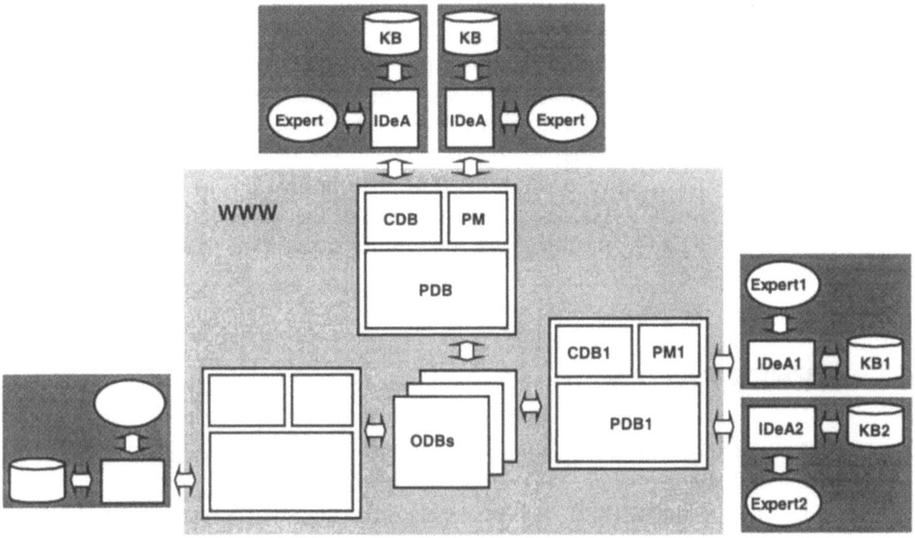

Figure 6. The overall schema of the P3 distributed collaborative design environment

5. SUMMARY AND CONCLUSIONS

As we approach the end of the 1990s it is becoming more and more obvious that 'good design' requires the continuous, synchronous (though not necessarily concurrent) involvement of many different experts throughout the design process. It is also becoming obvious that collaboration means much more than merely moving information from one participant in the design process to the next one: it requires the development of a shared-understanding among the participants. It is our belief that CAAD has a major role to play in facilitating 'true' collaboration, and engendering shared-understanding.

Our approach to facilitating effective collaboration can thus be considered a *combination* of existing approaches: the shared-product approach, the performance-evaluation approach, and the process-oriented approach. By combining them into one, unified whole, and by developing a networked environment where each module is designed to complement and enhance the other modules, we can overcome many of the individual limitations of the separate approaches. Specifically, our approach adds *valuation* and *deliberation* to the shared product approach, it adds the capacity of the human experts to *identify*, *extract*, *process* and *share* information to the performance modeling approach, and it adds *product description*, *evaluation*, and *contextuallity* to the deliberative process. The networking of all three modules into a unified network can help the experts arrive at a shared, high-level view of the emerging product, thereby overcoming their limited disciplinary worldviews. While our solution is technical in nature, we argue that its affect will be qualitative: it will facilitate *effective*, rather than merely *efficient*, collaboration.

Perhaps the most significant overall contribution presented by our approach is *continuity*: it does not revamp existing CAAD research, nor does it depend on scheduled breakthroughs. Rather, it draws on the respective strengths of individual research efforts of the past 50 years, and joins them into a unified new whole. This new whole clearly establishes the roles of the parts, while relieving each part from having to provide a holistic answer, on its own, to the original problem of helping designers to assess the quality, desirability, and the implications of their creations. It is the very maturity of CAAD research at the end of the 20th century that has made such joining of the parts possible, and, we argue, presents a viable and desirable direction for its future.

6. REFERENCES

Amor R., G. Augenbroe, J. Hosking, W. Rombouts, J. Grund, 1995, "Directions in modeling environments," *Automation in Construction* 4(3):173-187.

Armour G.C. and E.S. Buffa, 1968, "A heuristic algorithm and simulation approach to relative location of facilities," *Management Science* 9(2), 294-309.

Augenbroe G., 1995, "An overview of the COMBINE project," *Product and Process Modeling in Building Industry* (R.J. Scherer, ed.), A.A. Balkema, Rotterdam.

Axelrod R., 1984, *The evolution of cooperation*, Basic Books, New York.

Axelrod R., 1997, *The complexity of cooperation*, Princeton University Press, Princeton, NJ.

Benne B., 1998, *Web-based communication of design information for asynchronous design energy evaluation*, Master of Science Thesis, Department of Architecture, University of California, Berkeley.

Berger P.L. and T. Luckmann, 1967, The social construction of realit, *Anchor Books, New York.*

Carrara G., Y.E. Kalay and G. Novembri, 1994, "Knowledge-based computational support for architectural design," *Automation in Construction* 3(2-3), 123-142.

Cohen R.M., 1944, *A Preface to Logic*, Henry Holt & Co., New York.

Coyne R.D., M.A. Rosenman, A.D. Radford, M. Balachandran, and J.S. Gero, 1990, *Knowledge-based design systems*, Addison-Wesley, Reading MA.

Cross N., 1977, *The automated architect*, Pion Ltd., London, UK.

Cuff D., 1991, Architecture: the story of practice, *MIT Press, Cambridge, MA.*

Eastman C.M., 1994, "Computer integration of design and construction knowledge," *Knowledge-Based Computer-Aided Architectural Design* (G. Carrara & Y.E. Kalay, eds.), pp: 95-122, Elsevier Science Publishers, Amsterdam, The Netherlands.

Eastman C.M. and A. Siabiris, 1995, "A generic building product model incorporating building type information," *Automation in Construction* 3(4): 283-304.

Flemming U., 1994, "Case-based design in the SEED system," *Automation in Construction* 3(2,3): 123-133.

Galle P., 1995, "Towards integrated, 'intelligent', and compliant computer modeling of buildings," *Automation in Construction* 4(3):189-211.

Hacfoort E.J.H. and K.J Veldhuisen, 1992, "A building design and evaluation system," *Evaluating and Predicting Design Performance* (Y.E. Kalay ed.), pp. 195-211, Wiley Interscience, New York.

Hitchcock, R., 1996, *Improving life-cycle information management through documentation of project objectives and design rationale*, PhD dissertation, Department of Civil Engineering, University of California, Berkeley.

Hobbs R.W., 1996, "Leadership through collaboration," AIArchitect 3:11.
Jacobsen K., C. Eastman, and T. Sheng Jeng, 1997, "Information management in creative engineering design and capabilities of database transactions," Automation in Construction 7(1):55-69.
Jockusch, P.R.A., 1992, "How can we achieve a good design?" Evaluating and Predicting Design Performance (Y.E. Kalay, ed.), pp. 37-50. Wiley Interscience, New York.
Kalay, Y.E., L. Khemlani and J.W. Choi, 1998, "An integrated model to support collaborative multi-disciplinary design of buildings," Automation in Construction 7(2,3):177-188.
Kalay Y.E. and E. Shaviv, 1979, "A method for evaluating activities layout in dwelling units," Building and Environment 14(4):227-234.
Khedro T., M. Genesereth and P. Teicholz, 1993, "FCDA: a framework for collaborative distributed multidisciplinary design," AI in Collaborative Design (Gero & Maher, eds.), pp. 67-82, AAAI, Menlo Park, CA.
Khemlani L. and Y.E. Kalay, 1997, "An integrated computing environment for collaborative, multi-disciplinary building design," CAAD *Futures 1997* (Yunge, ed.), Munich, Germany.
Khemlani L., A. Timerman, B. Benne, and Y.E. Kalay, 1998, "Intelligent representation for computer-aided building design," *Automation in Construction* 8(1):49-71.
Kim I., T. Liebich and T. Maver, 1997, "Managing design data in an integrated CAAD environment: a product model approach," *Automation in Construction* 7(1):35-53.
Kuhn T., 1962, *Structures of scientific revolutions*, University of Chicago Press, Chicago.
Llavaneras G., 1996, *Fenestration agent*, Technical Report, Department of Architecture, University of California, Berkeley.
Mahdavi A.., P. Mathew, S. Kumar and N.H. Wong, 1997, "Bi-directional computational design support in the SEMPER environment," *Automation in Construction* 6(2): 353-373.
Mahdavi A. and G. Suter, 1998, "On the implications of design process views for the development of computational design support tolls." *Automation in Construction* 7(2,3):189-204.
Manning P. and S. Mattar, 1992, "A preliminary to development of expert systems for total design of entire buildings," *Evaluating and Predicting Design Performance* (Y.E. Kalay, ed.), pp. 215-238. Wiley Interscience, New York.
McCall R., G. Fischer and A. Morch, 1990, "Supporting reflection-in-action in the JANUS design environment," *The Electronic Design Studio* (McCullough et al, eds.), MIT Press, Cambridge, MA.
McCall R., 1986, "Issue-serve systems: a descriptive theory of design," *Design Methods and Theories*, 20(3).
McCall R., P. Bennett, E. Johnson, 1994, "An overview of the PHIDIAS II Hypercard system," *ACADIA'94* (A. Grin, M. Fraser eds.), pp. 63-76.
Mitchell W.J., 1977, *Computer-aided architectural design*, Van Nostrand Reinhold, New York.
Mohsini R., 1992, "On measuring project performance: some problems of aggregation," *Evaluation and Prediction in Design* (Y.E. Kalay, ed.), pp.239-250, John Wiley & Sons, New York, NY.
Musso, A. and H. Rittel, 1967, *Measuring the Performance of Buildings*, Technical Report. Washington University, St. Louis.
Oxman R. (1990). "Prior knowledge in design: a dynamic knowledge-based model of design and creativity." *Design Methods* 11(1):17-28.
Papamichael K., J. LaPorta, H. Chauvert, 1998, "Building Design Advisor: automated integration of multiple simulation tools," *Automation in Construction* 6(4): 341-352.

Pohl J. and L. Myers, 1994, "A distributed cooperative model for architectural design," *Knowledge-Based Computer-Aided Architectural Design* (G. Carrara & Y.E. Kalay, eds.), pp. 205-242, Elsevier Science Publishers, Amsterdam, The Netherlands.

Protzen J.P., and H. Dehlinger, 1972, *Debate and argumentation in planning: an inquiry into appropriate rules and procedures*, Technical report, Institute of Urban and Regional Development, University of California, Berkeley.

Pruitt D.G., 1981, *Negotiation behavior*, Academic Press, New York.

Raiffa H., 1982, *The art and science of negotiation*, Harvard University Press, Cambridge, MA.

Rittel H.W. and M.M. Webber, 1984, "Planning problems are wicked problems," *Developments in Design Methodology* (Cross, ed.), John Wiley, New York, NY.

Shaviv E. and G. Shaviv, 1977, "A Model for Predicting Thermal Performance of Buildings," *WP ASDM-8*, Faculty of Architecture and Town Planning, Technion, Haifa, Israel.

Shaviv E., A. Yezioro, I.G. Capeluto, U.G. Peleg and Y.E. Kalay, 1996, "Simulation and knowledge-based computer-aided architectural design (CAAD) systems for passive and low energy architecture," *Energy and Buildings* 23: 257-269.

Shibley R.G. and L. Schneekloth, 1988, "Risking collaboration: professional dilemmas in evaluation and design," *Journal of Architecture and Planning Research* 5(4):304-320.

Simon H.A., 1969, *The sciences of the artificial*, MIT Press, Cambridge, MA.

Steinfeld E. and Y.E. Kalay, 1990, "The impact of computer-aided design on representation in architecture," *ARCC Conference on Representation and Simulation in Architectural Research and Design*, Minneapolis, Minnesota.

Sun M. and S.R. Lockley, 1997, "Data exchange system for an integrated building design system," *Automation in Construction* 6(2):147-155.

Sycara K., 1988, "Resolving goal conflicts via negotiations," *Ninth International Joint Conference on Artificial Intelligence*, pp. 245-250, Los Angeles, CA.

Timerman A., 1998, *The ALPHA Project Object Database*, Master of Science Thesis, Department of Architecture, University of California, Berkeley.

Valkenburg R.C., 1998, "Shared-understanding as a condition for team design," *Automation in Construction* 7(2,3):111-121.

Wiezel A. and R. Becker, 1992, "Integration of performance evaluation in computer-aided design," *Evaluating and Predicting Design Performance* (Y.E. Kalay ed.), pp. 171-181, Wiley Interscience, New York.

Yaski Y., 1981, *A Consistent Database for an Integrated CAAD System*, PhD Dissertation, Carnegie Mellon University.

Exploiting the internet to improve collaboration between users and design team

The case of the new Computer Laboratory at the University of Cambridge

Paul Richens, Michael Trinder

The Martin Centre for Architectural and Urban Studies, University of Cambridge Department of Architecture, UK

Key words: collaborative design, internet, virtual reality

Abstract: Cambridge University and Microsoft are building a shared computer research and teaching laboratory on a green-field site to the west of the city. The clients wished to use internet based communication between themselves and their architects, including email, a web site and virtual reality. We explain how this is to be achieved, and describe experiences during the first six months of a two year project. Particularly successful has been the use of games software (Quake II) for 3d presentation of the emerging building design.

1. INTRODUCTION

Cambridge is an ancient University, squeezed into a small market town which even now has only 120,000 inhabitants. Until 100 years ago, the colleges and University buildings were all to be found within five minutes walk of the market place. Then, as the sciences began to assume their present importance there came an explosion in University building but still in, or immediately adjacent to, the historic centre. The original Botanic Garden was displaced a mile southward, and the site occupied by the Cavendish Laboratory. Downing College sold an undeveloped part of its site, and the combined area was rapidly covered with new science buildings - museums, libraries, teaching laboratories and research spaces. The development pressure was phenomenal: within fifty years the originally gracious layout of courtyards was choked by a sediment of infill buildings, animal houses, gas stores, bicycle sheds, and "temporary" huts crammed into the courtyards, on rooftops, anywhere. By the 1950's the pressure became intolerable and significant new developments were made further away - Chemistry to the south, and Veterinary Medicine pioneering in farmland a mile away to the west.

Less fortunate were the Computer Laboratory and Zoology Department, which were rehoused in situ, in a fragment of a megastructure known as the 'Arup Tower'. This was intended by the planners to be the first stage in a Buchanan inspired redevelopment of the city centre, with pedestrian circulation segregated to a high level system of decks and bridges. Only one other piece of this pattern was ever built - the courthouse on top of a car park across the street - but the bridges never appeared, and so the new building eventually came to be seen, like the skeleton prominently displayed in the Zoological part of its podium, as a beached whale.

The Cavendish Laboratory followed the Vets to the West, as did a scattering of other technological departments, and hi-tech businesses - including Schlumberger whose 'tent' (Hopkinson and Happold 1985) is the most distinguished building there. The building of the M11 motorway bypassing Cambridge to the West gave the site a definite boundary and identity and it became the long term strategy of the University to gradually decant the science and technology faculties, as they outwore their central buildings, to this largely vacant site.

A masterplan for this 'West Cambridge' site was drawn up by MacCormac Jamieson Prichard, presented to the University in 1997, and is currently being reviewed by the city planners. It shows the site, which extends from the western suburbs of Cambridge to the motorway, bounded on the north by the Bedford road, and to the south by the 'Coton footpath'; long used by academics as a place of recreation, but in fact the remnant of a still navigable medieval trackway connecting Cambridge to the villages of the west - Coton, Hardwick, Toft, Bourn. The plan shows a mix of high-tech industries along the north, accessed from the road, and University buildings to the south, reached on foot or by bicycle, from a covered 'colonnade' along the line of the Coton footpath. A central 'forum' is surrounded by shared large lecture halls, refectory, sports centre and other shared facilities.

Meanwhile the Computer Lab had expanded into parts of the old Cavendish, connected by a bridge at third floor level. By the 1990s the building was out of date and overcrowded. When the opportunity to rebuild arose, it was eagerly accepted by the University.

The proposed new Computer Laboratory is an interesting and complex project, which fits very well with the spirit of the masterplan. Funded partly by a donation from the William Gates III foundation, part of its floor space is to be leased on commercial terms to Microsoft for use as its major European research laboratory. It is the first building to be designed since the masterplan was published, and so is on the eastern edge nearest the city. As a mixed-use building it is on neither the northern road access, not the southern cycle way, but half way in-between.

1.1 Architectural expectations

Our first contact with the Computer Laboratory occurred in mid 97. At this point they had completed a written brief and were searching for an architect. The brief made some unusual demands of the architect:

> "The Laboratory and its principal benefactor are extremely enthusiastic about the use of computer modelling in the design and visualisation of the new building. This will have a number of benefits, including assisting the fund raising efforts of the Laboratory, allowing the Laboratory's members to gain an understanding of the design before it is realised, and publicly demonstrating the Laboratory's commitment to the use of computer oriented design techniques"

> "The appointed architect is expected to use 3D computer modelling as an integrated part of the design process. The data should be transferable to the Computer Laboratory in an agreed standard format, to enable our staff and students also to perform their own walkthrough and other experiments. Communication between the architect and the Laboratory should be by email, and also via a web site for the exchange of models and designs."

Aware that thorough-going 3D design was quite unusual in architecture, we replied that it was reasonable to ask the architect to use computer presentations, but design tools were personal to an architect, and should not be dictated by a client. Better choose an architect who can produce a good building, than one using a preconceived methodology.

The brief ended alarmingly:

> "It is crucial that an excessive burden is not placed on the academic staff of the Laboratory through meetings, requests for information etc. Information exchange via email could assist in achieving this. It is expected that those academic staff liasing with the architects should each average at most two hours a week on work related to the building"

This touched a nerve. Despite the endeavours of its building professionals, Cambridge University has a gift for getting bad buildings out of good architects - Stirling's disastrous History Faculty Library being the best known. Architects find the University and Colleges difficult and frustrating clients because they are faced, not with an enthusiastic client with whom they can build a personal relationship, but with a building committee, consisting of academics who are not really interested in what is going on, attend irregularly, and regard the whole operation as an irksome chore. At the end of a long campaign, the project architect may find himself the only person who has served continuously. This structure does not lend itself to responsible, consistent decision making.

There followed quite a lengthy email debate; our suggestion of the need for a more committed style of project management being countered by an enthusiasm for a 'standard box' of a building, which would minimise design risk and the need for consultation, and a horror of award-winning architects and award-winning buildings (the disliked Arup's Tower had won an RIBA Bronze Medal) .

The concern that the architect should be fully computerised was significant during the short-listing phase - for example one well-known name was rejected -

> ".. as a practice, they were not sufficiently wired. Incredibly, although all architects have workstations, email is not used inside the company at all, and email from clients has to go to a single address that is then processed by a secretary. We did not feel we could work with them."

In the end a limited competition was held. Most firms presented their proposals as HTML pages - which served chiefly to demonstrate that GIF is a hopeless way of presenting architectural plans. Some did a little animation or VRML. But good sense prevailed, and the winning architect – RMJM – was one who could demonstrate a workmanlike process, had built comparable buildings successfully, and (to our relief) could field a project architect to whom the head of the Laboratory took a strong personal liking.

Their use of CAD was quite normal and standard - the bulk of the design work is done in 2d on AutoCAD, with 3d Studio Max used for presentation. Workstations are less than 'one-per-desk', CAD is in the hands of specialists, and email though present, is not ubiquitous.

As it became apparent to the Computer Lab that the kind of intense electronic communication that the brief envisaged was not normal architectural practice (and perhaps more significant to the University authorities, could not be achieved without a fee uplift), they began to see it as a research opportunity, and asked us for proposals.

We had to move quickly, as the preliminary design was already underway. Fortunately, the project readily gripped the imagination of several industrial sponsors, who offered hardware and software contributions, and the EPSRC, who agreed to give accelerated consideration to an application for the staff costs. Getting agreement among the participants was rather more delicate (as will be discussed below). In addition to the Computer Laboratory and the architects RMJM, these now included Microsoft Research Ltd, Microsoft's facilities consultants, the University's Estate Management and Building Services (EMBS) division, who are the legal client, and a cost consultant.

1.2 The building

The brief asks for 10,000 m^2 of floor space. The total project budget is £20m, of which £12m is for the building, the rest being for equipment and relocation. As well as the embedded Microsoft laboratory, the building is to contain teaching space for several hundred undergraduate and graduate students, a library, machine rooms, research laboratories and a cafeteria. A central problem is to get the right degree of contact and separation between the three main categories of occupants - undergraduates, postgraduates and academic/research staff. The embedded laboratory is to preserve its separate identity, but facilitate interaction, both organised and accidental. The circulation space is to promote meetings and discussion "with chairs in cosy nooks and crannies .. lots of whiteboards scattered in public places .. public terminals .. cybercafe area." The building is expected to be environmentally friendly, with a minimum of air-conditioning "lots of natural light, openable windows and a nice view" while avoiding the overheating that has plagued some recent buildings. Technological requirements centre on the need to accommodate an intensive and ever-changing computer network. Amenities should include new cycle paths, cycle storage and showers, and "some kind of common room where the current tradition of morning coffee and afternoon tea for the whole Laboratory can be continued."

1.3 Research objectives

Our formal objective is to find ways of using the internet to achieve better buildings, by improving the consultation process between architect and eventual user. This means we need to engage people's interest in the project, keep them informed, facilitate discussion, and channel feedback to the architect.

The future occupants of this building - that is the staff and students of the Computing Laboratory, and the researchers employed by Microsoft - are ideal subject matter as they combine a high proficiency in electronic communication, with an unflattering opinion of architects and architecture. They see the first as a way of reducing the nuisance of having to deal with the second. Our project will succeed if it can use technological enthusiasm to seduce the computer scientists into a constructive relationship with the architectural project.

The first intention is to build a web server to provide on-line access to a comprehensive library of design information, linked to a bulletin board for gathering feedback. The second is to monitor the effectiveness of such as system, and study how it is used in practice to improve communication between client and design team.

Much of the material is expected to originate with the architects, using their in-house CAD systems (3D Studio Max and AutoCAD), but will need to be translated, extended, simplified, edited and interlinked to form a web library. Keeping the links working in the face of constant revision is likely to be a substantial problem.

The provision of 3d walkthrough we anticipated would be met in the early stages by generating VRML from 3D Studio, and later by bringing into play a viewer for very large models (million polygons upwards) which is the result of an earlier Martin Centre research project, and currently under commercial development.

Effectiveness is to be measured by regular web-based questionnaires. These will investigate changing perceptions of both the building, and the web-site.

2. PROGRESS

The project was eventually funded and started in July 1998, about six months after the architectural competition, so our immediate need was to catch up with progress already made. As it was also our first substantial web-based project, and our first research project to be based on Windows/NT, there was a somewhat steep learning curve to be climbed. We discovered that the Computer Laboratory, while fully wired, was largely using Linux on PCs, with Solaris on Suns, and deployed a variety of not quite fresh versions of Netscape. Even Lynx had some currency. On the other hand, the Microsoft end of our user-base would be using NT on Intel and the latest version of Internet Explorer.

This context argued for a very straightforward initial design for the site, avoiding even such mild elaborations as frames. It also became apparent that there were two levels to the project - the building and the site, as represented by the masterplan. The enjoyment of the Computer Laboratory would be substantially affected by progress (or the lack of it) on other parts of the site. Some issues that are mentioned in the brief - such as cycle ways - are outside RMJM's control, as they belong to the masterplan. Though responsibility is divided, both are important. So we conceived of two web-sites, an outer one presenting the masterplan and brief information as it became available on individual buildings, forming a context for the inner one, which would deal in great detail with the Computer Laboratory building itself. The outer site, we envisage, may last much longer than the inner, as the masterplan is likely to take 25 years to complete. It is also relevant to a much wider community than the inner.

The two sites achieve unity by using a similar style of header and footer on all pages, differentiated by a slight difference in colouring. The same design principles apply to each site: a straightforward tree structure - home page plus two levels of index to reach the content pages, page length restricted to a couple of screens full, sparing use of cross and off-site links. The pages were initially made fixed width (by use of tables), but freed after adverse comment from the Computer Lab. The consequence is that the fixed-width page headers and footers now look a little silly, and will have to be replaced at some time with navigational side-bars.

2.1 Outer site

This site was implemented second, largely using material from MacCormac's masterplan, and did not become live until February 1999. However, it is best discussed first, as it sets the context for the Computer Laboratory building.

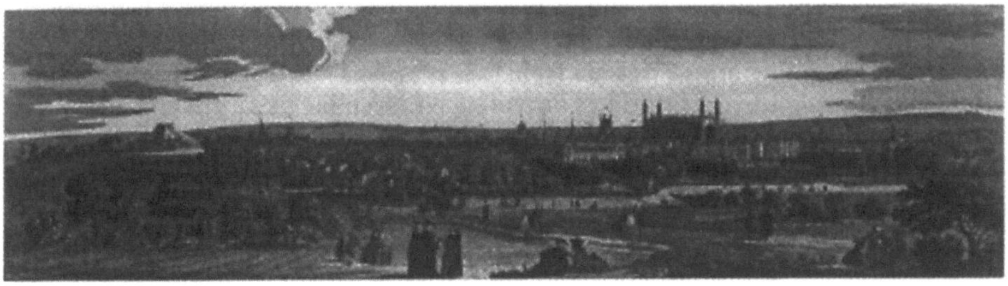

*Figure 1.*West Cambridge in 1743

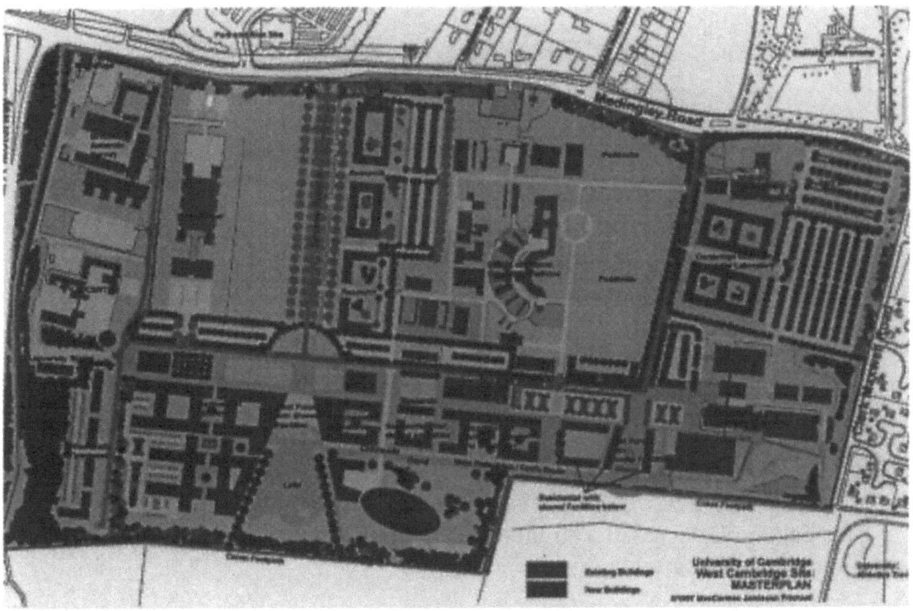

Figure 2. Masterplan used as entry image map to the outer site

The contents page shows a simplified version of the masterplan as an image map (Fig 2). Buildings and features under the cursor identify themselves, and clicking will (eventually) take you to the appropriate inner site. Also accessible from this page are:

- **Proposal** - full text of MacCormac's proposals for the site
- **Summary** - a non-technical summary of the official proposal
- **Diagram** - a variety of downloadable version of the site plan
- **Maps** - plan-form contextual material collected by ourselves. Most revealing are the aerial photographs (on which the masterplan and car and cycle access routes

can be superimposed), which clearly show the scale of the site (larger than the historic core of Cambridge) and its distance from current University facilities.
- **Views** - site photography and other scenography, including an old print showing the skyline of Cambridge as seen from the West in 1743 (Fig 1). A large part of this view survives, and is seen as providing an essential visual connection between the new site and the historic centre.
- **Diary** - history of events affecting the western development of Cambridge.

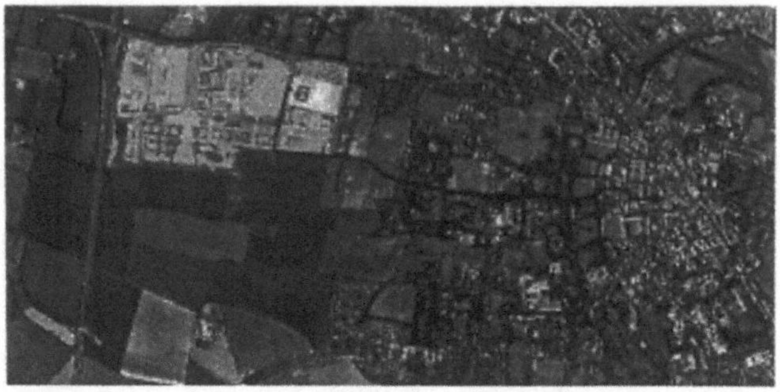

Figure 3. Aerial photograph with masterplan superimposed, and cycle route indicated. Demonstrates the size of the West site compared to the ancient city, and its relative isolation

2.2 Inner site

This page gives access to all the consultation drawings issued by the architects, and a variety of background documents.

2.2.1 Project Documentation

Textual information, starting with the original competition brief, and the quantified brief produced by RMJM in the early months of the project. This will eventually expand to full "room data" description of the building, giving usage, floor area, equipment, servicing needs and so on, for every room. Additional textual information, such as meeting notes and specialised reports, will be placed here.

2.2.2 Building proposals

Goes to an index of each major revision of the drawings. Each revision is represented by a page forming an index to all the drawings and models available.

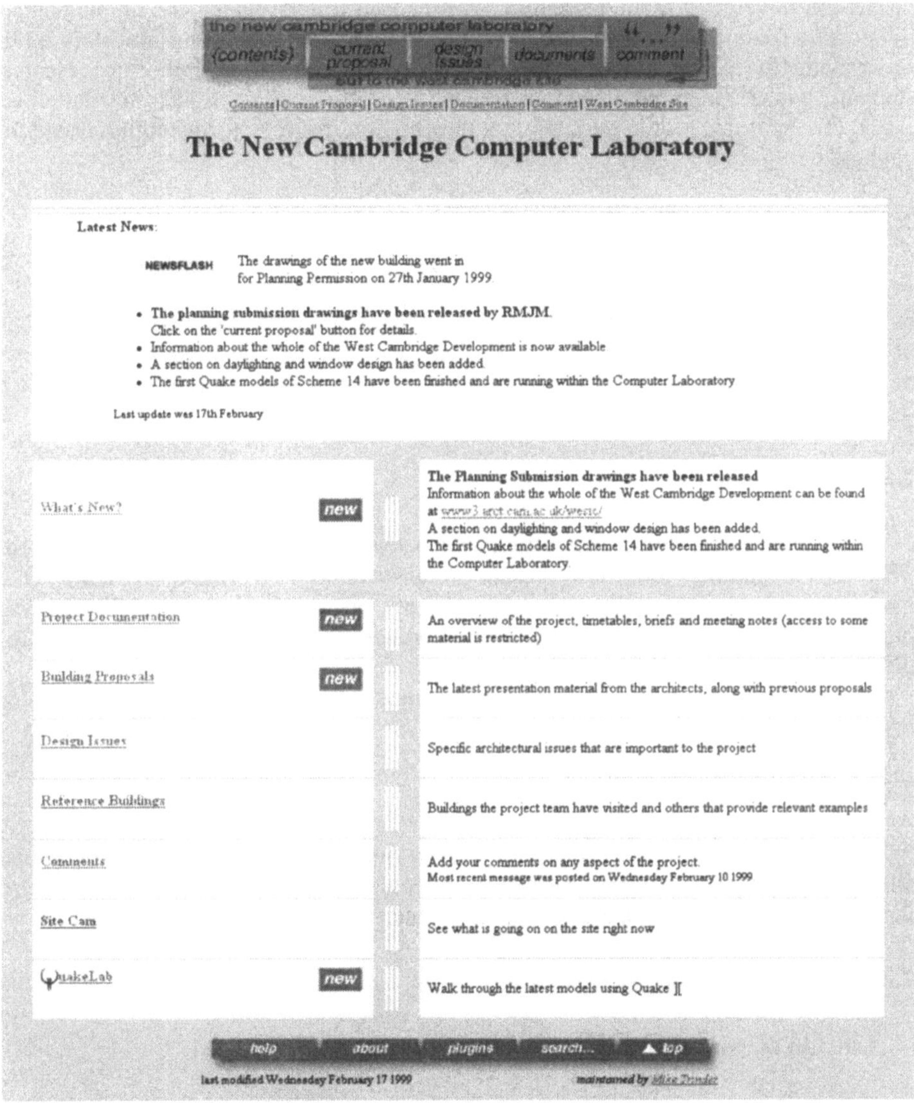

Figure 4. Contents page, inner site

2.2.3 Design issues

Essays prepared by ourselves to clarify architectural issues that have arisen in the discussion groups, and generally drawing on other areas of research in the Martin Centre. For example, the earliest proposals from RMJM addressed the 'environmentally friendly' requirements of the brief by employing a shallow plan form, E-W orientation, and night-cooled hollow plank floors. We contributed a paper on green architecture, explaining the general principles behind these and other practical approaches to low-energy building in the British climate. Later a discussion developed around suitable widths for corridors, which we supplemented by posting photographs of measured widths in the existing building. Recently quite a passionate discussion has developed about daylight and views. One of our research students, who is finishing a PhD on subjective reactions to daylight, contributed an essay on the topic, with background information from a forthcoming book on daylight being written by two colleagues.

2.2.4 Reference buildings

Initially we intended this page for buildings visited by the design team, or used by RMJM as references. Later we discovered that Bill Gates is supporting a number of parallel projects in American universities, and that they have web-sites in some degree parallel to ours, so we extended the section to provide links to web-sites related to interesting Computer Laboratories.

2.2.5 Comments

Each page has a comment button, which leads to a web-based bulletin board. This uses a slightly modified version of a standard Perl script WebBBS. One modification is to capture the identity of the page on which the comment button was pressed, with the idea of using this information later to analyse the relative impact of different documents.

The presentation of the BBS is conventional, with messages viewable in date or inverse thread order.

2.2.6 Site Cam

When construction starts, this will link to video camera views from one or two adjacent buildings. The London-based architects are looking forward with relish to the possibility of seeing what is going on on-site, whenever they wish.

2.2.7 Quake Lab

Links to the virtual reality pages described below.

2.3 Formats

Scanned material and computer renderings are normally posted as JPEG images, in a choice of resolutions, and are straightforward. More problematical is the presentation of CAD drawings originating in AutoCAD (for the building) or Microstation (for the masterplan). The predominance of Linux at the Computer Lab means that the many of the standard plugins are not available, and that Java based extensions have to be sought.

AutoCAD drawings were initially posted in the DWF format, together with a Java reader called CADViewer Light from Arnona (www.cadview.com). Even this failed initially, as some instances of Netscape at the Computer Lab were too old to run Java successfully. A forced upgrade to Netscape 4.5, courtesy of the Computer Lab sysops, fixed the problem. The freeware version of CADViewer provides reasonable performance in panning and zooming (thought not as lively as the WHIP plugin available on NT), but does not allow printing, which almost immediately became a problem. The solution was to convert the AutoCAD material to the Adobe Acrobat PDF format, which is available on all platforms, and seems entirely suitable, at least for drawings of the complexity we have encountered so far. The black-on-white format is preferable to the vivid colour-on-black that you get from DWF; proper fonts come naturally, panning and zooming on screen is easy, and drawings are easily printed at a definite scale.

Generating PDFs is not so easy, and we have yet to establish a foolproof workflow. In principle you open the DWG in an amenable CAD system (we have been using microGDS), and print it at the required scale to a postscript printer, checking the 'print to file' box. The postscript file can then be processed through the Acrobat Distiller, to generate PDF. The main difficulty is that most CAD systems seem to require you to have a driver and physical plotter on your network of the size required of the PDF image, before you can generate a postscript file.

We use a similar route out of Microstation, with a detour through Adobe Illustrator to tune up labelling, fonts and fills. To make an image map we then export as EPSF and render to GIF in PhotoShop. We have used video formats such as QuickTime and animated GIF to a limited extent to display moving images, such as sun studies. With interactive controls, it is easy to see how the pattern of sunlight and shadow varies with the time of day, and season of the year.

Apple's QTVR has long impressed us as an ideal web format for architectural presentation. The PC equivalent, LivePicture, has been used by RMJM to extraordinarily good effect. Their method is to construct an outline geometrical model in 3D Studio, and render it as a LivePicture panorama. The cylindrical texture map is printed, and used as a guide for hand drawing the interior architecture, which is then scanned, coloured, and substituted for the original. When viewed with a LivePicture viewer, you get a lightweight, charming and completely effective interactive display. There is a freeware Java viewer available from www.livepicture.com

VRML viewers such as Cosmo are available for all the platforms of interest, but their capacity is not impressive. They seem to be suitable for small models, such as a single room, but not for the overall building. We are currently investigating various methods of generating radiosity lit VRML models of rooms where the quality of daylight is of interest.

2.4 Quake

Quake II is a bloodthirsty shoot-em-up computer game based on an extremely effective 3d rendering engine, which will run on most platforms, including Linux. A good deal of the game can be downloaded in source code, and both the scenery ('levels') and the 3D animated characters ('monsters') are replaceable, making the game something of an hacker's delight but also something that we thought might be peculiarly appropriate for the Computer Lab web site.

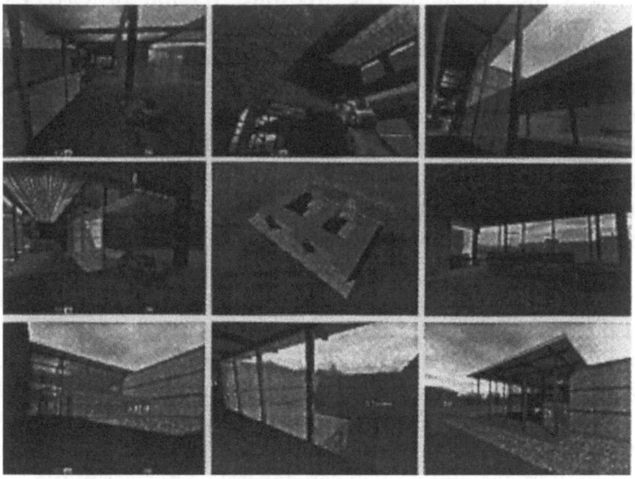

*Figure 5.*The Computer Laboratory proposal in Quake II

There are around half a dozen shareware level editors for Quake; we chose one called Qoole, which with a CD of useful resources cost us $120. The runtime system is contained on the ordinary Quake II disc, around £30 from any games store. Qoole provides an object oriented modelling environment in which to construct the architecture of the set, texture it, and locate lights. If you are developing a game, you will also locate and parameterise monsters, weapons, sensors and all the apparatus of gameplay. When completed, the level is handed over to a three-stage compiler, which produces a playable level. The first stage constructs a BSP (binary space partition) tree, using the bounding planes of architectural objects to subdivide empty space into nested convex subspaces. The virtue of the BSP tree is that it can be traversed in priority order, as seen from any viewpoint, leading to a very fast rendering algorithm. The second stage appears to work out intervisibility between subspaces, creating chains of spaces that can be seen from a particular location, as an optimisation of the basic algorithm. Without this visibility information Quake II can still render a level, although at much lower frame rates. The third calculates the lighting, on the radiosity principle, exploiting the visibility calculations that have already taken place.

Compiling a level of any complexity requires considerable resources. We use our web-server, which has twin 333Mhz Pentium II processors and 256Mbyte of memory, and can compile the structure, shell and core of the building (no internal partitions) in about 4 hours. Using much less memory triggers an orgy of paging whilst calculating the visibility; with only 128Mb of memory, compiling does not complete in a weekend.

The level editor prefers orthogonal geometry, and a modular grid, which turns out to be based on the size of the texture tile. It is a little difficult to establish the scale at which you are working - the basic yardstick is the eye-height of the viewer which seems to be about 50 units. We standardised on 64 units being 2 metres. You do not get manuals with this sort of software; you have to learn by doing, or joining an internet discussion. One tip we picked up from an experienced level builder (who happened to be visiting the Computer Lab), was to flag lights and other elements which do not enclose space as 'detail' so that they are ignored in generating the BSP tree. Unfortunately, flagging the columns as 'detail' speeded up the BSP and visibility calculations to around 1.5 hours, but reduced the frame rates considerably.

Our first level took about 2 weeks to complete, mostly learning how to do it. It comprised the floors, roof, columns, external and core walls, staircases, and lifts. We omitted windows, partitions and furniture. Originally, we used a lurid science-fiction landscape from the original game as a backdrop to the building, but it proved simple to switch this to digicam shots of the surrounding buildings on the West Cambridge Site, combined with a painted sky. As at that point no architectural commitment had been made to materials and textures, we used texture tiles from the original game throughout. These are engaging in a grungy sort of way but obviously unreal, so the model is readily understood to be about the spatial arrangement of the building, not its surface appearance. The lighting compiler was set so that the sun shone from where it was painted on the sky and cast shadows. The interior lighting, based on skylight through the windows, reflected sunlight, and interior area sources, is excellent. This is global illumination calculated not for a room but for the complete interior, within a few hours, and with no more fuss than specifying the number of interreflections to be considered and a general ambient lighting level to stop totally black shadows.

Considered as a VR system, the Quake engine provides very fast software rendering and smooth movement, full screen (640x480) on a 250MHz PII, with clash detection and gravity. Cheap 3D hardware reduces the processor requirement considerably. Precomputed lighting gives a huge lift in architectural quality. Moving objects are easily programmed - we have doors that open on proximity, and a functional elevator. The controls are easy to use, with the mouse used to 'look around' and the keyboard to walk. Best of all, the model can be populated with as much ease as you can plant trees in a normal CAD environment. The people are admittedly rather ugly, and rather liable to shoot you unless programmed otherwise, but they move around, and start to define the architectural scale like nothing else can. If the game is switched to 'deathmatch' several people can join in each from his own workstation, and seeing the others in the guise of monsters. With a little attention to the 'skins' of these monsters, we shall soon be able to let Geoff Cohen (the architect, in London) conduct Bill Gates (the sponsor, in Seattle) on a virtual tour, watched by anyone who cares to join in. (Provided, that is, that they surrender their guns at the door).

3. ASSESSMENT

3.1 Management Issues

At the time of writing, the project has been running only six months of its allotted two year span, but we have already learned a great deal. We assumed when writing our proposal, that the project was about openness and communication, but discovered from day one, that it was just as much about privacy and control.

The situation was an unfamiliar one to all the participants (except perhaps Microsoft), and each member of the building Project Team, had its own quite proper reservations. The architect quite naturally, did not want to lose control of the process of issuing information, or of the appearance of what was released. So we agreed with RMJM that material would first be mounted on a private web site, and not released to anyone else until they had approved it. EMBS did not want to lose control over the overall process, so we agreed that material would not be made visible to the Computer Lab until the Project Team had decided that it was appropriate to release it for consultation. They were also concerned that we should not put an extra load on the architect which might cause a fee uplift. Everyone realised that feedback from the site would have to be managed in some way - the architect could not be expected to deal with hundreds of emails from all over the organisation. It was agreed that a co-ordinator (seconded from EMBS to the Computer Lab) would moderate the bulletin board, and pull together issues that ought to be addressed. In practice, major issues which have arisen on the web site have been discussed for a while, and then brought to a Project Team meeting, usually by the head of department.

In the end we proposed five levels of privacy for the web site, with carefully monitored transitions between them. In no case would the webmaster move material from one level to another without express instructions. The levels were:

a) Private to the Martin Centre and RMJM, for material under development
b) Private to the Martin Centre, and the Project Team.
c) Released for consultation, available to all members of the Computer Lab, University administration, and Microsoft.
d) Available to the whole University, and Microsoft
e) Available to the world

This structure is designed to reassure rather than for ease of administration, and has caused greater delays than we would like to the flow of information. We are hopeful that, as confidence grows, there will be some simplification.

The feedback bulletin board was initially conceived of as being available to the whole of the Computer Lab, staff and students, but this wide exposure was found to inhibit discussion. At the head of department's request, a second inner discussion group of senior academic staff was established. Again, at the user's request this is configured as an email listserve, rather than using web forms, and is backed up by a mail archive on a secure server in the Computer Laboratory so that membership of the group can be validated through their login id's on the Computer Lab UNIX network, information not readily available to our web server. These changes have been effective in stimulating a much more directed and vigorous discussion.

Another case of privacy promoting discussion occurred when printed plans that were hung up on a publicly accessible whiteboard. They soon accumulated a variety of comments and redrawings, which we photographed and published. We offered to establish a webcam, to keep the whiteboard constantly in view. This was declined, on the basis that people wouldn't add their suggestions if they thought that they were being watched.

3.2 Design issues

The usefulness of the site is best judged by the quality of the discussion it develops, and its influence on the building itself. The opening messages were not too encouraging - being about ladies' lavatories (Curiously the discussion on the parallel site at MIT opened on the same issue). Green principles were the first serious issue, to which we responded by posting a summary of the design principles involved on the 'Design Issues' pages. At that point RMJM were proposing a system of night-cooled hollow concrete slabs called 'termodeck', which was unfamiliar, and required some explanation. Eventually RMJM organised a visit to a similarly equipped building at the University of East Anglia (which then appeared on the "Reference Buildings" page), and the issue became quiescent. Later still, when they had measured the actual heat gains from computer equipment, they dropped this proposal in favour of 'chilled beam' cooling.

Provision for cyclists and catering have been continuing themes, perhaps not yet resolved to everybody's satisfaction. But the first major debate occurred when the Scheme 3 plans were released. This showed the building organised as a 'finger' plan; four ranges of research accommodation cross connected by a range containing the teaching and communal spaces, lecture halls and library. This was criticised on a number of grounds, such as difficulty of circulation for researchers, who would have to exit and reenter the secure areas to visit a lab in a different range, poor views out, overlooking from the lab opposite. The discussion culminated in a meeting, where the architect was encouraged to look at a courtyard alternative. In fact he looked at many more, which were not published, until Scheme 14 was reached, which had a figure-of-eight plan, with two courtyards. This dealt with the principle issues, and discussion moved on from the overall building form to detailed dispositions. Offices which were shown as 2 by 5 metres were felt to be too long and thin and were changed to a squarer form, a computer room was moved from the sunny south facade to the north, and the library which partly occupied a courtyard was relocated within a range.

Concerns have recently been raised about the wiring strategy, for both power and IT. This is receiving a good deal of design attention, and the issue for us is to find a way of publishing the result, so that everyone can see that it is resolved.

Lately, since the plan has settled, discussion has focused on the size of windows, which some feel are too small for adequate daylight, or views, or both. This is more difficult to resolve, as the window width is conditioned by the planning module, and the height by the structural system, though the interconnection between these issues has not surfaced until now. As usual, when a serious issue arises, the head of department has summarised the discussion to the architect, and asked him to consider it further. There the matter stands, at present.

3.3 Web site issues

We had some initial problems with comprehensibility of the DWF plans, where the vivid colours of the lines puzzled everybody. They appeared to be coding for materials, but enquiry revealed that they simply denoted the line thickness that would be used if the drawing was plotted. It is a point in favour of the PDF files that they show the lines as various thicknesses of black, as intended. The scale of drawings also puzzled people - although the structural grid spacing was dimensioned 6000, it was not appreciated that this signified a dimension in millimetres. During the discussion on windows, and on many other occasions, we would like to have been able to take measurements off the web plans and sections. None of the viewers allow for this elementary function. We also find the ability to print all or some of any drawing to be essential. It is a major drawback of the freeware Java DWF viewer that it does not allow for this.

On the 3d side, the Quake version has been widely appreciated, both by the Computer Lab and the architect. Significant suggestions for improving it include finding a way to label spaces, so you know which room, range and floor you are on, and provision of a keyplan for orientation, and possibly for 'jumping' your viewpoint.

A deeper problem has been raised by the Computer Lab. They have a growing archive of correspondence, reports and minutes of decisions, which they would like to hold electronically, in the form of a cross-referenced, indexed and searchable archive. We are currently considering how best to accommodate this.

Another issue for the future concerns the topology of our web site. At present it is a shallow tree, with large drawings or documents as the leaves. We envisage that eventually we will have multiple representations, such as plan, schedules and VR for many or possibly all rooms in the building. If these are fully cross-linked, the topology will become a dense mesh, which would be flying in the face of most web usability studies, which suggest that shallow trees are desirable for comprehensibility.

Finally, we must remark on the most fundamental of our tools, HTML. We started off using an editor (Adobe PageMill), in the hope that we could avoid coding HTML directly. This was rapidly found to be an illusion. Testing our pages in Netscape on Linux, PC and Macintosh, and Internet Explorer on the latter two, we found inconsistencies and failures that could only be disentangled by examining and modifying the HTML. From that, it was but a short step to composing HTML directly. We might still use an editor for the initial formatting of a complex table, but from then on it will be maintained by direct editing the HTML. And HTML itself is a monster. As a graphic design tool, it makes a point of failing to control layout and appearance, which is what a graphic designer cares about more than anything. As a language, it reeks of the bad old days, when the GOTO was considered essential and spaghetti code an art form. Its lack of defined semantics guarantees inconsistent implementation, an opportunity joyfully seized upon by the browser vendors and enlarged by their incompatible extensions. It defies modularisation and maintainability, lacking even the elementary notion of a subroutine. If there was ever a need for the standards organisations to save us consumers from the vendors, this is it.

4. ACKNOWLEDGEMENTS

This project was inspired by Professor Mike Gordon of the Computer Laboratory, and made possible by the continued enthusiasm of the whole project team, but especially Derek McAuley (Microsoft Research), Richard Rhodes (Estate Management and Building Services), Professor Robin Milner (head of the Computer Laboratory), Geoff Cohen and Ray Stock (RMJM London Ltd). Liz Pride (MacCormac Jamieson Prichard) has been invaluable in providing masterplan information. Sebastian Macmillan at the Martin Centre is working with us on measuring the effectiveness of the site. The project is funded by a Grant from the Engineering and Physical Science Research Council under the Innovative Manufacturing Initiative, with industrial support from Microsoft Research, Bentley Microsystems, Informatix Software International, and Lightwork Design Ltd.

Urban Texture Analysis with Image Processing Techniques

Carlo Ratti and Paul Richens
The Martin Centre for Architectural and Urban Studies, University of Cambridge. UK

Key words: urban texture, image processing, urban environmental analysis, urban morphology, cellular automata

Abstract: A new paradigm for investigating the environmental consequences of urban texture is proposed. Using raster-based models and software algorithms derived from image processing, efficient methods of measuring geometric parameters and predicting radiation exchange are developed. The possibilities of generating synthetic urban textures, and integrating cellular automata, are explored. Results suggest the possibility of a raster-based urban model to inform planning and design.

1. INTRODUCTION

Over the past decades, many computer tools have been developed to assist in the environmental design of individual buildings. Heat, light, sound and especially energy consumption can be analysed in many different packages.

This is not generally true for urban design, especially at the medium scale. Although it is widely assumed that urban texture - the pattern of streets, building heights, open spaces and so on – will determine environmental quality both in the buildings and outside – tools for investigating the connections are sparse. The need for medium-scale understanding is confirmed by Givoni (1989): "The outdoor temperature, wind speed and solar radiation to which an individual building is exposed is not the regional 'synoptic' climate, but the local microclimate as modified by the 'structure' of the city, mainly of the neighbourhood where the building is located".

Apart from Hillier's (1996) Space-Syntax techniques for investigating urban configuration, there is a lack of tools to connect urban texture to the resulting urban quality. This is partly due to difficulties in describing the urban environment. The representation of the city needs "a geometry of order on many scales, a geometry of organised complexities" - state Batty and Lonely (1994), subsequently arguing for the replacement of traditional Euclidean geometry with Fractal geometry. Extending architectural analysis software to the urban scale fails because of the problems of building precise geometric models, and the difficulties of processing, as the time is likely to depend on the square of the number of geometric elements.

Our interest is in using very simple raster models of urban form. By borrowing techniques from image processing, we have been able to develop a range of promising algorithms, whose processing time is independent of geometric complexity, and relates linearly to the area under investigation

1.1 Relation to previous work

The use of image processing for urban analysis was proposed by Richens (1997), who implemented a number of simple algorithms using the public domain Macintosh software NIH Image (available on the Internet at http://rsb.info.nih.gov/nih-image/). This was used by Steemers et al. (1997) for analysis of directional porosity in cities, which they related to movement of pollution. Though NIH Image is very effective, it is restricted to a single byte of information at each pixel, and its programming language, a kind of truncated Pascal, is inadequate. In the present work we extend our image processing environment to include Matlab, a well known package for matrix manipulation which includes an Image Processing Toolbox and begin to consider further issues, such as the problem of urban texture synthesis, and the use of cellular automaton simulations of diffusion-like processes.

2. IMAGE PROCESSING BASICS

The simplest possible representation of urban texture is a figure-ground map such as that shown in Figure 1. If scanned, the resulting binary image tells us, for each pixel, whether it is built on, or open ground. If height information is included we obtain what geographers call a Digital Elevation Model (DEM), which is an image where each pixel has a grey-level proportional to the level of the urban surface. A DEM of an urban area in central London is shown in Figure 2.

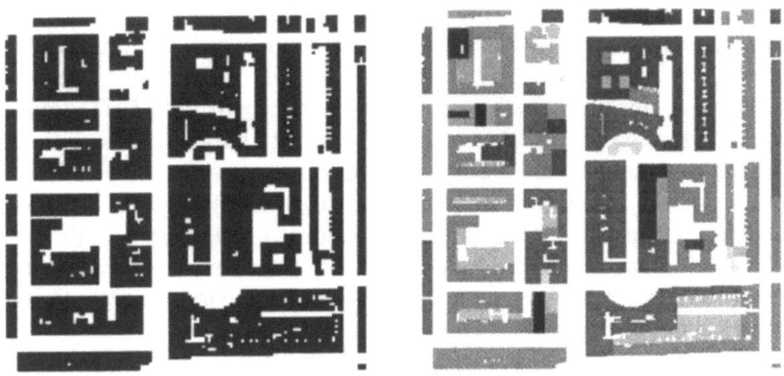

Figure 1,2. Urban ground map showing built and unbuilt areas (1) and urban Digital Elevation Model – DEM (2) for a selected case-study in central London

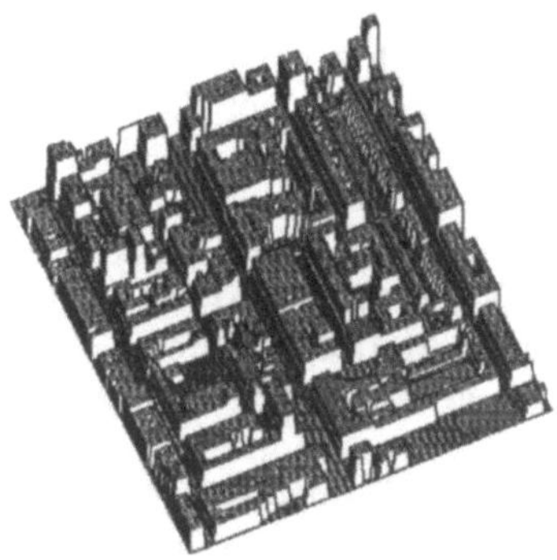

Figure 3. Axonometric view of the central London DEM

Urban form represented as a DEM can be analysed with image processing techniques using a simple package like NIH Image. Many of the ideas in Image should be familiar from popular programs such as Adobe Photoshop. The difference is one of emphasis: Photoshop is concerned mainly with visual appearance, whereas Image concentrates on measurement and analysis.

The basic data structure is the image, a 2D array of pixels, each storing a single byte (8 bits), typically understood as an integer in the range 0..255. Closely associated with the image, is its LUT (Look up table), which translates each of these integers into a colour that is displayed on the screen. The LUT can be freely changed, changing the appearance of the image, without damaging the underlying data stored in each pixel. Image and LUT can be written to disk as a unit, typically as a TIFF file. Images usually originate with a scanner; video frames can be captured directly if the computer has an AV board. Images can be originated or edited on-screen using the usual paint-box tools for line-drawing, selection, fill, brushing and erasing.

A large part of the functionality in NIH is aimed at image enhancement, using two major techniques. The first is LUT modification, which can alter the brightness and contrast, or even apply false colour to the pixel values. For instance, thresholding or density slicing can be performed by LUT modification. Thresholding turns the grey-scale image into a black and white one, with interactive control of the threshold value. Density slicing turns a narrow range of pixel values into another colour, again with interactive control. Dragging the slice through the range gives a precise way of exploring the values in the image. Both thresholding and density slicing are performed by LUT modification, but it is possible to force the changes into the image itself, which then becomes a binary image with values of 0 (false, white) and 255 (true, black) only.

The second method of image enhancement is by applying filters. The fundamental idea behind image filters is the kernel, a small window or region of pixels (say 3 by 3 or 5 by 5). The filter operates by centring the kernel on each pixel in the image in turn, and performing some computation on the values seen under the kernel. The central pixel is replaced by the result of the computation. So for example the Max filter replaces each pixel by the maximum value in its 3 by 3 neighbourhood, and the Median filter replaces each pixel by the median (central) value of its neighbourhood (which is a popular way of removing noise from an image).

Convolution filters work by forming a weighted average of the pixels under the kernel; the kernel cells specify the weights. If the weights are all positive, the result is some sort of smoothing or blurring. More sophisticated kernels, with some negative weights, can sharpen an image, differentiate it, or detect edges.

NIH Image provides some all-important features for doing arithmetic on complete images, for example by adding or multiplying corresponding pixels, or taking the maximum or minimum values. The result is a complete new image. In general these operations are computed at higher precision, scaled and offset by a stated amount, and then clipped to the 0..255 range allowed in each 8-bit pixel

3. BUILT FORM STATISTICS

Given a simply constructed DEM, Image was able to measure most of the usual built form statistics. Histograms of building height, arbitrary cross-sections, and axonometric representations (Figure 3) are standard commands. A very simple macro to measure the whole image by summing the pixel values (with suitable calibration corrections) gives the built volume. We obtain the built area by thresholding the DEM at $z > 0$, and counting. This works because the ground profile in the sample areas is essentially flat, and at zero throughout. If this were not the case we could use a background removal function to find ground level, which could be subtracted from the image before measuring areas and volumes. By using thresholding to extract an image at any particular level, we can repeat area measurements at any horizontal cross-section.

Measuring external wall area (or the thresholded equivalent, perimeter length), proved more challenging. There is a command to extract edge pixels from the image, which seems a good start. However, summing them and multiplying by the scale does not necessarily give the right answer. Walls aligned to the axes measure correctly, but walls at 45 degrees measure short, by a factor of $\sqrt{2}$.

A more accurate answer can be obtained by using a standard image-processing technique, the Sobel edge-detector. The image is convolved separately with two kernels:

$$\begin{matrix} 1 & 2 & 1 \\ 0 & 0 & 0 \\ -1 & -2 & -1 \end{matrix} \qquad \begin{matrix} -1 & 0 & 1 \\ -2 & 0 & 2 \\ -1 & 0 & 1 \end{matrix}$$

The first constructs the derivative in the y direction f_y, the second the derivative in the x direction f_x. These derivatives are equal to eight times the projected area of the wall in each direction. The total area is therefore obtained by forming the square-root of the sum of the squares of the two derivatives. If the convolutions are carried out separately, it is also possible to use the results to classify edges by their orientation computed as $atan(f_y/f_x)$.

Deep-plan areas – those at a certain distance from the perimeter - which are significant consumers of energy (as they must have artificial light and ventilation), can be displayed using density-slicing to extract the level at which to work, and then applying a standard image erosion function which strips away edge pixels to leave only the deep core.

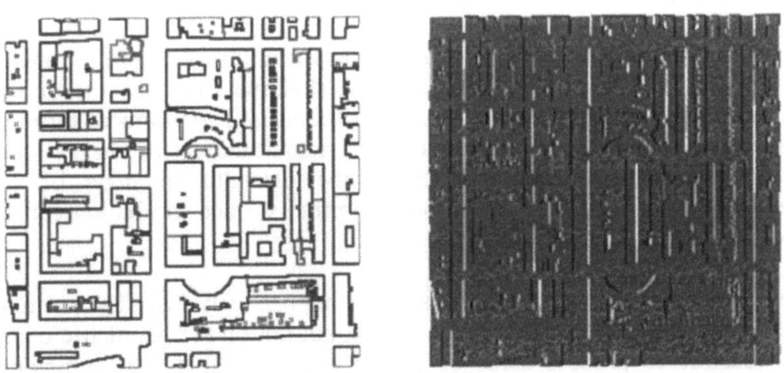

Figure 4,5. Edge (4) and east/west façade (5) detection on the central London DEM

4. URBAN ENVIRONMENTAL ANALYSIS

4.1 Shadow casting algorithm

Daylight and sunlight are fundamental constituents of the microclimate. Surprisingly effective algorithms can be devised, using image processing macros, to compute them from DEMs.

Our central algorithm (Richens, 1997) is one to compute shadows for an arbitrary angle of lighting. The approach taken is to compute shadow volumes as a DEM, that is, the upper surface of the volume of air that is in shadow. We start by defining the three components of the vector pointing towards the sun. Then we compute the components of an opposite vector, scaled so that the larger of the x and y components is just 1 pixel, and the z component is adjusted to the image calibration. If we translate the DEM by the x and y components, and simultaneously reduce its height by subtracting the z component, we get part of the shadow volume. If we continue translating and lowering by multiples of this vector, and take the maximum of this volume with that previously calculated, we build up the whole shadow volume. The process can be stopped when all levels are zero, or the translation has shifted the volume right off the image. This is easily encoded as an image processing macro. The results are shown in Figure 6.

To reduce the shadow volume to an actual map of shadows on the roofs and ground of the city, the original DEM is subtracted from the shadow volume. Pixels with negative or zero value are in light, positive values are in shade.

This calculation computes shadows for an arbitrary lighting angle; it is straightforward to add a procedure to calculate shadows from the sun for any given latitude, time of year, and time of day, using the usual astronomical formulae. The next stage is to calculate solar irradiation, the actual amount of energy received at each pixel, from the sun, taking into account the solar altitude, angles of incidence, and shadowing. Computing the clear-beam normal surface irradiation can follow a standard method – we used that in Page (1986).

These algorithms are simple to write, and impressively fast. The running speed depends on the number of pixels in the image, but not on the geometric complexity. So it is feasible to process acres of city at a time, something unthinkable with traditional geometric models.

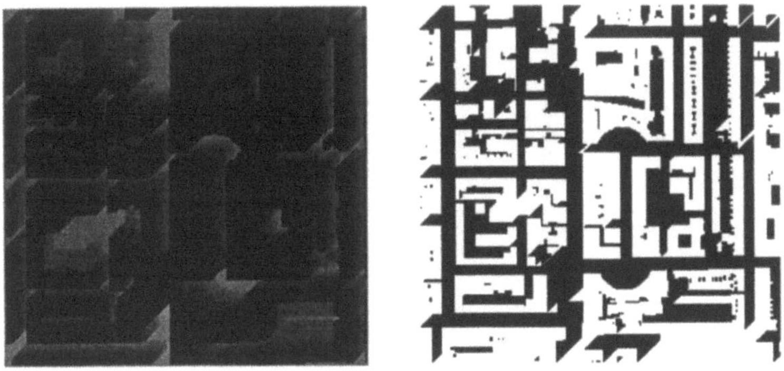

Figure 6,7. Shadow volume (6) and shadow results (7) on the central London DEM

4.2 View factors

The view factor is a geometric parameter that is related to many urban environmental processes. Also called shape factor, form factor or configuration factor, it was originally introduced in heat transfer theory to model radiant exchange between diffuse surfaces. Later, it has been successfully applied to urban climatology (Oke, 1988).

In particular, the view factor from the city to the sky, which represents the openness of the urban texture to the sky, takes part in three major environmental processes: the loss of long wave radiation from city to sky, responsible for cooling down of the urban surface during the night; the absorption of solar short wave radiation causing daytime heating; and the illumination received on a horizontal plane from an overcast sky, which determines daylight potential.

Several methods have been suggested in urban climatology for deriving the view factor from the city to the sky. Glenn and Watson (1984) derived an analytical expression to determine the sky view factor for different urban canyon arrangements. Steyn (1980) suggested a graphical method based on fish-eye lens photographs.

We calculate the view factor by repeated application of the shadow algorithm. We simply compute the shadows for a large number of light sources, distributed over the sky, and for each pixel count the number of times they are in light. So if we use 255 samples, any pixel whose count is 255 can see all the sky and has a sky view factor 1, while a count of 0 means that it cannot see the sky at all.

To get meaningful results, it is necessary to distribute the sample points over the sky in the correct manner. If a uniform distribution is used, then what is measured is not the view factor but the solid-angle of sky visible from each point. However, if we distribute our points evenly over a unit circle in the horizontal plane, and then project up to a unit hemisphere, we obtain a cosine-weighted distribution, which is the correct distribution for computing the view factor from the city to the sky.

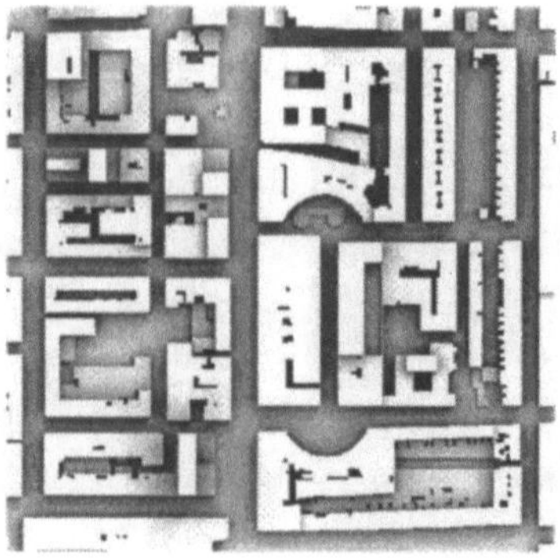

Figure 8. Sky view factors on the central London DEM

4.3 Informing urban design: an example

An example of how the above techniques could be used to inform urban design is given here. A more complete review can be found in Steemers and Ratti (1998, 1999).

The Martin Centre (University of Cambridge, UK) and Cambridge Architectural Research (Cambridge, UK) acted as research consultants for an urban development in Greenwich (London) by the Richard Rogers Partnership. The image processing analysis of DEMs was used to assess the environmental potential of different design schemes.

Most of the analysis focused on the comparison of daylight and sunlight conditions on six urban options (Figure 9). For each option images with shadowing contours were produced, showing on each pixel the number of hours of shadow (Figure 10). A simplified parameter, the mean shadow density, was determined for each image. This parameter can vary by approximately 19% in winter and 21% in summer, following the geometry of the development. Furthermore, the summer maximum does not coincide necessarily with the winter one. It was therefore possible to minimise shadowing in wintertime, when the sun is beneficial, while maximising shadowing in summertime.

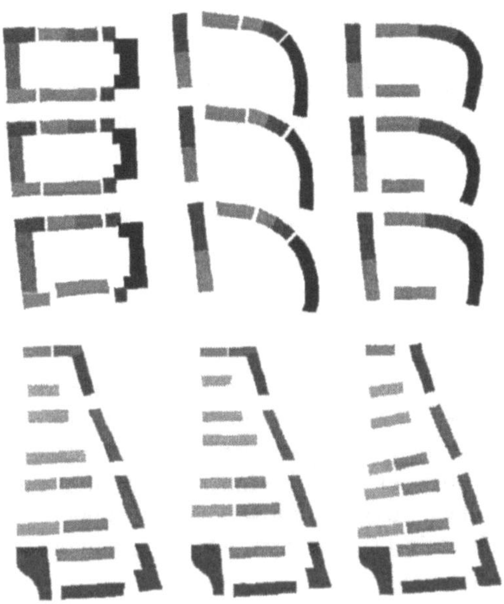

Figure 9. Sketches of development options for the Greenwich peninsula (Richard Rogers Architects Limited)

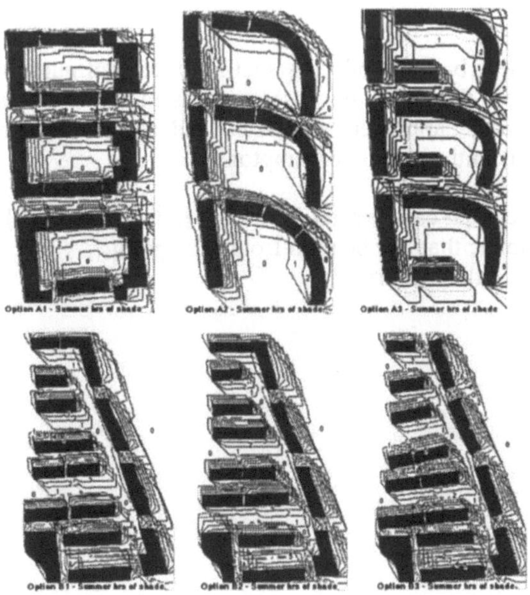

Figure10. Hours of shadows during summer for each development option presented above

5. MORPHOLOGICAL ANALYSIS

The simple image processing capability of NIH Image has limitations. The value associated with each pixel is a single byte, and the macro language, a kind of debased Pascal, is inflexible. For more advanced work, therefore, we have used Matlab, which gives us a double-precision variable at each pixel, and a complete matrix-oriented programming language. We find this to be much more flexible, though sometimes slower, and much more memory intensive, than Image.

5.1 Urban Ground view factor

Peponis et al. (1998) proposed an extension to the theory of Space Syntax (Hillier and Hanson 1984; Hillier 1996) analysing the visible relations between the surfaces that bound and subdivide space. The study of plan configuration according to the covisibility of surfaces, and the introduction of parameters such as visual connectivity would, they suggest, allow a more subtle investigation of spatial configuration than Hillier's original Space Syntax, which is mainly based on the topology of circulation spaces.

We have found DEM techniques highly adapted to the analysis of these measures. We defined a parameter called the urban ground view factor and computed its value on each pixel by modifying the above macro for the sky view factor (4.2 above). The urban ground view factor is a measure of the extension of the view of the ground from each urban pixel and directly relates to the strong visibility polygon (O'Rourke, 1987, Peponis et al., 1998). Results on the selected London case-study are shown in Figure 11.

In addition to the urban ground view factor, other configurational characteristics can be calculated from the DEM. For instance, it is easy to write Matlab functions to compute the following parameters suggested and applied by Batty (1999) to the analysis of the town centre in Wolverhampton:

- the mean distance that can be seen from each point, called visibility;
- the furthest distance that can be seen from each point;
- the angle at which this occurs.

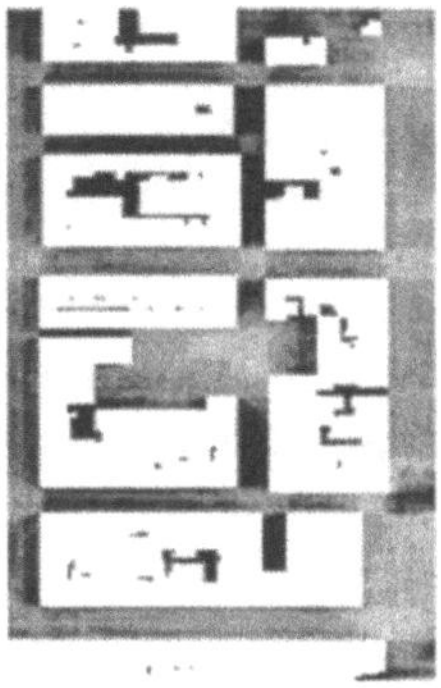

Figure11. Urban ground view factors on a selected portion of the central London DEM

5.2 Synthetic cities

> "But right there between prediction and game playing, between science and science fiction, lies the realm of sharpened intuition, informed speculation, and educated guess".

Here Coucleis (1997) refers to the significance of computer simulation of urban growth, using cellular automata. Many other workers have tried to find algorithms for growing city-like structures. Fractal forgery has been used at the very large, almost regional, scale (Batty, 1994). Multi-agent algorithms have been used to investigate the patterns of urban growth and to generate "fictitious cities" (Cecchini and Viola, 1992).

Our own approach, which aims at synthesising textural details down to a scale of about 1m, corresponding to the resolution of our DEMs, is derived from recent advances in texture replication in the field of computer graphics. This work is at a preliminary stage, but is not without promise.

In image synthesis, textures (such as brickwork) are often represented by a sample tile, perhaps a few hundred pixels square. If a large brick wall is to be imaged, then the tile is repeated. Unfortunately, the eye is very receptive to the regularity of the resulting pattern, and it does not look like a brick wall any more. The problem is to synthesise a larger area of brickwork, which "looks the same" as the original sample, but does not involve exact repetitions.

This problem eluded solution until the algorithm of Bonet (1997), which improves on the work of Heeger and Bergen (1995). Both approaches use a Laplacian pyramid to analyse the sample texture at different scales. This is constructed, roughly speaking, by repeatedly blurring the image, and halving its resolution. The Laplacian pyramid (Figure 12) consists of the differences between successive blurred images, and is thus in some sense a kind of spectral analysis of the image, dividing it up into different visual frequency ranges.

Bonet's algorithm uses this analysis to generate a new pyramid, corresponding to the output texture, by a kind of constrained random sampling. The input pyramid provides joint feature occurrences across multiple scales, and is used as a probability density estimator to constrain sampling. So the output texture contains recognisable fragments of the input texture, placed in similar but not identical contexts.

We implemented a version of this algorithm in Matlab, to operate on urban textures represented by DEMs. Given a small piece of city, it synthesises a larger piece. The result, shown in Figures 13 and 14, is surprisingly good, given that the algorithm has no knowledge of urban processes. It preserves the general structure and orientation of the street pattern, and the mix of building heights. Features up to the size of a single block can probably be found in the original image, large assemblies probably not. It is remarkably, though not totally, free of unurban conditions like buildings which block major streets.

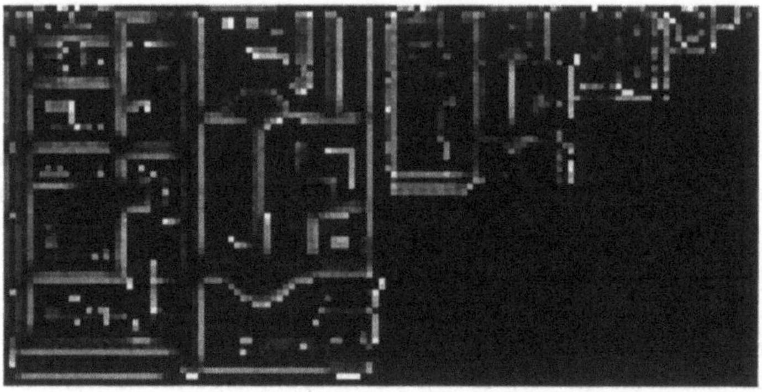

Figure 12. Laplacian pyramid of the central London DEM

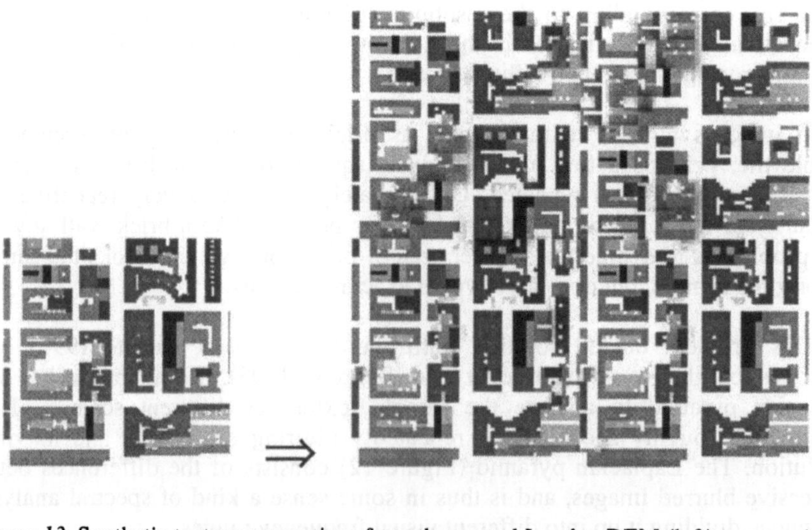

Figure 13. Synthetic texture generation: from the central London DEM (left) a new urban portion with similar textural characteristics is produced (right)

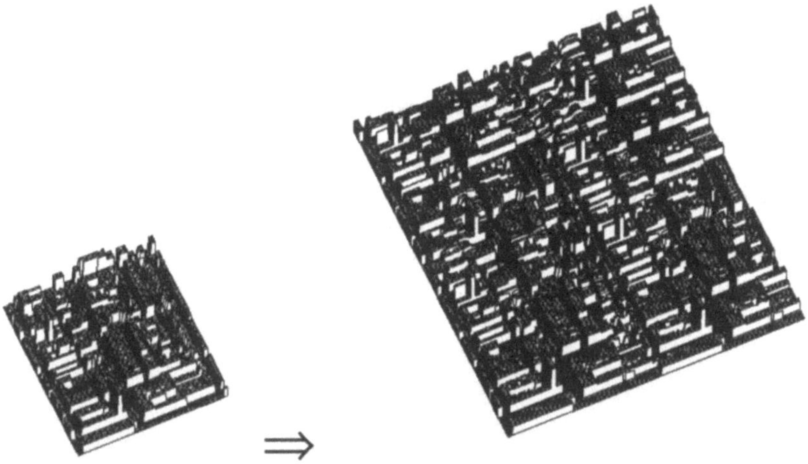

Figure 14. Axonometric view of Figure 8

6. DEM AND CELLULAR AUTOMATA

‘Cellular automata are mathematical idealisations of physical systems in which space and time are discrete, and physical quantities take on a finite set of discrete values. A cellular automaton consists of a regular uniform lattice (or 'array'), usually infinite in extent, with a discrete variable at each site ('cell'). The state of a cellular automaton is completely specified by the values of the variables at each site. A cellular automaton evolves in discrete time steps, with the value of the variable at one site being affected by the values of variables at sites in its 'neighbourhood' on the previous time step. The neighbourhood of a site is typically taken to be the site itself and all immediately adjacent sites. The variables at each site are updated simultaneously ('synchronously'), based on the values of the variables in their neighbourhood at the preceding time step, and according to a definite set of 'local rules" (Wolfram, 1983).

Although CA have mainly been used in urban studies to simulate spatial dynamics (Batty et al., 1997), we will consider here their applications in modelling small-scale processes within the urban texture.

The physics of pollution dispersal, wind flow and noise diffusion can be studied by CA. CA have also been very successful in modelling complex situations for which traditional computer techniques are not applicable, such as flows in porous media, erosion and transport problems, etc. (Chopard and Droz, 1998).

In addition to the modelling of physical processes, CA have been used to simulate the local action of individual actors or agents. Batty (1999) applied CA to model how pedestrians move in town centres and buildings. More generally, agent-based models may constitute the base of built space analysis complementary to Space Syntax.

CA techniques are highly compatible with DEMs. The DEM, a raster map, is the natural format to input urban geometry into a CA. Operation on cells can simply be viewed as operations on pixels. Image processing and CA can therefore be integrated into a single body of analysis techniques to work with DEMs. As an example, a CA macro has been written in Matlab to simulate diffusion of pollutants within the urban texture. Results are shown in Figure 15: a steady source is positioned at the bottom of the image, while pollutant particles migrate upwards.

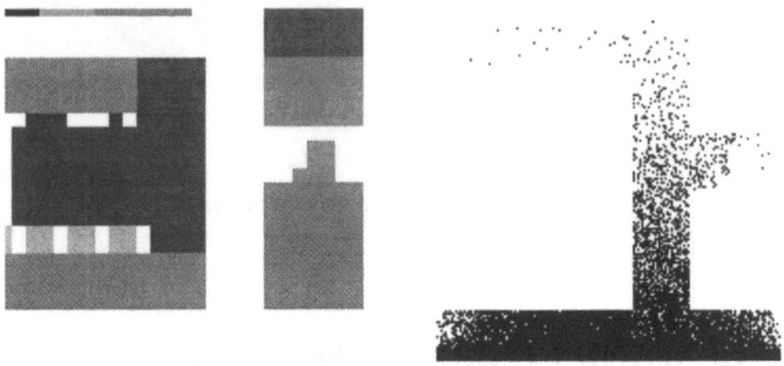

Figure 25. Modelling diffusion on a selected portion of the central London case-study (left) using cellular automata (right)

7. CONCLUSIONS AND FUTURE WORK

DEMs have proved to be an unexpectedly effective format for representing the surface of the city. The innovative use of image processing and CA techniques allows the computation of a wide range of environmental and morphological parameters in urban areas.

A great deal more could be done to extend the modelling capability, using multiple images, rather than simply a DEM. If the ground surface is complicated, separate ground and building DEMs would be necessary. Additional channels could be used to classify surface types (vegetation, water, roads, buildings), and to provide more information about buildings, of which the number of stories would be most valuable, followed by information about glazing, albedo, and materials generally.

Raster urban data should be increasingly available at low cost in the forthcoming years. DEM can already be produced with semi-automatic digital photogrammetry and Synthetic Aperture Radar (SAR) techniques. Additional surface data in urban areas can easily be obtained from remote sensing. The analysis of DEMs could constitute the basis for an integrated image-based urban model to inform planning and design.

8. ACKNOWLEDGEMENTS

The research was partly funded by the European Commission through Project PRECIS (Assessing the Potential for Renewable Energy in Cities). Acknowledgements go to Dr. Nick Baker for his suggestions and many informative discussions; to Dr. Koen Steemers for supporting the application of image processing techniques to urban design; to the PRECIS team for its productive research environment.

9. REFERENCES

Batty M, Longley P, 1994, *Fractal Cities – A Geometry of Form and Function* (Academic Press, London)

Batty M, Coucleis H, Eichen M, 1997, "Urban Systems as Cellular Automata", *Environment and Planning B: Planning and Design,* 24 159-164

Batty M, 1999, "Visual fields", *http://www.casa.ucl.ac.uk/agent-homepage/visual-fields.htm*

Bonet J S, 1997, "Multiresolution Sampling Procedure for Analysis and Synthesis of Texture Images", *Proceedings Computer Graphics ACM SIGGRAPH,* 361-368

Cecchini A, Viola F, 1992, "Ficties (fictitious cities): a Simulation for the Creation of Cities", paper presented at the *International Seminar on Cellular Automata for Regional Analysis,* Venice, Italy; copy available from A. Cecchini, Laboratorio sulla Simulazione Stratema, Istituto Universitario di Architettura di Venezia, Ca' Tron 5, Croce 1957, 30135 Venice, Italy

Chopard B, Droz M, 1998, *Cellular Automata Modelling of Physical Systems* (Cambridge University Press, Cambridge)

Coucleis H, 1997, "From Cellular Automata to Urban Models: New-Principles for Model Development and Implementation", *Environment and Planning B: Planning and Design,* 24 165-174

Glenn T J, Watson I D, 1984, "The Determination of View-Factors in Urban Canyons", *Journal of Climate and Applied Meteorology,* 23 329-335

Givoni B, 1989, *Urban Design in Different Climates* (World Meteorological Organisation WMO-TD n. 346, Geneva)

Heeger D J, Bergen J R, 1995, ". Pyramid based texture analysis/synthesis", *Proceedings Computer Graphics ACM SIGGRAPH* 229-238

Hillier B, 1996, *Space is the Machine* (Cambridge University Press, Cambridge)

Hillier B., Hanson J., 1984, *The Social Logic of Space* (Cambridge University Press, Cambridge)

Jain A K, 1989, *Fundamentals of Digital Image Processing* (Prentice-Hall, London)

Johnson G T, Watson I D, 1984, "The Determination of View-Factors in Urban Canyons", *Journal of Climate and Applied Meteorology,* 23 329-335

Oke T, 1988, "Street Design and Urban Canopy Layer Climate", *Energy and Buildings,* 11 103-113

O'Rourke J, 1987, *Art Gallery Theorems and Algorithms* (Oxford University Press, Oxford)

Page J K (ed), 1986, *Prediction of Solar Radiation on Inclined Surfaces* (Reidel Publishing Company, Dordrecht)

Peponis J, Wineman J, Rashid M, Bafna S, Kim S H, 1998, "Describing Plan Configuration According to the Covisibility of Surfaces", *Environment and Planning B: Planning and Design,* 25 693-708

Richens P, 1997, "Image Processing for Urban Scale Environmental Modelling", *Proceedings of the International Conference Building Simulation '97 - Prague*, 163-171

Steemers K, Baker N, Crowther D, Dubiel J, Nikolopoulou M, Ratti C, 1997, "City Texture and Microclimate", *Urban Design Studies,* 3 25-49

Steemers K, Ratti C, 1998, "Nuovi Strumenti per la Progettazione Bioclimatica Urbana", *Housing* 9 98-100

Steemers K, Ratti C, 1999, "Informing Urban Bioclimatic Design", *To appear in the proceedings of the International Conference Architecture and Engineering - Plymouth*

Steyn D G, 1980, "The Calculation of View Factors from Fish-Eye Lens Photographs", *Atmosphere-Ocean* 18 254-258

Wolfram S, 1983, "The statistical mechanics of cellular automata", *Review of Modern Physics,* 55 601-643

A Web-centric CAD System for Collaborative Design

Raymond McCall
University of Colorado, Denver

Key words: net-centric, Web-centric, hypertext, hypermedia, CAD, HyperCAD, knowledge-based critics, collaboration, executable content, argumentative design, IBIS, PHI, design rationale

Abstract: Web-PHIDIAS is a hypermedia-based, intelligent CAD system that delivers both CAD functionality and design information to anyone with Web access. This system is above all designed to facilitate collaborative architectural design. It provides both private (individual) and collaborative (group) drawing and text authoring spaces, with a variety of types of authoring and viewing privileges for groups. This enables a single designer to work in privacy on one piece of a design and later "publish" it to a supervisor or a group. It also enables a group to work in privacy and later publish its work to a different or larger group. This notion of "levels of privacy/publication" is a crucial but too-often missing component of collaborative design systems. With Web-PHIDIAS, all drawings are stored in a central repository accessible from the group server. This means that they are accessible from anywhere in the world to any viewer who has the required viewing and/or authoring privileges. This enables designers to access and modify stored drawings while travelling or when out on the site, even if the site is in another country. It also enables them to create new drawings and store them in the central repository from anywhere in the world. Web-PHIDIAS consists of an interactive Web-based client that serves as an interface to the PHIDIAS hypermedia server. This client, which is implemented in Java, provides basic, 2D graphical editing functionality and as well as display of 3D views. It also provides access to multimedia information useful for whatever design task is at hand. This information includes text and graphical descriptions of design precedents as well as various issues in design of a particular type of building.

1. INTRODUCTION

1.1 Goals of our work

Our work has for more than two decades been aimed at the creation of hypertext systems that support an argumentative approach to design [Rittel 1972]. This approach treats difference of opinion as a positive, democratic force that can improve the quality of things designed by helping designers to explore a wider range of design issues, solution options and evaluation criteria. The argumentative approach is especially well suited to group design processes, both in the sense of collaboration in design teams and participation by clients and users in design projects.

To support the argumentative approach, Rittel originally proposed IBIS (Issue-Based Information Systems) [Kunz and Rittel 1970], a method—not a computer system—for eliciting and organizing design rationale. Our systems support PHI (Procedural Hierarchy of Issues), a method that attempts to improve on IBIS by making it more general and more scalable [McCall 1991]. PHI, like IBIS, organizes information around the deliberation of design questions, referred to as *issues*. Various *answers* are proposed to these issues, and *arguments* are then given on the answers and on other arguments. Where PHI differs from IBIS is above all in the way issues are linked together. While IBIS used a variety of inter-issue relationships, PHI uses only relationships that indicate ways in which the answering of one issue depends on the answers given to other issues. By concentrating on these *dependency relationships* between issues, PHI creates more orderly networks of issue-based discussion. This in turn enables the creation and management of issue networks that can easily be ten to a hundred times the size of typical IBIS networks.

We have sought to create software systems that support the argumentative approach. Our systems have included a range of technologies, including hypermedia, CAD graphics and knowledge-based computation; but they have not included the sort of support for collaborative design that the argumentative approach calls for. Until recently, creating such support was limited by the limited nature of networking technology for computing. In past few years we did develop a multi-user prototype that worked in local-area networks [McCall Argumentative Agents article in Construction Automation]. But this system did not support the work of highly distributed teams or participation-at-a-distance by users or clients.

In recent years the explosive growth of Internet technologies has dramatically changed the potential for computer supported collaboration. In particular, the World Wide Web, Java, and other tools for creating Web-based client-server applications enable us to support group design in a far more full-blown manner. This article describes the first prototypes of Web-PHIDIAS, a software system that uses such tools to support collaborative design. We begin by briefly describing our prior work on hypermedia-based CAD. We then describe four stages in which we modified our system to support the delivery of design information and group design over the Web. This includes a description of a computational mechanism that we have devised to support group work by providing different contexts for collaborative communication. Finally we describe our plans for future prototypes.

1.2 Our previous prototypes

1.2.1 Our first hypertext systems

Our work began with the creation of single-user, text-only hypertext systems to support argumentative discussion in design, using PHI as a more comprehensive and scalable version of IBIS. Our first systems, PROTOCOL [McCall 1979] and MIKROPLIS [McCall, Mistrik and Schuler 1983; McCall 1989] were intended to support the creation of project-specific databases of issue-based design rationale—or *issuebases*—from scratch. MIKROPLIS, in particular, was designed to eliminate most of the extensive and tedious "secretarial" work of recording and updating design rationale in an organized and retrievable fashion. Up to that time, it appeared that this "secretarial" work was so great that it was preventing effective implementation of both IBIS and PHI.

1.2.2 Domain-oriented issue bases

Extensive experience with use of MIKROPLIS by end users revealed that reducing the secretarial work of issue-based documentation was not enough to make it practical to develop project issuebases from scratch. MIKROPLIS did make it possible to develop issuebases many times the size of pervious issuebases; nevertheless there still remained far too much conceptual and editorial work to justify the creation of issuebases for all but the most important projects.

We therefore adopted the strategy of constructing *domain-oriented issuebases*, i.e., collections of issues, answers and arguments that recur frequently in a given problem domain—for example, the design of a particular type of building. Here, the enormous work of creating an issuebase could be justified, because a domain-oriented issuebases could be used in many projects—potentially all the projects in a given domain. Such an issuebase can "jump start" the creation of project-specific issuebases by enabling designers to create the rationale for a given project simply by adding comments and judgments to the generalized issuebase for the domain—a process we call *differential reasoning*. Thus, for example, a generalized issuebase for the domain of house design could easily be tailored to create an issuebase for design of a particular house on a particular site.

Domain-oriented issuebases have another advantage. They can provide designers with specialized information for the problem domain. They do this by serving as a sort of souped-up FAQ (Frequently Asked Questions) that provides useful information about a specialized domain. Over the years we have built extensive issuebases for a number of domains, including kitchen design, house design, layout of computer networks in buildings, and NASA's design of space-based habitats—plus one very large issuebase on health care policy.

1.2.3 Integrating CAD graphics: JANUS

By 1984 it had become clear that our text-based information systems would not be sufficient for architectural design. Support was also needed for graphical design and for ways to provide architects with information as they designed with graphics. Our next system, JANUS [McCall, Fischer and Morch CAADF 1989] presented the designer with two different faces (thus the name JANUS), one for CAD-graphic form development and the other for argumentative reflection on design decisions. To connect the graphical and argumentative aspects of the system, JANUS used knowledge-based critics that critiqued partially developed graphic solutions using rationale drawn from a domain-oriented issuebase. Schoen's theory of "reflection-in-action" [Schoen 1983] provided a useful theoretical framework for JANUS and all our subsequent systems. This theory sees design as a continual alternation between two types of thinking, which Schoen refers to as *knowing-in-action* and *reflection-in-action.* We interpreted these respectively as *form making* (using CAD graphics) and *argumentative deliberation* (using PHI-based hypermedia).

1.2.4 Intelligent HyperCAD

JANUS employed a heterogeneous system architecture, with three, loosely coupled major subsystems: a CAD system, a hypertext system and a knowledge-based system. Added on to these were a number of minor subsystems. As we added new functionality to JANUS, it tended to "grow like Topsy," becoming progressively more complex and, thus, more difficult to debug and extend. We therefore devised a simpler and more integral system architecture that we dubbed *Intelligent HyperCAD.* This offered the same basic threefold functionality of JANUS—i.e., CAD graphics, hypermedia browsing and knowledge-based computation—but did so using a common, computational substrate of fine-grained hypermedia. This architecture became the basis of PHIDIAS [McCall, Bennett and Johnson 1994]. In this system, all data, information and knowledge were represented as hyper-nodes and hyper-links. Complex graphical objects were represented as composite hyper-nodes, and knowledge-based computation was done by using the hypermedia network as a semantic network. A typical screen from PHIDIAS is shown in Figure 1.

Because of its integral architecture, PHIDIAS proved far easier to extend than JANUS. In particular, it became much easier to inter-relate and combine the major types of functionality to create new functionality. This greatly facilitated our efforts to create the context mechanism, described below, that our latest prototype uses to support collaborative communication.

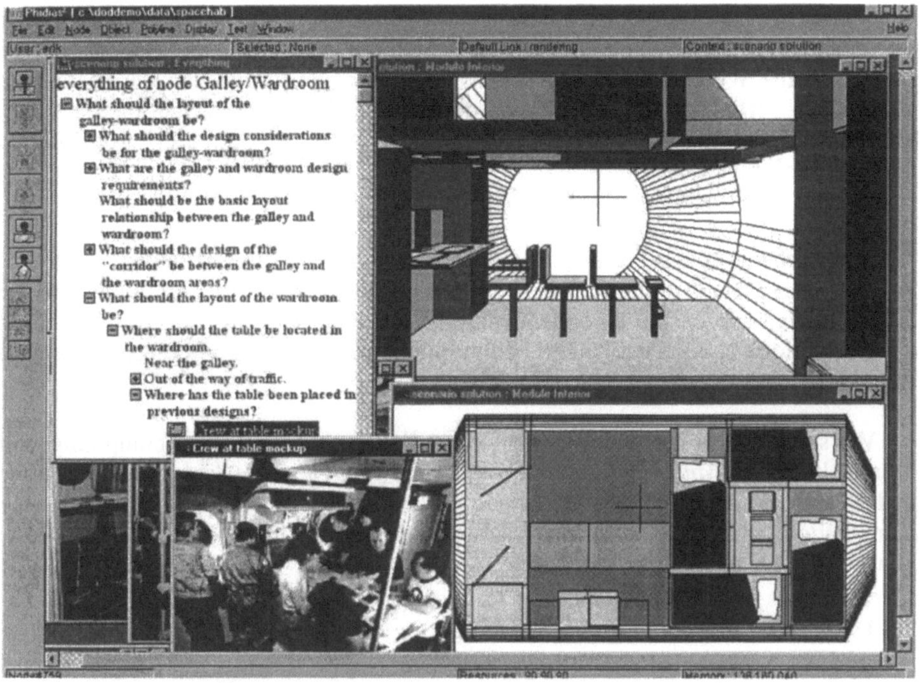

Figure 1. This screen of the stand-alone version of PHIDIAS shows its capabilities for displaying text, raster images and vector graphics. Text and vector graphics could also be created and edited with the system. Additional capabilities of PHIDIAS included knowledge-based critics, voice annotation and video display.

2. WEB-PHIDIAS

2.1 Significance of the Web for our project

The growth of the Internet and, in particular, of the World Wide Web is of enormous significance for our work. For one thing, the "gospel" of hypermedia that we and others had been preaching for years is now widely understood and accepted. As the Web first began to take shape, hypertext pioneer Ted Nelson—who coined the terms hypertext and hypermedia in the 1960s—was, as he put it, "promoted from 'mad man' to 'visionary.'" By the beginning of 1995 it was already clear that his then 30-year-old notion of a world-wide network of hyper-linked information on computers was not merely a visionary's futuristic dream; it was an explosively emerging reality. At the end of the century we find that hypermedia, in the form of the World Wide Web, has become a major cultural and economic force. The influence of the Web will continue to grow over the first decade of the 21st Century.

The success the Web made it easy for us (for the first time) to persuade people of the value of the hypermedia concepts on which our prototypes were based. At the same time the Web's success also caused many people to question the need for a system like PHIDIAS that has its own approach to hypermedia—an approach that differs from the Web in a number of ways. "Were we seriously proposing to compete with the Web?" they asked. Some argued that while the Web proved our point about hypermedia, it also effectively put us out of business.

For us, the issue was never how to compete with the Web, but rather how to use it—and other aspects of the Internet—to give a boost to our efforts to implement the argumentative approach. To understand how the Web helps us achieve our goals it is useful to look at what networked computing is likely to offer in the way of general future functionality.

We subscribe to the widely repeated prediction that networked computing will ultimately offer us *anything, anytime, anywhere* (AAA), where "anything" should be understood to mean 1) any information, 2) any communication, and 3) any computation. Let us look at what each of these means and then relate this to the goals of the argumentative approach.

2.1.1 Any information

This means that whatever information we need we will be able to find it whenever we want and from wherever we happen to be. Access to a vast amount of high-quality information has emerged as one of the first and most obvious things that the Web offers us. The bandwidth limitations we now experience will disappear over the next decade. The quantity and quality of information available on the Web will increase greatly. It is not inconceivable that all documented human knowledge will be accessible online within a decade.

Significance for the PHIDIAS project: A goal of the argumentative approach is to improve design–e.g., improve the quality of buildings—by providing designers with useful information. To do this, PHIDAS could use the Web/Internet as a vehicle for distributing information. Secondly, PHIDIAS could be use information that is on the Web to augment its own information. Among other things, PHIDIAS can serve as a means for finding and organizing the design-related information that is distributed on the Web. Perhaps the hallmark of information on the Web is that is comes from a plurality of points of view; so the fit between PHIDIAS and the Web is in this respect very good.

2.1.2 Any communication

This means that we will be able to communicate with any person we want to, whenever we want to and from wherever we want. Internet-based communications is headed in this direction with email, newsgroups, interactive "chat", instant messaging, and so forth. As wireless communication—including satellite communication—and various portable computing devices proliferate over the next decade, it will become increasingly possible to communicate with any person in any location and at any time—at least asynchronously. Communication for collaborative work will be ubiquitous—so much so that the notion of *groupware* will likely fade away as all software becomes *group-aware*, i.e., comes with collaborative communication capabilities built in.

Significance for the PHIDIAS project: Collaborative design communication is the central idea of the argumentative approach. Documenting and managing such communication are what IBIS and PHI were created for. Documenting design rationale has proved very difficult to do in the past, because it required extra work from designers. But it can happen easily in the future by being made a side-effect of routing design communication through computers; once the text, voice, drawings and video used for computer-mediated communication are in digital form, we only need to keep persistent records of them. PHIDIAS can use this fact to better capture and distribute design rationale. Collaborative design using networked computers is going to create highly detailed records of design thinking for the first time in history. We will need systems like PHIDIAS to manage the deluge of data that will inevitably result.

2.1.3 Any computation

This has two related meanings: one for hardware and one for software. On the hardware side, it means that whatever hardware resources we need to do computation will be available to us wherever and whenever we need them. The computer that we use to access the Internet will not be the only one we use for computation. We already see this trend in the movement toward Web-based client-server systems. These appeared first in the form of database systems that manage large Web sites. Next came "personalized" Web sites with selected news, email and calendar functionality—all with content tailored to the individual users and having "pages" that are computed "on the fly" at display time. A host of new technologies is waiting in the wings to give users additional access to the computational capabilities of other computers, including CORBA, RMI, ASP, JSP, and servlets. In addition, currently experimental uses of Internet-networked computers as distributed supercomputers could also accelerate dramatically—turning Sun's motto, "The network is the computer" into a reality.

On the software side, "any computation" mostly refers to *executable content*, i.e., content of Web pages that is not merely text and pictures, but is interactive software. Whether implemented using Java, ActiveX, Dynamic HTML, or some as-yet-unimagined technology, the trend toward Web pages whose content executes on the user's (client side) computer seems certain to accelerate. Limitations of current computers and network technologies now impede use of executable content, but the advances of the next decade will change this situation dramatically. Ultimately users will have *software on demand*; any type of software the user might want would be available on the World Wide Web for immediate use without any complex installation process. "Software" might well become a type of *service* rather than a type of *product*. The notion of going to a store to buy software in a bubble-pack and then struggling to install it on a hard disk might in the near future be considered a bizarre ritual from the Stone Ages of computing.

In summary, nearly any type of software and hardware capabilities that any user could need will become available for use over the Internet. The limitations of the desktop computer and its installed software will in no way limit the computation that users can do.

Significance for the PHIDIAS project: The argumentative approach is most easily and strongly supported when collaborative design is supported. The best way to enable collaboration in design by a wide range of people is to eliminate distance as a barrier to participation. This can be done by providing an interactive interface to a collaborative design system that is available over the Internet—e.g., as executable content in a Web page. Such an interface must provide communication with other participants in the design project—regardless of their number or location; and this can best be done by means of a central server that acts as a switchboard for all communications. Using a "thin client" strategy relieves the client computers of computational burdens and therefore increases the types—and thus the number—of Internet devices that can be used as means for participation in design. This server-based, thin-client strategy exploits both the hardware and software aspects of "any computation."

Using the server-and-thin-client strategy, the PHIDIAS functionality—including CAD graphic editing and display, knowledge-based critiquing, information retrieval (with task-based indexing)—can be made available to every desirable participant in a given design project—from anywhere in the world they happen to be, from a wide range of Internet-aware devices, and at any time of night or day that they wish to participate. It will possible to use a library computer, a Web-enabled television or even a PDA to access project drawings, modify them, store them in the central PHIDIAS database and send them to any of the members of the design team or any clients or users. Collaborative design will be possible anytime and anywhere.

2.2 Four stages of implementing Web-PHIDIAS

2.2.1 Stage 1: Putting NASA's MSIS online using Web-PHIDIAS

Our first opportunity for exploring the transition to a Web-based version of PHIDIAS came from our work with NASA. Two years ago, people at the Flight Crew Support Division of the Johnson Space Center became convinced that they needed to put their guidelines and requirements for design of space-based habitats on a Web site, where they would be available to NASA contractors and others who might be interested. These guidelines and requirements are collected in a document known as the Man-Systems Integration Standards (MSIS) and currently take up somewhat less than a thousand pages of paper per version, with a separate version of being created for each new habitat. Thus, the generalized MSIS version is labelled NASA STD-3000, while the specific version for the International Space Station is labelled SSP 50005.

The MSIS was originally created and is still maintained using a desktop publishing program. Since this software can automatically produce an HTML version of any document it creates, it might seem to be a simple task to produce a reasonable Web site containing the MSIS. In fact, for a variety of technical reasons, the NASA staff found it to be impossible. Since no commercial software seemed to offer the needed functionality, the PHIDIAS team—i.e., the author and several of his former students—was therefore hired to create a Web-based MSIS using PHIDIAS. We succeeded by using an approach of first translating the MSIS data into PHIDIAS' proprietary format and then using PHIDIAS as a hypermedia-database engine that created Web pages "on the fly." Creating a manageable and malleable Web site for long, hierarchically constructed documents such as the MSIS is not a simple task, but it is one for which PHIDIAS is especially well suited. The HTML pages produced by PHIDIAS are simple, but the task of managing MSIS information, which is hyper-linked both hierarchically and cross-referentially, is not. Storing information as separate Web pages is not a good approach when that information is updated frequently. The PHIDIAS approach stores information on the level of the paragraph. Figure 2 shows a screen image of the Web-PHIDIAS version of the MSIS.

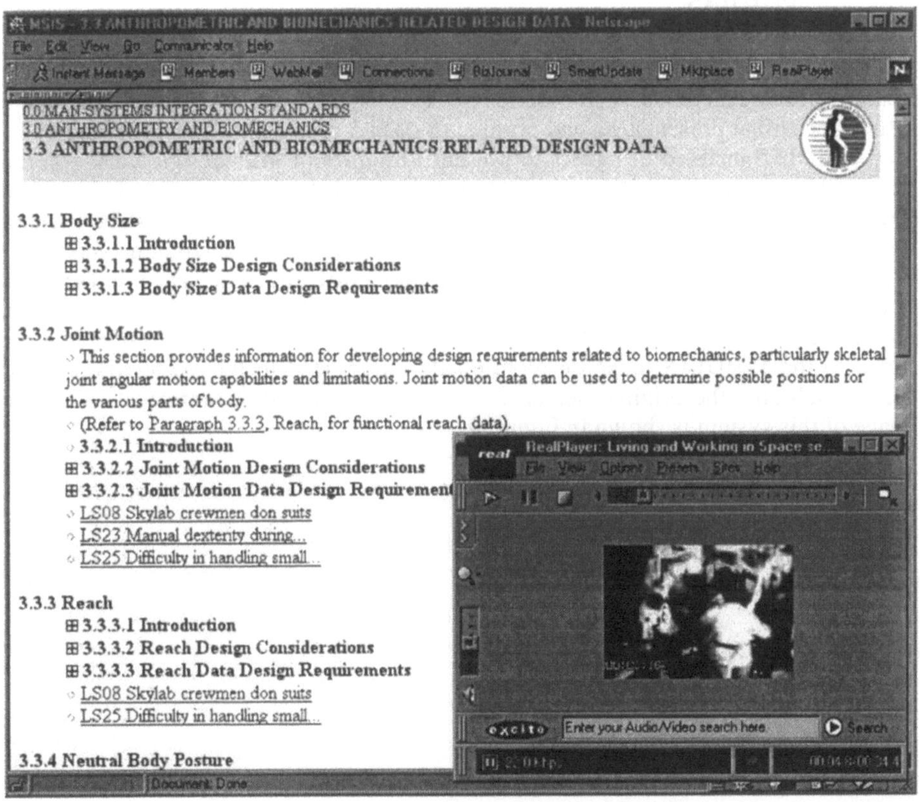

Figure 2. A page from the Web-based MSIS is shown here with one of its many streaming video clips. The Web pages are generated on demand by Web-PHIDIAS from its internal model of the MSIS document. The "plus buttons" represent hierarchical links. The underlined "Paragraph 3.3.3" is a cross-reference link. The remaining links are video clips.

For the MSIS project, we ignored the PHIDIAS user interface and instead used the PHIDIAS hypermedia database engine as a system for managing hyper-linked information. Every database management system uses some language for retrieving data. We used LINQ (Language for Inference, Navigation and Query), the functional (i.e., applicative) hypermedia language that has been the basis of our prototypes since the early 1980s [McCall et al. 1990]. What the Web-MSIS project did was enable us to exploit the inherently language-based approach that we have been using for many years to create a client-server system that would function over the Web. In particular, clicking on a link displayed on a Web-MSIS page sends a LINQ query to the PHIDIAS database system, which then retrieves the data specified in the LINQ query and displays it using a specified HTML page-template. The irony here is that our decision in the early 1980s to use a language-based approach to hypermedia became the basis for creating the Web-based functionality of our current prototypes.

2.2.2 Stage 2: Presenting and critiquing design proposals online with Web-PHIDIAS

In the next stage of our work we moved a significant step closer to our goal of creating a fully functional Web-based HyperCAD system. In this stage we created a system for online presentation and critique of design proposals. This new version of Web-PHIDIAS enabled designers to present examples of their design work—in the form of VRML models—over the Web together with the rationale for their design choices. Each VRML model was represented as a node in the PHIDIAS system. Rationale was stored in linked collections of nodes—as in the stand-alone version of PHIDIAS. Rationale for a design proposal was presented in a Java interface that enabled users at remote locations to call up the rationale and then make comments in response to the design and/or the rationale. These comments were then immediately added to the PHIDIAS database of rationale and posted to the Web, where other users—including the original designer—could respond to the comments. A screen image of this system is shown in Figure 3.

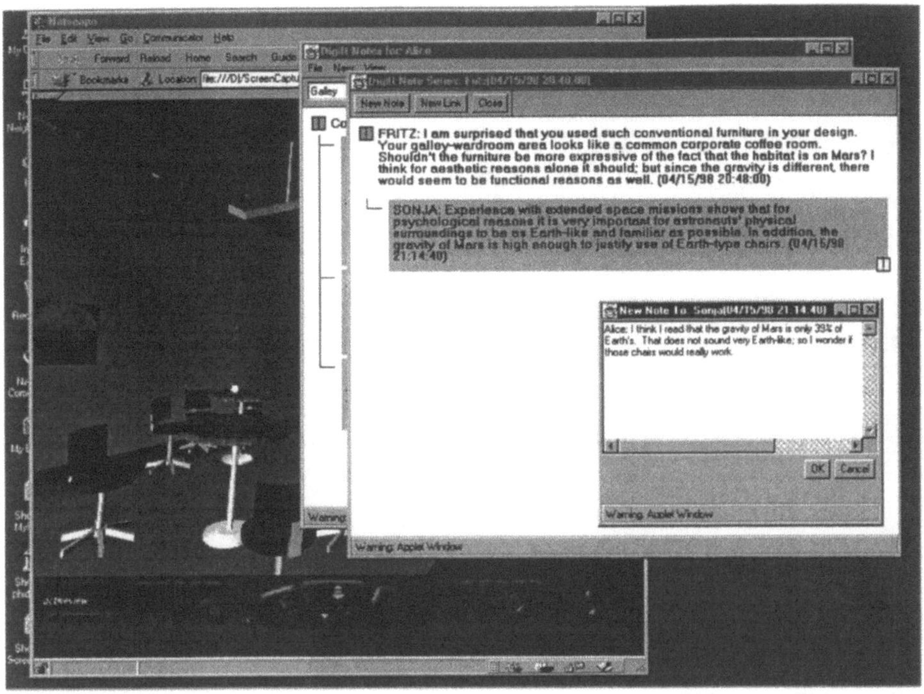

Figure 3. Here we see that a student (Sonja Holmes) has created a VRML model of her design for a Mars Habitat and then displayed this in a Web page created by Web-PHIDIAS. Fritz, a visitor to the Web site, has made a comment on Sonja's design and she has responded. A second person, named Alice, is shown here using the authoring interface to create another comment on Sonja's design. The interface for display and authoring of comments is written in Java. Once entered, comments are stored in the PHIDIAS database, which then displays them on the Web.

2.2.3 Stage 3: Collaborative design online with Web-PHIDIAS

The third stage in creating the Web-PHIDIAS system was to create a Java interface that enabled users to utilize more of the functionality of the server-side PHIDIAS system. In particular, we created 1) an interface for the display of rationale in an expandable/ collapsible outline format, 2) an interface for display and editing of 2D representations of 3D forms in plan, elevation and section, 3) 3D perspective representation of 3D forms, and 4) interfaces for the display of additional media types, such as raster images and streaming video. A screen image of this first real HyperCAD prototype interface is shown in Figure 4.

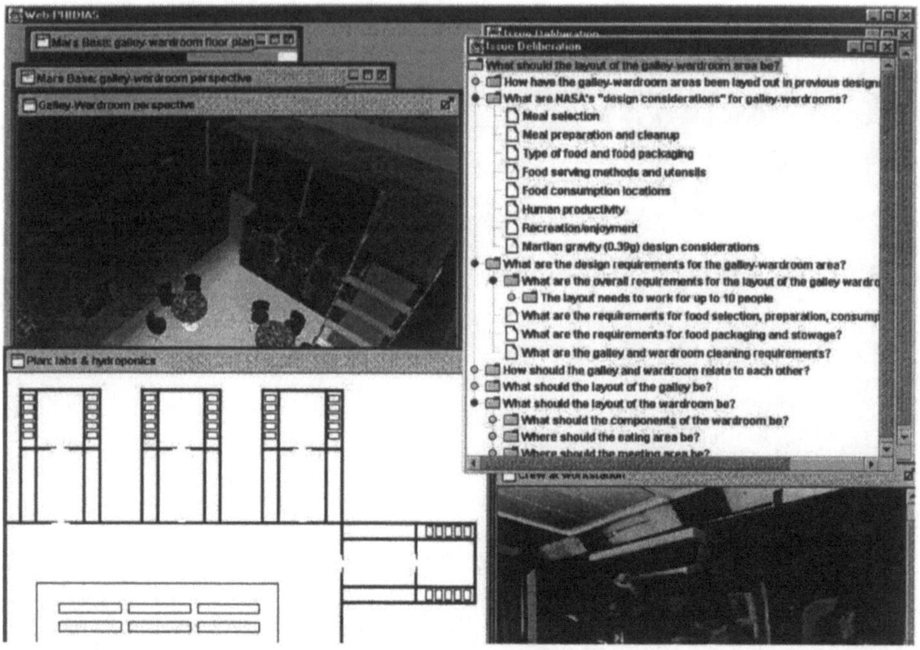

Figure 4. The Web-PHIDIAS Java interface enables use of the inherent functionality of the PHIDIAS hypermedia database. This functionality includes capabilities for display and editing of vector graphics and hierarchically structured design rationale as well as display of a variety of media types including raster images and streaming video.

2.2.4 Stage 4: Supporting collaboration with virtual copies of hypermedia networks

The object-oriented functionality of PHIDIAS provides inheritance of nodes, links and even entire hypermedia networks. This has many uses, but the most important for this paper is the support it provides for collaboration. In particular, it is the functionality for inheritance of hypermedia networks that provides the main means for managing collaborative communication in Web-PHIDIAS.

In Web-PHIDIAS, one hypermedia network inherits from another by creating a *virtual copy* of the network inherited from. To understand what a virtual copy of a network is, imagine a sheet of clear acetate laid over a graph representing the node-link structure of a hypermedia network. We could create a "new" network by drawing new links and nodes on the acetate and by "whiting out" old links and nodes on the acetate. To the casual observer it might appear that we had modified the original graphic image. But picking up the sheet of acetate reveals that we have merely created on the acetate a description of the differences between the original network and the "new" network. When the acetate containing this *differential description* is overlaid on the original image, the result is called a *virtual copy*. With PHIDIAS we say that the original network and the virtual copies of that network are each in a different "hypercontext." When it is used to support collaborative communication we call each hypercontext a collaboration context. (Though this functionality and its implementation strategy are unique to our system, they elaborate basic ideas about virtual copying of non-hypermedia networks in [Mittal, Bobrow and Kahn 1986]. Our own approach to virtual copying is described in [McCall 1993].)

The crucial point for understanding how hypercontexts—including virtual copies of hypermedia networks—can function as collaboration contexts is to see that each hypercontext is a view of the underlying hypermedia database. In any given hypercontext we see some nodes and links but not others. Thus, a hypercontext both enables and restricts our viewing of the database. From this, we can see that it is trivial to associate viewing privileges with hypercontexts: all we have to do is password-protect each hypercontext. It should also be clear that it is a simple matter to give more than one user viewing privileges for any given hypercontext and to give each user viewing privileges in more than one hypercontext.

We can in a similar manner assign various levels of authoring privileges to specific users in specific hypercontexts. (Here by "authoring" we are referring to the creation of both texts and CAD drawings.) The persons authorized to view a drawing or text can be assigned various levels of authoring privileges. They might have no authoring privileges. Or they might be entitled to add to a drawing or text but not edit it; this is like being allowed to draw on the acetate but not "white out" anything. They might be allowed to edit a virtual copy of the text but not the original; this is like drawing and "whiting out" on the acetate. Or they might have full authoring privileges, in which case they can alter the original; this is like drawing and "whiting out" on the original. It is only when we have assigned both viewing and authoring privileges to hypercontexts that we have created collaboration contexts.

Inheritance hierarchies of collaboration contexts can be used to support level of privacy/publication of information. This is very important for supporting collaborative work by larger groups. To see why, consider the following considerations for viewing and authoring privileges in a three-level collaboration of 1) individuals 2) workgroups and the whole design team, containing several workgroups:

There are notes and drawings that I create but for various reasons do not want to share with my workgroup. Similarly, there are things that I and my workgroup collaborators create but might not share with our boss or the larger design team of which we are part. On the other hand, things agreed to in the larger design team ought to be used in the work of the smaller design workgroup; and things that this group agree to ought to be seen and used in my individual work. But I should not be able to alter things agreed to by my work group; nor should this workgroup—or I—be able to alter things agreed to by the larger design team.

The crucial point here is that inheritance of collaborative contexts naturally supports the sorts of restrictions of viewing and authoring—such as those described above—required for multi-level collaboration. If a workgroup inherits its collaboration context from the context of the larger design team, then by default the workgroup can see and use (add to and combine) all the data in the team context but cannot alter it. At the same time, the other team members cannot see anything in the workgroup's collaboration context. When the workgroup is ready, it can publish a selected subset of its work by assigning to this subset viewing privileges for everyone in the team. The same kind of authoring/viewing relationships exist when an individual's context inherits from the workgroup's collaboration context.

Several things should be noted about the collaborative context mechanism. One is that its concept of inheritance hierarchy does not restrict its use to the strict tree structure for the structure of a large design team. It also enables formation of arbitrary ad hoc group structures—such as in so-called "ad-hocracies." A second point is that collaborative contexts still allow conflicts to arise whenever authoring privileges for a given data item are given to multiple users. To deal with such cases, our strategy is to adopt "node locking." In other words, if any person wants to alter a node, he or she "locks" the node until finished making the desired changes. This prevents anyone else from making changes while the node is locked. Further conflicts between users over changes to a given information node must be handled by means not described in this article—such as, for example, the "argumentative agents" that we described in [McCall and Johnson 1997].

3. CONCLUSION AND FUTURE WORK

We have implemented the first full prototype of Web-PHIDIAS, a Web-centric version of the PHIDIAS Intelligent HyperCAD System. This system has a Java client and uses PHIDIAS as a server-side hypermedia database that manages both information and communication for collaborative design. Web-PHIDIAS employs a special context mechanism that uses virtual copies of hypermedia networks to support complex, collaborative communication in potentially large design groups.

Web-PHIDIAS is a major step toward our goal of supporting the argumentative approach to design first proposed by Rittel. It does this by using the Internet and the Web to distribute and capture argumentative information, to distribute computational aids to design, and thus to eliminate distance as a barrier to participation in design projects. Web-PHIDIAS is the first system to demonstrate how the information, communication and computation capabilities of networked computing can be used to promote the pluralistic view of design inherent in the argumentative approach.

Much work remains to be done to flesh out and refine the Web-PHIDIAS system. Our current prototype is still fairly primitive both in its user interface and its failure to provide access to the full functional of the PHIDIAS server-side system. For example, our Java client does not yet make use of the PHIDIAS functionality for knowledge-based critiquing. Other functionality found in previous prototypes—e.g., "argumentative agents"—will have to be re-implemented, perhaps from scratch. We also plan to incorporate voice-based communication and collaborative sketching into future prototypes; and these will require the addition of entirely new code.

In conjunction with the creation of additional functionality, we plan to do a great deal of testing of our system with end users. For this purpose we plan to focus our work on more conventional architectural domains than space-based habitation. We have for a number of years worked on issuebases for the design of houses and this domain is likely to become the focus of our user testing with individuals and small groups. Testing with larger groups will probably require a more complex domain.

REFERENCES

Kunz, W. and Rittel, H. "Issues as Elements of Information Systems," Working Paper # 131, Center for Urban and Regional Development, University of California, Berkeley, 1970.

McCall, R. *On the Structure and Use of Systems in Design*, Dissertation, University of California, Berkeley, 1978, University Microfilms, 1979.

McCall, R.; Schaab.; Schuler, W. and Mistrik, I. "A Database Model as the Basis of a Workstation for the Designer," in *Representation and Exchange of Knowledge as a Basis of Information Processes*, H.J. Dietschmann (ed.), North-Holland, Amsterdam, 1984.

McCall, R., "MIKROPLIS: A Hypertext System for Design," *Design Studies*, Vol.10, No.4, 1989, Butterworth Scientific Ltd., Surrey, UK, pp. 228-238.

McCall, R.; Bennett, P.; d'Oronzio, P.' Ostwald, J.; Shipman, F. and Wallace, N. "PHIDIAS: Integrating CAD Graphics into Dynamic Hypertext," in *Hypertext: Concepts, Systems and Applications* (Proceedings of the 1990 European Conference on Hypertext: ECHT '90), Rizk, A.; Stretz, N. and Andre, J. (eds.), Cambridge University Press, 1990, pp. 152-165.

McCall, R.; Morch, A. and Fischer, G. "Supporting Reflection-in-Action in the Janus Design Environment" in *The Electronic Design Studio* (Proceedings of the CAAD Futures '89 Conference), Mitchell, w.; McCullough, M. and Purcell, P. (eds.), MIT Press, 1990, pp. 247-259.

McCall, R. "PHI: A Conceptual Foundation for Design Hypermedia," *Design Studies*, Vol.12, No.1, 1991, Butterworth Scientific Ltd., Surrey, UK, pp. 30-41.

McCall, R. "The Locks and Keys algorithm for virtual copying of networks," Working Paper 73, Center for Environmental Design Education and Research, University of Colorado, 1993.

McCall, R.; Bennett, P. and Johnson, E. "An Overview of the PHIDIAS II HyperCAD System," in *Reconnecting: ACADIA '94 (Proceedings of the 1994 Conference of The Association for Computer Aided Design in Architecture)*, A. Harfman and pp. 63-74.

McCall, R. and Johnson, E. "Argumentative Agents as Catalysts of Collaboration in Design," in *Proceedings of the 1996 Conference of the Association for Computer Aided Design in Architecture (ACADIA '96)*, Patricia McIntosh and Filiz Ozel (eds.), Tucson, AZ, 1996, pp. 155-163.

Mittal, S.; Bobrow, D.; and Kahn, K. "Virtual Copies: At the Boundary between Classes and Instances," *OOPSLA '86 Proceedings*, Association for Computing Machinery, September, 1986, pp. 159-166.

Rittel, H. "On the Planning Crisis: Systems Analysis of the 'First and Second Generations'" Bedriftsokonomen, No. 8, October 1972 (Norway).

Schoen, D. *The Reflective Practitioner*, Basic Books, 1983.

Immersive redlining and annotation of 3D design models on the Web

Thomas Jung[†], Ellen Yi-Luen Do, and Mark D. Gross
Sundance Lab for Computing in Design & Planning, University of Colorado, Boulder, CO, USA
[†] & C.R.A.I. Centre de Recherche en Architecture et Ingenierie, Nancy. France

Key words: VRML, immersive environment, virtual annotation, computer-aided design, building models

Abstract: The Web now enables people in different places to view three-dimensional models of buildings and places in a collaborative design discussion. Already design firms with offices around the world are exploiting this capability. In a typical application, design drawings and models are posted by one party for review by others, and a dialogue is carried out either synchronously using on line streamed video and audio, or asynchronously using email, chat room, and bulletin board software. However, most of these systems do not allow designers to embed annotations and proposed design changes in the three-dimensional design model under discussion. We present a working prototype of a system that has these capabilities and describe the configuration of Web technologies we used to construct it.

1. INTRODUCTION AND MOTIVATION

We are constructing a collaborative three-dimensional environment for design review, comment, and exploration of design alternatives. Our system offers two distinct mechanisms for collaborative review. In both mechanisms, a designer posts a model in VRML format for on line review. Reviewers can walk through and inspect the model using a standard Web browser that has VRML capability. In the first mechanism a reviewer annotates the design by using a Java applet to drop text notes and simple annotation symbols (spheres, arrows, call-out dots with identifiers) into the model. The annotations are objects in the VRML world, although they are stored and loaded as separate files from the base world that is being discussed. In the second mechanism, a reviewer explores, selects, and make changes to the objects in the model by interacting with a Java controlled VRML browser. The reviewer can change the location, colour and texture of the design objects (e.g., furniture, window, walls) in the VRML world. As with the first mechanism, the new objects with changes are stored in a separate file from the base model, and are loaded in to the shared VRML world by other reviewers.

The paper outlines the motivations for this research; it reviews related work on 3D interaction in Virtual Reality both in laboratory settings and over the Web, design rationale, and on-line collaborative design; and describes our working system for interactive immersive redlining and annotation of VRML models.

We are interested in building, using and studying systems that support a group of people participating in a design process. Collaborative design means not only the collaborations among different designers, but also between designers and their clients. Often a design is carried out as a conversation and negotiation process among architects and their clients. The process is not merely acceptance and rejection of presented alternatives, but a collaborative negotiation of individual decisions about dimensions, positions, and selection of components.

We believe a virtual environment that provides designers and clients the opportunity to negotiate through annotations and changes to the design alternatives inside the design artefact will be useful. This environment could serve as a record of what people agreed to do, their design and use rationale, and what actions they have agreed to perform. An environment that enhances this communication and encourages co-operation, overcoming the barriers of space and time constraints will be valued. For example, when people collaborate in daily life, from commissioning a design project, to several iterations of the review and redesign process, scheduling is usually the hardest part besides communication. A meeting usually requires the architects, clients, and other concerned parties to be present in the same location at the same time. This is not always convenient or even possible. Furthermore, clients often have difficulty understanding design presentations of floor plans and cut sections. They may easily be seduced or confused by graphic representations of the design (e.g., water colour painting, airbrush rendering, pencil sketches) and fail to understand the basic elements of the design. Therefore, a simulated environment of the design presented in a format that allows people to access, review and even change the design from a distance can help transcend the boundaries of time and space.

A shared world of virtual reality offers a more sophisticated, interactive, stimulating and engaging environment for people—architects and clients alike—to communicate. Therefore, we built a working prototype that facilitates this kind of communication and collaboration in a VRML environment. We call it the "Immersive Redlining and Annotation" project, or Redliner for short.

In the following, section 2 describes how architects and their clients (or domain experts and consultants) might use our Redliner environment to discuss and interact about the decisions of a house design. Section 3 follows with a description of the implementation of our working prototype. Section 4 reviews related work in the domain of VR and CSCW. Finally, section 5 concludes with discussion and future work directions.

2. SCENARIO

We present here a scenario showing how architects and clients can work and communicate in a shared virtual environment by leaving annotated objects and making changes in the VRML world of an architectural design.

Our story has three key players: the architect Bill who is hired by a couple, John and Mary, to design their new house. Bill works at his architecture firm in California, John and Mary live in Pennsylvania, and their new house is to be built in Georgia. John's profession as an executive in a successful Internet startup requires him to travel frequently around the country. All three have easy access to the World Wide Web (in the office, community library, or Internet cafes).

Our architect Bill, after initial discussions with his clients and the examination of the site proposed an initial design for the residence. He posted a Web address of a model of his newly designed house made in a 3D VRML format using the Redliner program. He calls and sends emails to John and Mary to inform them of his progress on the project and invites them to explore and review the design.

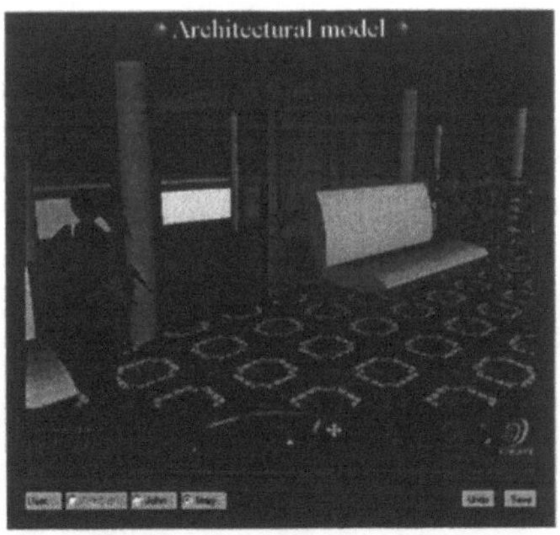

Figure 1. Mary parks her car, views the exterior of the building (left) and walks inside, finding herself in an entrance area with a tile floor (right). Note the annotation buttons for John, Mary, Architect, Undo, and Save.

Mary is the first to visit the virtual house on the site. She rolls up on the set in her virtual Mercedes (well, actually, the Cosmoplayer VRML browser plug-in for Netscape Navigator or Internet Explorer on PC or Mac). She sees the building, as it would appear from the driveway. The door opens as she approaches and the lights come up, bathing the interior of the entrance area with a rosy glow.

To her left, Bill (the architect) has made a place for the Possible Mondrian Painting they inherited from her parents, and he has illuminated it from above by a lamp that turns on when she looks at the painting on the wall (Figure 2, left). Looking around, she hears a waltz by Chopin playing from the piano downstairs (Figure 2, right). She feels that the space is a little dark, though, and she begins to think about what changes she would like to see.

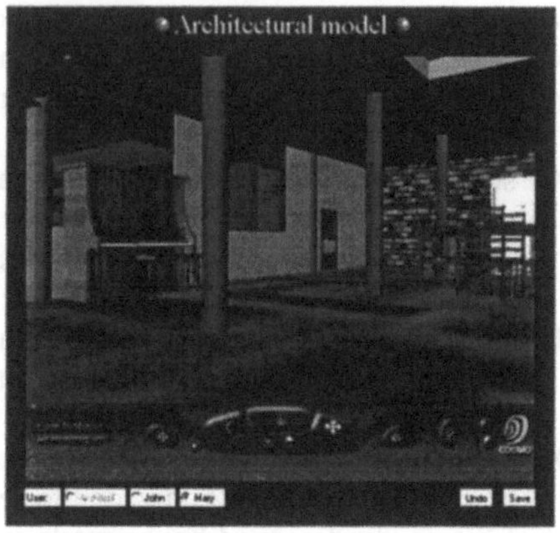

Figure 2. Mary looks around the entrance area. The light comes up over a painting (left) and music comes from the piano (right).

Mary decides to leave a note for Bill (Figure 3). She drops a green annotation marker (a small sphere, the VRML equivalent to a PostIt note) on the wall. A new green marker (numbered '1') sticks to the wall, and a text annotation window appears. She types, "The wall behind the piano looks too dark". She closes the annotation window and moves further into the house.

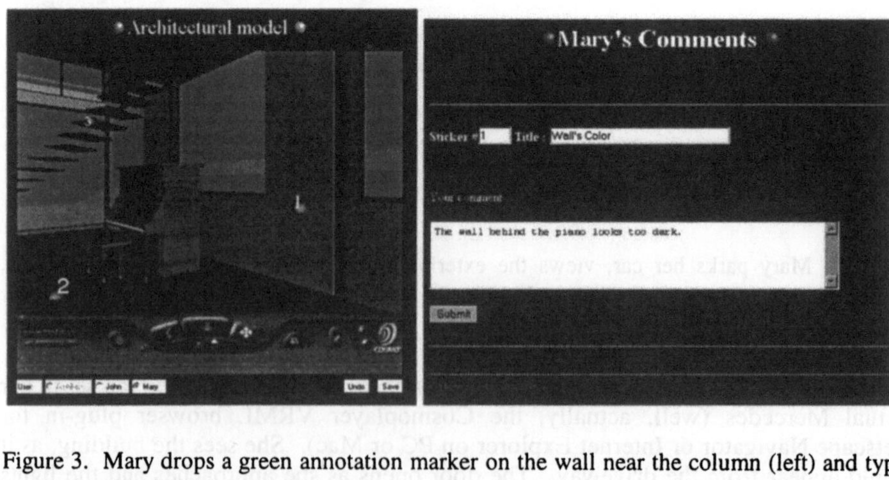

Figure 3. Mary drops a green annotation marker on the wall near the column (left) and types in a comment (right).

Later that day, John logs on from his hotel room in Seattle. He makes comments using red markers. Figure 4 shows comment #1, in which he asks about what appears to be a cut off column in the middle of the room.

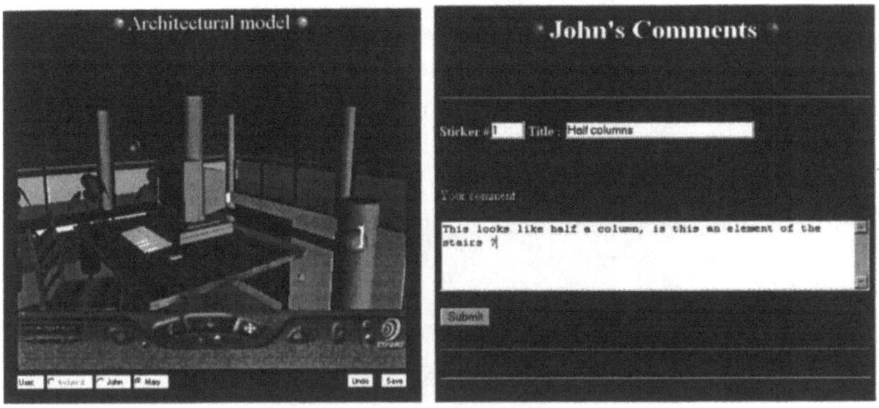

Figure 4. John drops a note (#1) on the cut-off column asking whether it is part of the stairs.

After both John and Mary have visited the virtual design, Bill returns to the world to review their comments (Figure 5, left). He responds to each of their comments (Figure 5, right).

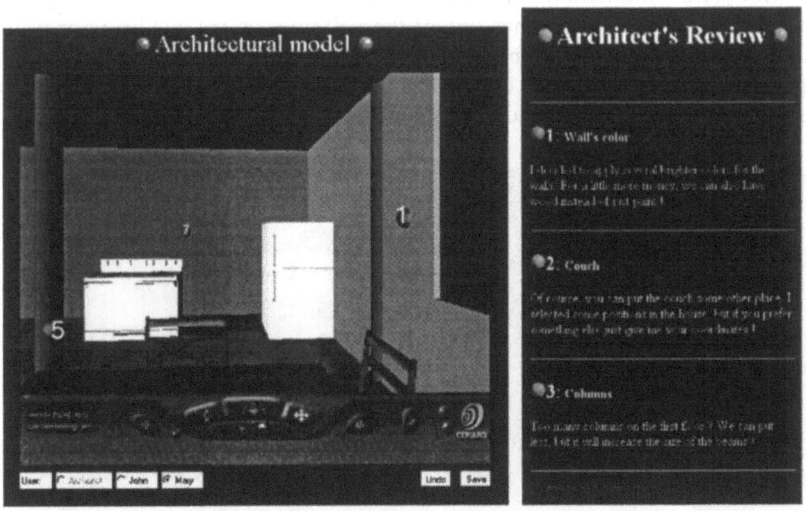

Figure 5. Left, Mary has left annotation markers inquiring about the wall materials and furniture arrangement; right, Bill responds to Mary's questions.

In some cases, he responds by creating several alternatives, for example, enabling Mary and John to view different colours and materials for the walls (Figure 6). He replaces their red and green annotation markers with new, yellow ones that link to a page of his design revisions in response to their comments.

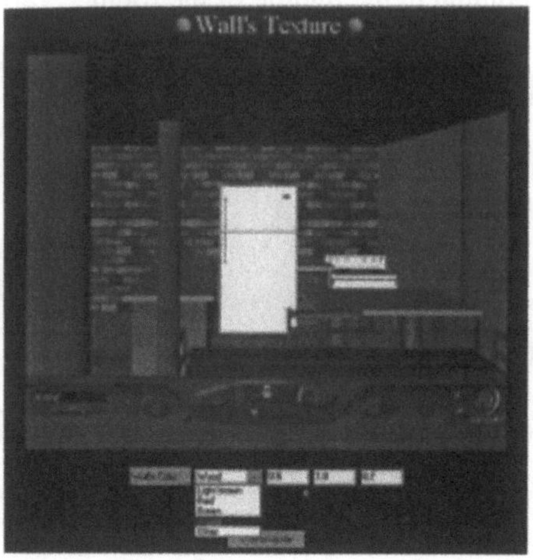

Figure 6. Bill has provided an ‘options window’ to try out different wall materials. Note the colour and materials choice menus and the colour co-ordinate menu on the bottom, as well as the comments button.

John returns the next day to review the revised design. He sees the design as it was before, with yellow markers in the locations where his and Mary’s red and green spheres were before. When he touches one of the yellow markers, his browser pops up a window showing Bill’s responses to their comments (Figure 5 right). This HTML page also links to a custom configured VRML browser that allows him to explore design alternatives Bill provided (Figure 6, 7).

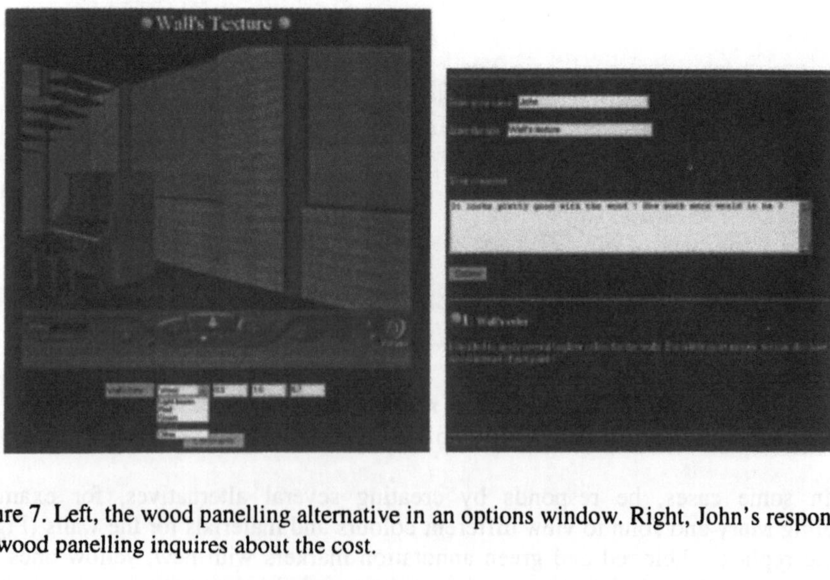

Figure 7. Left, the wood panelling alternative in an options window. Right, John’s response to the wood panelling inquires about the cost.

John tries the different wall material options (Figure 7 left). Lacking a subtle taste, he prefers the wood panelling and he leaves a note (Figure 7 right) for Bill and Mary, expressing his preference and asking how much this option would cost. (Each change in this design negotiation process creates a new .wrl file with annotation objects and links to the comment pages.)

Also, Bill has provided several options for the kitchen and dining area. The original design (Figure 8, left) leaves the kitchen open to the rest of the house. One option (Figure 8, right) encloses the kitchen with a partial wall and places the dining area outside the wall. As with the choice of wall materials, John and Mary can explore these wall alternatives and record their preferences for Bill to complete the design.

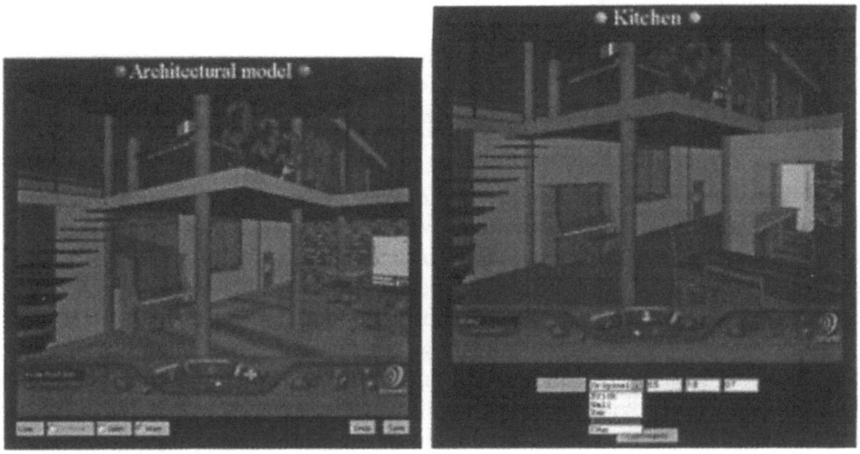

Figure 8. Left, original configuration with no walls; right, new walls define the kitchen and the dining table is moved outside the kitchen.

We constructed this scenario to demonstrate several of Redliner's features. Designers, clients, and consultants browse a VRML world that is stored on a server. Beyond just browsing the model, they can post their comments about the design, by leaving markers (coloured spheres) on or near the parts of the design they are commenting on. Others who visit the VRML model later can see these markers and read the posted comments. With a little more effort (this is the designer's job), objects in the model can be assigned optional positions, sizes, and colours, and an interface is provided for clients and consultants browsing the model to explore these variations and record their preferences.

3. IMPLEMENTATION OF THE WORKING PROTOTYPE

We describe briefly the technologies we explored and employed in this project. There are four levels of implementation: 1) basic VRML modelling and behaviour, 2) VRML communication to Java through Script Node;. 3) VRML communication to Java applets through EAI, and 4) CGI scripts in the Perl language to support input, saving, and processing of forms, new world files and file uploads.

The implementation of Redliner consists of several modules. First, VRML models are stored on the server, and a CGI program manages the storage and retrieval of posted text. Client side Java applets handle user mouse events in the VRML world. The Java applets communicate with the VRML world in several ways: via VRML's internal object behaviour language, Script Nodes (in an earlier version of Redliner), using the External Authoring Interface (EAI), and using the CGI Perl script.

3.1 Using VRML's Scene Description Language

The first level implementation of Redliner uses VRML's scene description language to specify object behaviours. VRML provides a set of object-oriented graphics objects augmented by hypermedia links with geometry rendering. It is a file interchange format[1]. We use VRML to specify object behaviours and their response to user events. Using ROUTEs, we connect 3D object nodes and fields to behaviour driven sensors and timing in the VRML world. For example, when a user enters a room, the ProximitySensor can activate and turn on a SpotLight (DirectionalLight, PointLight) node to provide radiating light rays. The other object nodes in the scene that are visible will be illuminated and material and texture are properly shaded. The Sound node places an audio clip (.aif, .wave) at a certain location; it can be rendered spatially (e.g., music comes from the right speaker) and can be triggered on and off. Similar examples of Sensors (e.g., TouchSensor, ProximitySensor, PlaneSensor) are embedded in objects like doors and windows, and the switches for a television or an elevator.

A WWWAnchor node can be placed in the world to link to a new world or a Web site (with a URL, Uniform Resource Locator specified). For example, a high rise building can be divided into separate worlds (.wrl files) and connected through the anchor buttons in an elevator. The reason to break the high rise building into separate worlds is because of memory and file size concerns. Anchors have nothing to do with sensors but behave like TouchSensors. This is the basic level to specify object node behaviours, such as doors that open, or a loop event that rotates a globe. To achieve our goal of greater interaction with the users and letting users control the manipulation of the scene requires the next level of implementation.

Figure 9 shows how a simple event of activating a viewpoint by a click works. For example, when the user clicks on an elevator button, the TouchSensor becomes active (its value becomes True). ROUTE receives the isActive value and turns the

[1] VRML became an ISO specification (International Specification ISO/IEC 14772-1, http://www.vrml.org/) in December 1997.

set-bind field of viewpoint (an x,y,z co-ordinate value, i.e., a vector of a 2nd floor location) to True. The user viewpoint is then assigned to this preset value.

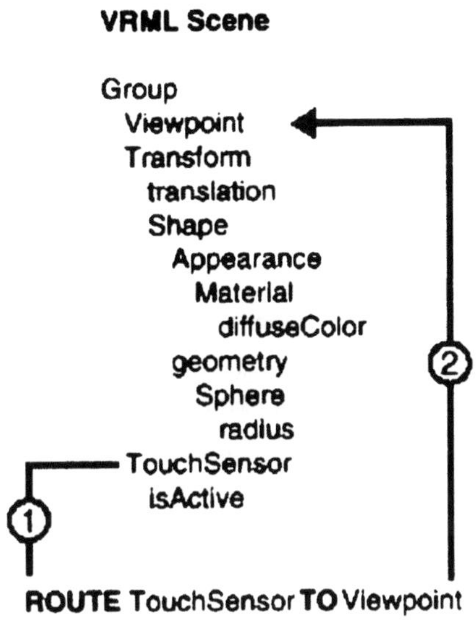

Figure 9. Simple events described in VRML

3.2 VRML to JAVA via Script Node

In Redliner's second level of interaction Java communicates to VRML through a Script node interface with APIs. The Applications Programming Interface (API) provides hooks from Java to any component of the object nodes in a scene. For example, one can use API scripts to access any field type in VRML. VRML provides ways to get and set the value of its own fields or get and set global information associated with the VRML browser. Script Nodes encapsulate Java (or JavaSscript, VRML script or any other language) functionality and connect (and convert) variables in the Java environment with VRML object field values. Script Nodes can contain a program, identify a change or a user action, receive events from other nodes, and send events to change others. For example, Figure 10 below shows our first prototype of touching an object to change its location according to user input. This approach is more interactive than the version above because a user can supply inputs based on specific, individual preferences.

For example, a sphere object that responds to a mouse click event by changing its colour, texture and location requires communication between a Java class or Java Script to access the object's fields and respond to events through Script Nodes. Figure 10 shows the process flow. First, when the TouchSensor isActive, its ROUTE sends an EventOut to the Script node. The Script node activates a JavaScript or Java class that takes user input and sends it as an EventOut (changePosition) back to the Script node. ROUTE takes this EventOut and sends its value to an exposedField that translates this object to a new location according to the x,y,z coordinate received by Java.

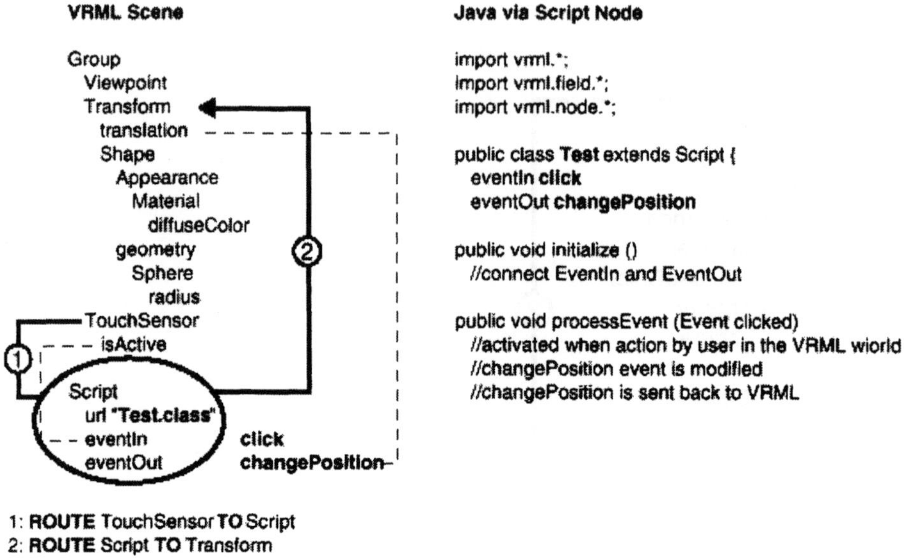

Figure 10. Communication between VRML and Java through a Script node.

3.3 JAVA to VRML via EAI

The above example works for simple interactions such as changing the colours of the object from red to blue or from one location to another. Basically, the Script Node provides a connection to Java from inside the VRML world. A more complicated modification of the scene from user input (e.g., type in RGB values or x,y,z co-ordinates) requires accessing several different Java classes and windows in order to make them interact with each other. For example, to accomplish the task of letting a user specify a change of view location, we need communication among a VRML browser, a Script Node containing a Java program and a Java window class with an interface for input. Any modification of the interaction requires changes to both the Java program and also the VRML file (fields of the Script nodes). Therefore, we investigated using the VRML External Authoring Interface (EAI). The EAI provides a Java interface that communicates with an external HTML Web browser. EAI applets can send information to and from VRML scenes embedded in an HTML page[2]. Using EAI, a Java applet can handle any new world from "outside" without going "inside" it to change its specification for events and fields.

[2] (EAI information can be found from the Web, www.vrml.org, and www.cosmosoftware.com/developer/eai_whitepaper.html).

The current version of Redliner has several modules. The first provides the functionality of adding annotation markers on the surface of objects as a person walking through the scene. A Java applet is always watching the VRML world and observes the user's actions (e.g., moving the cursor to a location, capturing mouse clicks). Once an event (e.g., a mouse-click) occurs, the program logs the location of the click, creates an annotation sphere in the VRML world, then passes these messages and values back to the VRML browser so the object appears with an ID number. The second module is the interface that enables the user to select design options and put in changes of location, texture, and colour through the type-in boxes.

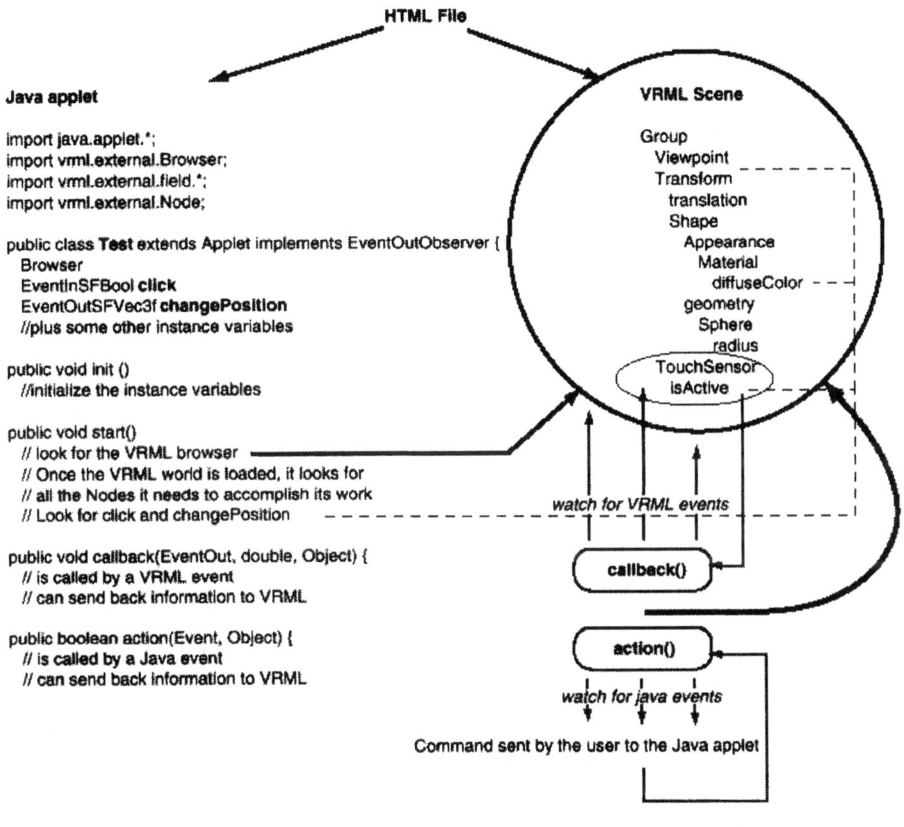

Figure 11. VRML and Java communicate via the Extended Authoring Interface (EAI).

The Java applet has four methods: 1) an **init** method, 2) a **start** method that looks for a VRML browser and special nodes as defined in the program, 3) an **action** method that waits for Java events, and 4) a **callback** method that implements the EventOutObserver class to watch for VRML events. For example, Figure 11 shows how a Java applet can control all the interaction of a classic VRML file. The Java applet and a VRML file are loaded at the same time in a HTML document. The first function of the applet (after initialising some instance variables) is to recognise the VRML world and some specific nodes that will be used and changed. This is done by the **start** method. Then the **callback** and **action** methods are called whenever something new happens. In Redliner, both these methods are used for dropping the annotation markers (spheres). The **action** method is used to recognize the user—John, Mary or Bill—and send to VRML the colour of the Note (red for John, green for Mary, yellow for Bill). The **callback** method, constantly watching for VRML event, is activated when the user clicks in the VRML world to drop a node. **Callback** records the position of the click and sends these co-ordinates back to the VRML files where a new sphere appears.

3.4 CGI Perl Access to Server to Facilitate Reviews

As designers and reviewers add annotations and make changes to the VRML world, these modifications must be stored so that other users can view them later. Saving new VRML object additions and the annotation texts is handled through a Perl CGI (Common Gateway Interface) script. A server side CGI script in Perl reads values (e.g., strings, numbers) set by a Web page form, does some processing on these values, and creates an output (e.g., result, generating a new HTML page) after closing the document. It also allows uploading a file, creating a log file back to the server.

In Redliner, a VRML browser and several buttons (to identify different users, activate comments and submission by Java applets and Perl scripts) are embedded in an HTML page. A user first identifies herself (Mary) by clicking on a button, then explores in the VRML world. She posts annotation nodes (green spheres) in the scene with her signature. When she clicks on the annotation markers a window pops up with a new HTML page that contains a form for her to enter remarks associated with the current location. The Perl script on the server handles this form. Comments entered in the form are written back to the page and redisplayed below the form. The new VRML (.wrl) file with the annotation objects are also saved as a new file back to the server. A later appendix to the comments at the same location can be entered through the same interface, and changes appended to the previous input in chronological order. The second HTML page where the users review and select design options, or enter a new specification of the locations and material of the design objects is handled in a similar way. This page has interfaces for selecting among the designer's proposed alternatives (colours: red, pink, orange). The page also offers type-in boxes for a user to specify a particular value (e.g., of location, or texture) if unsatisfied with the alternatives the designer provided. The result values and the world are saved back to the server and are ready for the next iteration of review.

3.5 Summary of Functionality

In summary, here is the basic functionality of our current Redliner prototype. Users can view a shared design model stored on a central server from their VRML-enabled Web browsers. They can add text annotation objects to the design model, which appear as small coloured spheres with numbers on them. The spheres are linked to text that is maintained in a log on the server, and the colours of the spheres indicate the authors' identities. The annotation objects (and all additions and changes made by users) are stored in separate VRML files on the server, which are loaded on top of the base model VRML file that contains the original design. User annotations are available for all to see and comment on.

Designers can post several alternatives for various elements in the design, allowing browsing users to change the colours, positions, and other parameters of the walls, windows, and floors, in the design. The users can record their preferences, which are saved back to the server as a VRML overlay file to the base model.

Designers and users can record preferred viewpoints in the model, thereby describing a path through the design. These viewpoints are stored on the server, and using a Java applet, users can select a navigation path and move through the model.

3.6 Optimisation of VRML models

In our demonstration project, we constructed the architectural model in Form•Z, a commonly used three-dimensional modelling application, and we used that program's export function to produce the VRML model. Form•Z (version 2.9.5) does not optimise the VRML files it produces, and therefore some effort was spent reducing the complexity of the VRML representation of the model by hand. Therefore, we are also developing a set of simple translation tools to help with these tasks. For example, one tool will convert a complex Form•Z model such as a cylinder with multiple facets to a single cylinder object node in VRML.

4. RELATED WORK

Visionary architects and designers have heralded the potential of virtual reality and immersive environments. Bertol's book, Designing Digital Space: an architect's guide to virtual reality (Bertol 1997) offers visions of possible applications, combined with more technical descriptions of recent virtual reality research. Donath and Regenbrecht (Donath and Regenbrecht 1996) have explored sketching in three-dimensional virtual reality spaces, using hand gestures to sweep out planes and solids and to manipulate the virtual objects in a space. Davidson and Campbell (Davidson and Campbell 1996) developed and experimented with Greenspace 2, a prototype virtual environment for architectural design review. Virtual reality systems have also been used to help architects assess the performance of their designs (Pinet 1997). Campbell (Campbell 1998) describes a real world test application, more pragmatically oriented, in which a design firm used VRML models to communicate construction information (e.g., dimensions) about the design of a stair with a fabricator. In the model, each object in the VRML was defined as an anchor for call-out notes and links to the specification document.

Web PHIDIAS, a system with aims somewhat similar to ours, has been under construction in our laboratory for the past two years. The principal differences between Redliner and Web PHIDIAS are as follows. First, Web PHIDIAS, which derives from the PHIDIAS system that has been under development for over a decade, is based on an issue-based information system (IBIS) approach, in which comments are structured in a hierarchy of design rationale (issues, answers, and responses). Our Redliner comment system is at present more loosely structured;

users simply add to comment logs associated with the coloured spheres. Second, at least in the version described in (McCall, Holmes, Voeller and Johnson 1998) the objects in the VRML world had to be explicitly prepared to allow designers to attach a hyperlink to the PHIDIAS issue base. In Redliner, designers and clients add new annotation objects to the existing VRML world, attaching coloured spheres to the design elements in the model as they browse through the world scene. Third, we are developing a tool kit that makes it easy for a designer to modify VRML objects in the model so that users can move them, change their colours, positions, or other properties.

Virtual Reality (VR) was defined originally as an advancement of technology that engages human senses (e.g., sight, hearing, touch) through input and output devices. As Heim pointed out, VR has "three I's" – immersion, interactivity, and information intensity (Heim 1998). Immersion is usually achieved through a head mounted display (HMD) or a CAVE environment (Cruz-Neira, Sandin, DeFanti, Kenyon et al. 1992). A desktop virtual reality ("fish tank VR") usually is equipped with stereoscopic glasses and manoeuvring device (e.g., data glove, joy stick or space mouse). Dille argues that VRML is an information management tool for the Internet that provides a capability for simulation and participation from multiple users (Dille 1996).

VR applications have included simulation of site (e.g., walkthrough, flight, rides) and object manipulation (e.g., product assembly, work operation). The Committee on Virtual Reality Research and Development's report to the US federal government agencies (National_Research_Council 1995) recommends that four areas show the most promise for synthetic environments (that is, virtual environments, teleoperation, and augmented reality). They are: 1) design, manufacturing, and marketing; 2) medicine and health care; 3) hazardous operations; and 4) training.

Examples of VR projects abound. In the category of design and manufacturing systems, simulation tools have been built to support configuration lighting control (Rygol, Dorbie, Worden, Ghee et al. 1995) and to support testing and evaluation of robot welding work cell design in virtual prototyping (Bickel 1998). Medical applications include a dynamic simulation connected with a force feedback arm to position a drug molecule in potential binding site (Ouh-young, Beard and Brooks 1989), remote surgery (Green, Hill and Satava 1991; Satava 1992), and a medical virtual workbench (Lawton, Poston and Serra 1995). Training systems for hazardous operations include flight simulators (Furness 1986; Vince 1995), Steve (Soar Training Expert for Virtual Environments) (Johnson and Rickel 1997), and Jack, a simulated human figure to carry out control and repair tasks (Badler, Phillips and Webber 1993), and remote control for atomic bomb factory waste disposal (TRW Space and Defense Systems, for the US Department of Energy).

A relevant arena of virtual environments is Computer Supported Collaborative Work (CSCW). This work focuses on managing and visualising information. For example, SeeSoft (Eick, Streffen and Sumner 1992) uses arrangements and colours to represent large bodies (millions of lines) of source code. Cone Trees (Robertson, Card and Mackinlay 1991; Robertson, Card and Mackinlay 1993), Perspective Wall (Mackinlay, Card and Robertson 1991) and Web Forager (Card, Robertson and York

1996) at Xerox PARC use physical metaphors such as tree, walls, and books to display hierarchically structured data as objects in a three dimensional virtual world.

Current VRML development follows the larger trends of VR development and has augmented VRML with extensions (see, for examples, the proceedings of the VRML 98 symposium). The human simulation system Jack (now distributed by Transom Technologies, Inc.) has been incorporated into the OpenWorlds (Eick, Streffen and Sumner 1992) toolkit as a base for creating applications of human performance—a responsive actor in a story, or the simulation and analysis of an astronaut's body movement in space. A Showroomkit (Dauner, Landauer and Fraunhofer 1998) is under development to support adding predefined exhibition objects into a showroom. Another project explores integrating geographic information (Abernathy 1998) such as topography and GPS (Global Positioning System) information into VRML. A psychology application uses VRML with objects of different colors and shapes to test people's spatial cognition (Givaty, vanVeen, Christou and Bulthoff 1998) in a 3D world and records the memory test on a 2D plan view.

5. DISCUSSION AND FUTURE WORK

We have constructed a working prototype to demonstrate the potential of interactively, asynchronously annotating a three-dimensional architectural design model. Our scenario involves an architect and two clients interacting around the design of a house. The same techniques could be used to support design critiques in a 'virtual design studio' project conducted at several architecture schools around the world, or in a more professional context in which architects, consulting engineers, and other experts collaborate.

Sun released a 3D toolkit for Java called Java3D in 1998. Java3D is an API for creating and controlling 3D objects and behaviours[3]. Java3D programs are saved as Java bytecodes, not as modelling format. It's a useful approach but implementation is tedious. It requires the programmer to understand the structure of any imported object in order to apply behavioural control in the proper places. A VRML-Java3D working group is exploring the interoperability between Java 3D and VRML[4].

Our Redliner prototype is constructed from a diverse collection of current Web technologies including VRML scene description language and CGI scripting. These particular technologies are already being replaced by newer approaches; for example, Java3D is likely to replace VRML in the near future. We are less interested in the specific technologies than in the opportunities for a collaborative discussion around a three-dimensional design model.

We look forward to a number of enhancements and extensions to the current Redliner prototype. Some of these are described below:

[3] http://java.sun.com/products/java-media/3d/

[4] http://www.vrml.org/WorkingGroups/vrml-java3d/

We adopt a conservative (and inexpensive) approach to virtual reality: we do not employ head-mounted displays, CAVE (VR room) technology, haptic feedback, or immersive audio (we do not have access to this hardware). An obvious enhancement to the current system would simply place our Redliner system in a hardware immersive virtual reality environment. Along similar lines, our VRML models do not include constraints, such as fixed viewpoints at eye level, and interference of solids; any object can be placed in any location, and viewers can fly around and through the model. These immediate and relatively simple enhancements are needed.

We have built our system on an asynchronous model, but it could also be used in a synchronous mode. We would then need to construct avatars to represent the various designers as they move through the model. It might prove valuable to embed the Redliner system in a larger online conferencing system with text chat, whiteboard, and audio-video conferencing.

Currently, designers interact with the building model by posting annotations, or by manipulating objects that have been pre-programmed (by the designer) to allow for variation. An interesting way (for a designer) to interact with a virtual world might be to sketch on top of the VRML model (in a kind of transparent window over the browser), to add elements to the model or to change the positions or attributes of existing elements. We have built an early prototype of this capability but additional work is required.

At present our VRML models are simply geometric; their elements are geometric surfaces and solids. A logical enhancement would link our Redliner system with online catalogues of building products so that the designer could access dimensionally accurate design elements that would store their cost, material, dimensional, and performance attributes. Reviewers could not only manipulate pre-built building components, but they could also select new ones from the product libraries and add them to the model. As with our current system, the 'base model' would be stored separately from the reviewers' annotations and additions.

The current Redliner system just logs text typed into the annotation screens, associating it with each coloured sphere. The text could also be automatically emailed to the designer(s) who are responsible for the design elements under discussion, or the logs could be structured as for example in the PHIDIAS system of McCall and his colleagues.

6. ACKNOWLEDGEMENTS

This research was supported in part by the National Science Foundation under Grant No. IIS-98-19856. The views contained in this material are those of the authors and do not necessarily reflect the views of the National Science Foundation. The authors also acknowledge support from the Chancellor's office at the University of Colorado at Denver, and a grant from the Undergraduate Research Opportunity Program (UROP) for Jimmy Davidson. Jimmy Davidson constructed the VRML building model and Dongqiu Qian and Peter Kappus helped with Java and Perl debugging. John and Mary's house is modelled loosely on a design by Bennett Neiman.

7. REFERENCES

Abernathy, M. (1998). Integrating Geographic Information in VRML Models. *VRML 98 – Third Symposium on the Virtual Reality Modeling Language*. S. N. Spencer. Monterey, CA, ACM: 107-114.

Badler, N. I., C. B. Phillips and B. L. Webber (1993). *Simulating Humans: Computer Graphics, Animation, and Control*, Oxford University Press.

Bertol, D. (1997). *Designing Digital Space: an Architect's Guide to Virtual Reality*. New York, Wiley.

Bickel, D. (1998). Virtual Reality for Industrial Applications. *3D Realtime Simulation and VR-Tools in the Manufacturing Industry*. F. Dai. Berlin, Springer: 123-138.

Campbell, D. A. (1998). Architectural Construction Documents on the Web: VRML as a Case Study. *Digital Design Studios: Do Computers Make a Difference?* T. Seebohm and S. V. Wyk. Quebec City, Canada, ACADIA 98: 266-275.

Card, S. K., G. G. Robertson and W. York (1996). The WebBook and the WebForager: An information workspace for the World Wide Web. *CHI '96 ACM Conference on HUman Factors in Software*, ACM: 111-116.

Cruz-Neira, C., D. J. Sandin, T. A. DeFanti, R. V. Kenyon, et al. (1992). "The CAVE: Audio Visual Experience Automatic Virtual Environment." *Communications of the ACM* **35**(6): 65-72.

Dauner, J., E. Landauer and I. Fraunhofer (1998). 3D Product Presentation Online: The Virtual Design Exhibition. *VRML 98 – Third Symposium on the Virtual Reality Modeling Language*. S. N. Spencer. Monterey, CA, ACM: 57-62.

Davidson, J. N. and D. A. Campbell (1996). Collaborative Design in Virtual Space: Greenspace II: A Shared Environment for Design Review. *Proc. 1996 National Conference Association of Computer Aided Design in Architecture (ACADIA '96): "Collaboration, Reasoning, and Pedagogy"*. P. McIntosh and F. Ozel. Tucson, AZ, 165-179.

Dille, E. (1996). *Exploring Moving Worlds*. Research Triangle Park, NC, Ventana.

Donath, D. and H. Regenbrecht (1996). Using Virtual Reality Aided Design Techniques for Three Dimensional Architectural Sketching. *in Proc. 1996 National Conference Association of Computer Aided Design in Architecture (ACADIA '96): "Collaboration, Reasoning, and Pedagogy"*. P. McIntosh and F. Ozel. Tucson, AZ, ACADIA: 199-212.

Eick, S. G., J. L. Streffen and E. E. Sumner (1992). "SeeSoft – A tool for visualizing software." *IEEE Transactions on Software Engineering* **18**: 957-968.

Furness, T. A. (1986). The supercockpit and its human factors challenges. *30th Annual Meeting of Human Factors Society*. Dayton, OH: 48-52.

Givaty, G., H. A. vanVeen, C. Christou and H. H. Bulthoff (1998). Tele-Experiment – Experiments on Spatial Cognition using VRML-based Multimedia. *VRML 98 – Third Symposium on the Virtual Reality Modeling Language*. S. N. Spencer. Monterey, CA, ACM: 101-105.

Green, P., J. H. Hill and R. M. Satava (1991). "Telepresence: Dextrous procedures in a virtual operating field." *Surgical Endoscopy*(57): 192.

Heim, M. (1998). *Virtual Realism*. New York, Oxford, Oxford University Press.

Johnson, W. L. and J. Rickel (1997). "Steve: An Animated Pedagogical Agent for Procedural Training in Virtual Environments." *SIGART Bulletin* **8**(1-4): 16-21.

Lawton, W., T. Poston and L. Serra (1995). Time-lag reduction in a medical virtual workbench. *Virtual Reality Applications*. R. A. Earnshaw, J. A. Vince and H. Jones. London, New York, Academic Press.

Mackinlay, J. D., S. Card and G. Robertson (1991). Perspective Wall: Detail and Context Smoothly Integrated. *ACM SIGCHI '91 Conference on Human Factors in Computing Systems*. New Orleans, LA, ACM: 173-179.

McCall, R., S. Holmes, J. Voeller and E. Johnson (1998). World Wide Presentation and Critique of Design Proposals with Web-PHIDIAS. *Digital Design Studios: Do Computers Make A Difference? ACADIA 98*. T. Seebohm and S. V. Wyk. Quebec City, Canada, ACADIA:Association for Computer-Aided Design in Architecture: 254-265.

National_Research_Council (1995). *Virtual Reality – scientific and technological challenges*. Washington D.C., National Academy Press.

Ouh-young, M., D. Beard and F. Brooks (1989). Force display performs better than visual display in a simple 6D docking task. *IEEE Robotics and Automation*: 1462-1466.

Pinet, C. (1997). Design Evaluation Based on Virtual Representation of Spaces. ACADIA 97, Association of Computer Aided Design in Architecture: 111-120.

Robertson, G. G., S. K. Card and J. D. Mackinlay (1991). Cone Trees: Animated 3D Visualization of Hierarchical Information. *CHI 91*, ACM: 189-194.

Robertson, G. G., S. K. Card and J. D. Mackinlay (1993). "Information visualization using 3D interactive animation." *Communications of the ACM* **36**(4): 57-71.

Rygol, M., A. Dorbie, A. Worden, S. Ghee, et al. (1995). Tools and Metaphors for User Interaction in Virtual Environments. *Virtual Reality Applications*. R. A. Earnshaw, J. A. Vince and H. Jones. London, New York, Academic Press: 149-161.

Satava, R. M. (1992). "Robotics, telepresence and virtual reality: a critical analysis of the future of surgery." *Minmally Invasive Therapy* **1**: 357-363.

Vince, J. (1995). Virtual Reality Techniques in Flight Simulation. *Virtual Reality Systems*. R. A. Earnshaw, M. A. Gigante and H. Jones. London, New York, Academic Press: 135-141.

How do distributed design organizations act together to create a meaningful design?

Towards a process model for design coordination

Jeffrey Huang
Harvard University, Graduate School of Design, USA

Key words: Distributed Design, Coordination Theory, Process Modeling, Process Redesign, Collaborative Design Environments.

Abstract: This paper describes the results of a longer-term research project that looked at CAAD as an enabler of completely new collaborative processes rather than as a support for existing collaborative processes. In order to question existing design processes and invent new collaborative processes systematically, we applied a process modeling methodology that employed recent developments in coordination theory. The methodology contained four steps: (1) Decomposition, (2) Dependency Analysis, (3) Process Substitution, and (4) Recomposition. In this paper, we describe how this approach was used to redesign a sample collaborative design process in building design, and present the resulting process coordination model. We describe the implications of this model for the development of collaborative environments, and illustrate its practical application in a case study. We conclude by reiterating the contributions made.

1. INTRODUCTION

With the popularization of the Internet and the World Wide Web, we observe a shift of interest in the use of computers in architectural design (CAAD): from a stand-alone machine to support individual design to a networked device for coordinating distributed design activities [Schmitt 1996, Junge 1997, Jeng and Eastman 1997]. Recent research efforts in collaborative design have focused on solving technical problems of collaborative design, such as drawing exchange standards, product models and application sharing, and on the support of teaching design collaboration among designers within the same domain, such as in "virtual design studios" [Wojtowicz 1994, Kolarevic et al. 1997]. In most of these examples, CAAD has been applied to facilitate and improve existing collaborative processes.

This paper describes the results of a three year long research project (1995-98) that took a slightly different approach. Our research focused on inter-domain design collaboration, among architectural designers, mechanical, structural and electrical engineers, manufacturers, owners and local authorities, and aimed at questioning and redesigning the existing collaborative processes among these participants by the use of web-based CAAD technologies. Thus, what distinguishes our approach from other research initiatives is our perspective. We looked at CAAD as an enabler of completely new collaborative processes (which in turn may drive the development of appropriate CAAD systems), instead of as a support for existing collaborative processes.

In order to question existing design processes and invent new collaborative processes systematically that use web-based CAAD systems, we applied a process modeling methodology that employed recent developments in coordination theory [Malone and Crowston 1994]. In this paper, we describe how this approach was used to redesign a sample collaborative design process in building design, and present the resulting process coordination model.

The remainder of this paper is organized as follows. In the next section, we describe the process handbook methodology and coordination theory which form the basis for our research. Section 3 then presents the coordination model for a sample building design process. Next, in section 4, we describe in detail the implications of this model for redesigning collaborative design process and for the development of collaborative environments. We use the design of a window as a microcosm. Finally, we reiterate the contributions made.

2. PROCESS HANDBOOK METHODOLOGY

We started our investigation by looking at other industries, including examples from the aerospace, automobile and professional service industries, and studied how advanced firms in these industries deploy information systems to support distributed design and problem-solving efforts. In all examples we studied, a similar pattern emerged; the improvements did not come simply from the computerization of existing "production" activities, but from a fundamental change in and redesign of the "coordination" activities: what is in-between activities, and how activities performed by different participants were coordinated.

We then contrasted this with the way information technology has been applied to building design so far. In the building industry, information systems have been mostly used to automate existing processes: information technology is applied in design firms to expedite the existing drafting processes, and document production processes. In contrast, the examples in the other industries have used IOS to invent new processes, which in turn led to the development of appropriate technologies. In the examples we studied, information technology was an *enabler* of new organizational processes, which in turn were *drivers* of new technology.

From this comparison, it became clear that in order to leverage the potential of Internet-based technologies, it would be necessary to challenge and fundamentally rethink existing processes in building design, in parallel with considering the potential of enabling information systems.

2.1 Approach

To challenge and redesign the existing processes in building design systematically, we used and adapted the process handbook methodology [Malone et. al. 1993], a methodology currently under development at the Center of Coordination Sciences, MIT Sloan School of Management.

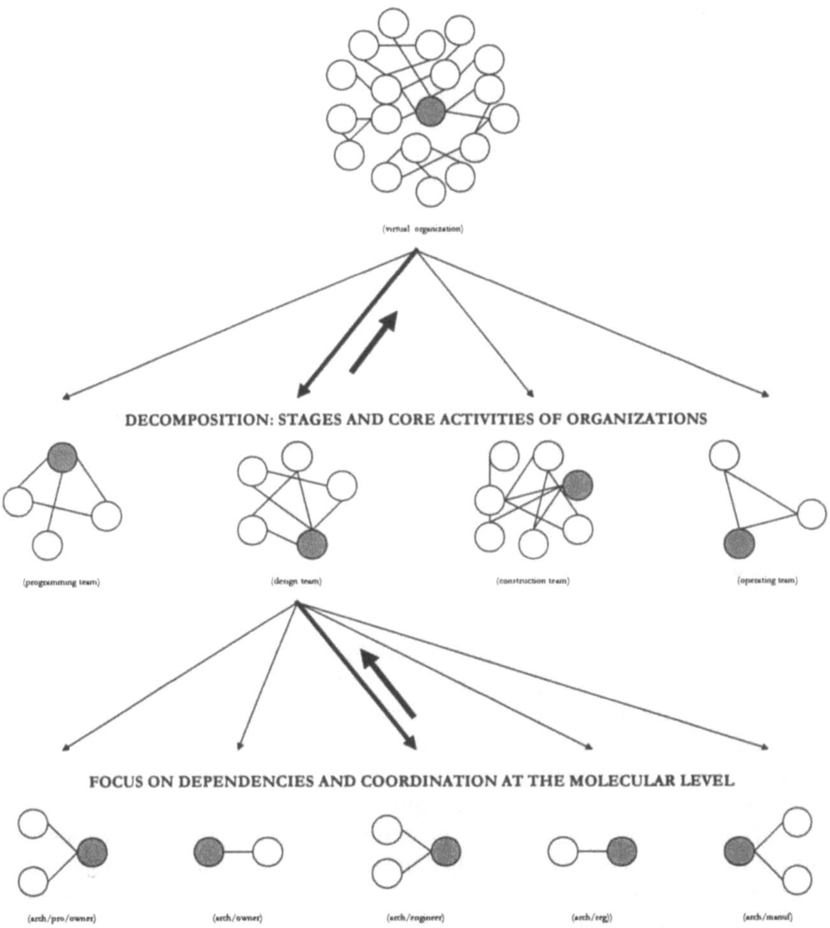

Figure 1. Schematic representation of the proposed coordination approach to process redesign.

Simply put, the proposed approach contained the following steps: (1) decompose the complex organizational processes into manageable subprocesses, (2) investigate the interorganizational dependencies among those subprocesses, (3) replace the existing coordination processes (which are used to manage these dependencies) with alternative coordination mechanisms that use IOS, and (4) recompose the processes into a new design organization.

Figure 1 illustrates the approach schematically. At the top is the complex design organization composed of multiple processes performed by different autonomous participants. The first decomposition (in the middle) disentangles this web of connections into phases and participants, and a further decomposition (bottom) makes explicit the dependencies between the interorganizational processes and the coordination mechanisms --the "glue" that binds together the processes at the molecular level.

Based on this explicit process model, alternative coordination mechanisms that use new information systems could be deployed to replace the existing ones. The recomposition of the new processes (the arrows pointing upwards) resulted in a new design organization.

This approach involved interpretations of, and generalizations from, field interview data. In order to provide a more rigorous framework, while maintaining closeness to practice, it was therefore important to use an appropriate process modeling language. In the following, we describe the process representation and taxonomy we have used in more detail.

2.2 Data Modeling

Recent trends in organizational studies and management, including total quality management (TQM), and Business Process Reengineering (BPR) have stressed the importance of a process view of organizations. As a consequence, several process modeling tools have become available. For this research, we reviewed several modeling methodologies for their compatibility with our purposes, including methodologies originating from academic research, business process modeling and database design. These included Yourdan's dataflow diagrams [Yourdan 1989], the Process Handbook [Malone et al. 1993], IDEFo [DeWitte et al. 1997], Martin's OOAD [Martin 1992], Chen's ER model [Chen 1976], and Booch's OOD [Booch 1994].

We selected the process handbook methodology (PH), developed at the Center for Coordination Sciences (CCS) at MIT [Malone et al. 1993], as the most appropriate modeling tool. The advantage of the PH methodology derives from its two key concepts of "dependency analysis" and "specialization", which are explained in more detail below.

Essentially, the process handbook supports not only the modeling of the core activites of an organization, as other modeling tools do, but offers an orthogonal view of the processes, i.e., a view of the "processes between the processes". In other words, it provides a representational logic for analyzing the dependencies between processes, and for finding alternative coordination mechanisms for managing them.

Since our focus was on *inter*-organizational processes and on the potential implications of new coordination technologies, the process handbook methodology was most appropriate.

The process handbook methodology is based on coordination theory [Malone and Crowston 1994], which is defined as "a body of principles about how the activities of separate actors can be coordinated". Coordination then is defined as "managing dependencies". Based on this theory, the process handbook offers a language for (1) modeling organizational processes in activity hierarchies comprised of *decompositions* and *specializations*, and (2) categorizing the *dependencies* among these activities, and evaluating the *coordination* mechanisms required to manage these dependencies.

2.3 Decomposition and Specialization

In the process handbook methodology, processes are modeled using activity hierarchies. Activities are broken down into *decompositions* and *specializations*. Decomposition breaks an activity into its smaller components (subactivities), while specializations consists of the various ways in which the activity can be accomplished.

Using the concept of specialization, the processes are arranged in a hierarchical network with very generic processes at the top and increasingly specialized processes at lower levels. Specialized activities inherit the properties of the (more generic) parent activity.

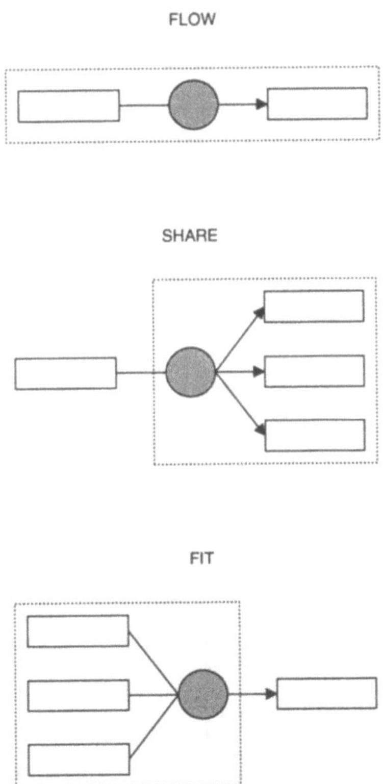

Figure 2. The three basic types of dependency: Flow, Share, and Fit

2.4 Dependency Analysis

Once an activity hierarchy is established, the dependencies between the activities can be analyzed. According to coordination theory [Malone et al. 1993], there are three basic dependency types: flow, share and fit. These three dependency types denote the relationship between two or more activities with respect to a resource:

- A *flow* dependency arises when a first activity produces a resource that is used (flows to) by a second activity. This type of dependency occurs almost all the time and is the focus of existing process mapping techniques (such as flow charts).
- A *sharing* dependency arises when multiple activities use (share) the same resource. For example, a particular machine that is used by different actors for different activities must be shared. Typical resources that need to be shared include time, money, space, machines and the manager's attention.
- Finally, a *fit* dependency occurs when multiple activities concurrently produce a single resource. A typical example is concurrent design in which several designers and engineers work together on the same design. Their activities must fit together in order to produce a good design.

Dependencies are managed by coordination processes.

2.5 Process Taxonomy

In order to ensure a certain degree of generality, and make certain that the proposed data model is shareable, a certain rigor was necessary in naming the processes. The process handbook offers a taxonomy for classifying activities in a specialization tree, which proved to be a useful point of departure.

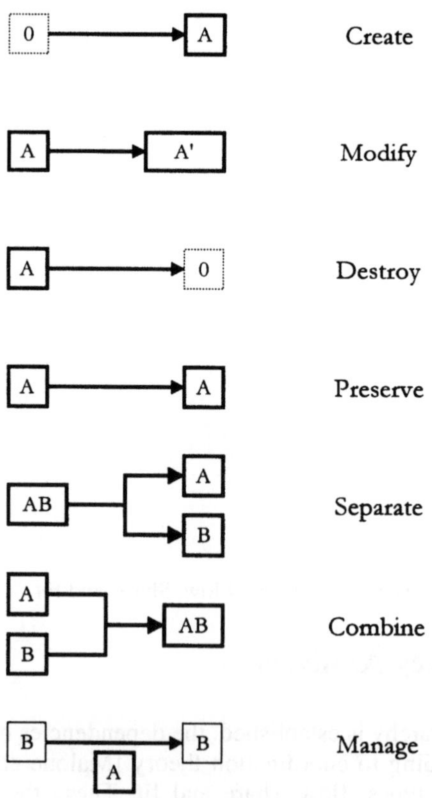

Figure 3. Logic of the top-level activities

The top level of the specialization tree consists of seven verbs: "create," "modify," "destroy," "preserve," "separate," "combine" and "manage." These seven activities are the first specialization of the root verb "act." They derive logically from the basic possibilities or effects a verb can have in relationship to a resource or another verb. "Create," "modify," "preserve," "destroy," "separate," and "combine" act on a resource, while the verb "manage" acts "perpendicularly" on the other activities, in the sense that it manages their dependencies. These seven top-level activities provide a framework for classifying all other existing organizational processes. Any organizational process can be modeled as a specialization of one of these seven basic verbs.

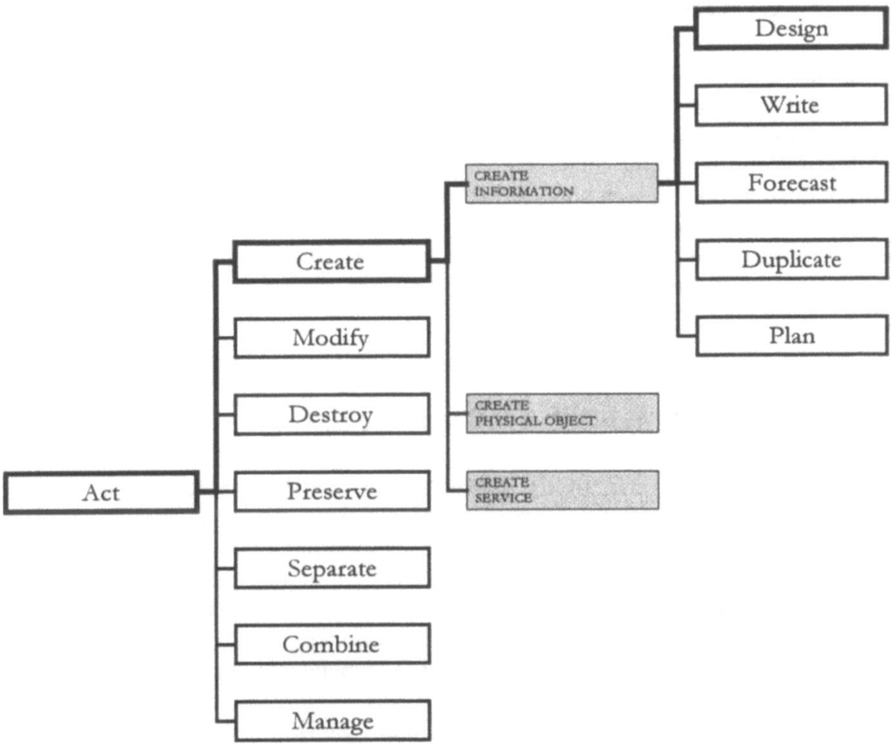

Figure 4. Design as a Specialization of "Create"

The logic of the relationship between verbs and resources, which forms the rationale for the classification, is the following. "Create" contains the specialization of verbs that transform a non-resource (0) into a resource (A). "Modify" contains the verbs that transform a resource A into resource A'. "Preserve" contains the specialization of verbs that keep a resource (A) as it is. The specializations of "Destroy" transform a resource (A) into a non resource (0). "Separate" contains the verbs that transform two or more combined resources (AB) into separated resources (A) and (B), "Combine" contains the activities that put two or more previously separated resources (A) and (B) together (AB). Finally, "manage" contains the verbs that manage the dependencies between the activities.

Design, for example, by its very nature a creative process, is a specialization of "create information." Similar to other activities under the same specialization branch "create information," design reduces entropy or randomness of information. Drawing from a definition by Bakos and Pollalis [1987], design is an activity to "impose meaningful order."

This taxonomy and classification scheme gives process analysis a point of reference and a common denominator, which make it possible to share the process model with participants in the building industry, and learn from the best practices in other fields.

3. COORDINATION MODEL

In this section, we describe how the process handbook methodology outlined above was used to analyze the inter-organizational dependencies in distributed design organizations and develop a coordination model. Our coordination model was organized according to the three-stage model of design described by [Pollalis 1994]: concept design, final design and production design. Here we present only a fraction of the coordination model we have developed. We document the major dependencies in the concept design phase, a sample dependency, and the perceived alternative coordination processes. The complete model is described in [Huang 1997]

3.1 Dependencies in Concept Design

Figure 6 shows a map of the dependencies that arise between the activities performed by the individual organizations during concept design. The core activities are aligned vertically within the box of the corresponding organization that performs the activity. For example the owner organization to the very left performs the activities *identify needs, assemble design team (architect), assemble design team (engineer), select site, approve program, select financing,* and *approve concept.* The lines and circles indicate the existence of a dependency between two or more processes.

We identified seven generic types of interorganizational dependencies which occur during concept design. Dependencies arise: (A) between assemble team/sell service, (B) between identify needs/define requirements, (C) between select site/provide land, (D) between select site/define requirements/complete survey and approve program, (E) between select financing and provide financing, (F) between select financing/develop concept/study financial feasibility/study technical feasibility and approve concept, and (G) between obtain approvals and approve concept.

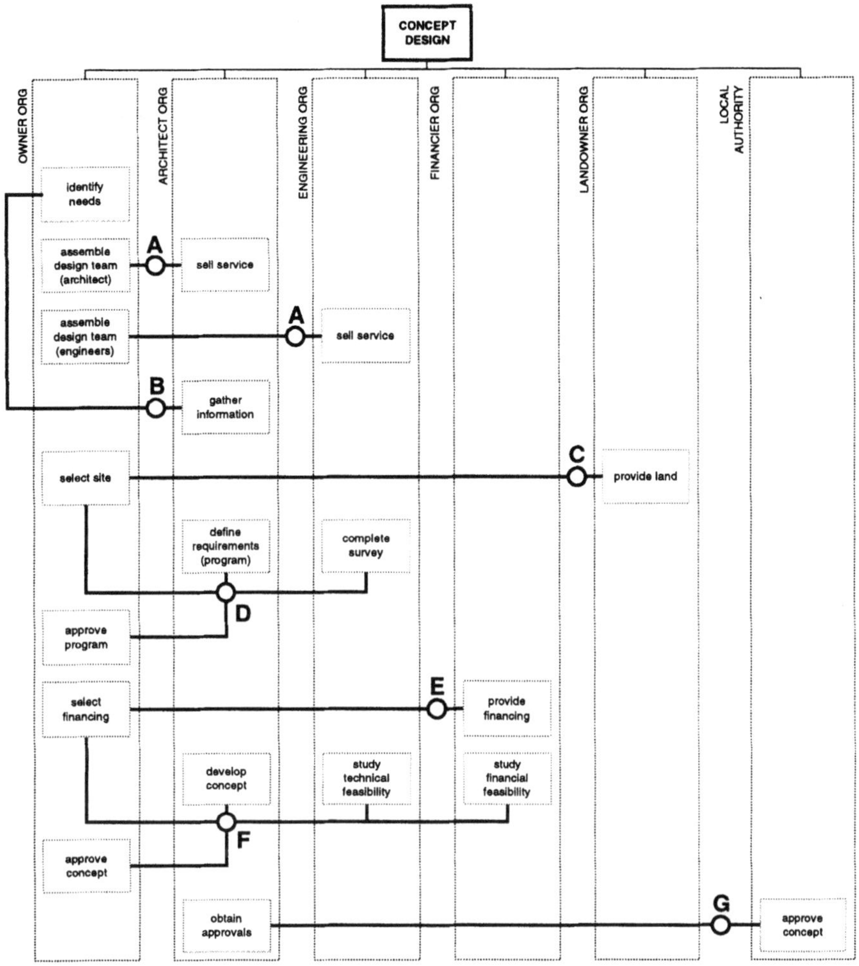

Figure 5. Inter-organizational dependencies between processes performed during concept design

Each of these dependencies can be managed differently. In the following, we will describe the sample dependency "A" and the alternative coordination mechanisms used to manage the dependency.

3.2 Dependency "A": Assemble Design Team/Sell Service

At the beginning of concept design, the number of organizations that might be involved in the project is almost unlimited. Any design and engineering organization with the appropriate knowledge and available resources may choose to participate in a particular project, and sell its services to do so. Type A dependency exists between these *sell service* processes performed by the candidate firms and the *assemble design team* process performed by the owner.

The dependency between the owner's *assemble design team* process, and the designer's *sell service*, is a *flow/share* dependency. The resource, an assignment or contract for a particular project, *flows* from the owner's activities to the architect's activities. This dependency must be managed in order to ensure that the right contract is awarded to the right organization. Because many candidate organizations apply for the same assignment, there is also a *share* dependency among the *sell service* activities performed by them: resources (the job to be done, and, in particular, the compensation for the service) are limited and must be *shared* by the appropriate organizations.

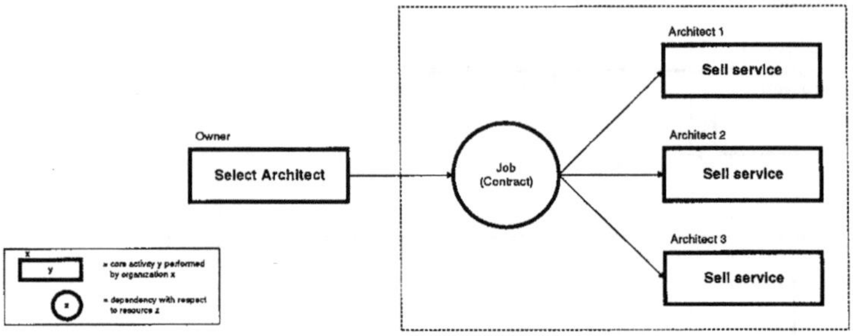

Figure 6. The dependency between the owner's selection and the architect's sell service process

This flow/share dependency has the following characteristics. The knowledge exchanged is of tacit nature. Consider, for example, the design of a new corporate headquarters for a high technology firm. Typically, the owner organization is not in a position to state explicitly its precise needs in advance. This is not surprising as, in reality, part of the design process is finding and formulating the building needs of the owner organization. On the other hand, it is difficult for the architectural organization to show ex ante (before the contract) what kind of design it will develop. The architect organization must therefore be able to show its tacit knowledge of designing (its competencies, abilities, and values) through explicit documentation of previous work, appropriate subcontracting of competent specialists to complement the team, and references and recommendations from past customers. Further, it must quickly articulate the intended approach to the particular project in question through sketches, proposals, and oral presentation.

Typical coordination mechanisms to manage this flow/share dependency between team selection and sell service include: *manage by competition*, *manage by prequalification* (short-listing), *manage by direct assignment* based on prior experience and reference, and *manage by using an in-house architect*. Alternative coordination mechanisms to manage this flow/share dependency that use shared information systems include: *manage using knowledge network, manage by posting on a bulletin board, manage by on-line broker, manage by a bargain finder agent*,.

By modeling the dependencies that exist and capturing alternative coordination mechanisms to manage them in a systematic "process handbook", it is then possible to redesign complete design processes by replacing existing coordination processes with alternative ones that are appropriate for a given context.

4. PROCESS REDESIGN: AN EXAMPLE

In this section, we will show, in a concrete example, how the proposed coordination model was used to redesign existing processes in design systematically. The process we have selected as an example is window design, a microcosm of the larger design process in which a significant portion of the complexity of a typical cooperative design effort is already represented. The process involves the participation of several autonomous design organizations, whose efforts need to be coordinated in order for a meaningful design to emerge.

4.1 Existing Process of Window Design

Let us first consider the process of designing a window in detail. Who is involved? What are the typical processes each participating organization performs? What are the dependencies and typical coordination processes?

The participants in the process of window design are the architect, the engineer, the general contractor, the subcontractor and the manufacturer. Together they form the core design organization, with each participant contributing its specific knowledge to the process. Figure 7. depicts the window design organization as a nexus of distributed design knowledge.

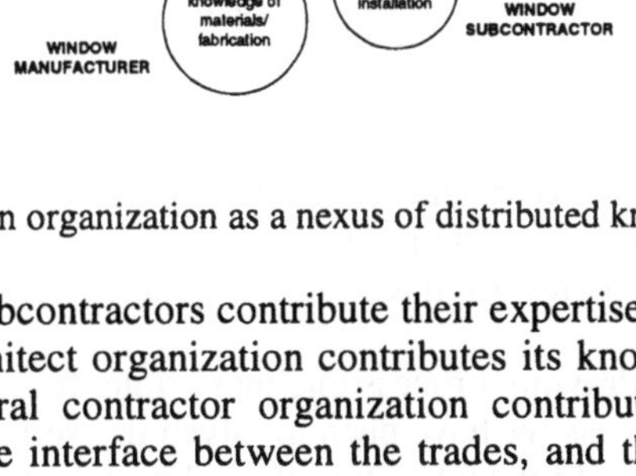

Figure 7. The window design organization as a nexus of distributed knowledge

Manufacturers and subcontractors contribute their expertise concerning materials and installation. The architect organization contributes its knowledge of the overall design intent. The general contractor organization contributes its knowledge of constructibility and of the interface between the trades, and the structural engineer organization contributes its knowledge of the structural system.

4.2 Problems of the Existing Process

The existing process of designing a window is considered very inefficient by almost all participants we have interviewed. It is inefficient in terms of time, costs and quality. The typical duration of the process described above ranges from 6 - 8 weeks. Of these 6 - 8 weeks, however, only a fraction is effectively spent on advancing the design. Most of the time, the design documents, which form the basis for the advancement of design, are in *transition* from one process to another, "in-between the processes." They are at the copy institution, in the post-office, in the transportation truck and plane, or sitting in the backlog of a participant, waiting to be processed. According to one of the participants, all of the design work could be done in less than three days, instead of the 6-8 weeks, if all participants were collocated in one place.

The process is also relatively costly for what it actually achieves. The direct labor costs (actual man-hours spent) aside, there are the costs of reproducing and of materials utilized. Consider the costs of photo-copying the shopdrawings. Each shopdrawing is photo-copied an average of thirty to forty times during the shopdrawing process. For each minor revision, the appropriate shopdrawing needed to be copied five to six times, in order to be distributed to all participants.

More importantly, however, there are design quality issues. Because of the inefficient nature of the process, the designer is far removed from the actual window makers. This alienation makes the designer increasingly lose control of the manufactured product, which, in the end, impacts on the quality of the window, and leads to a deterioration of design quality. The process, as it is set up today, does not foster an effective use of manufacturer knowledge. Design suggestions have to travel too far a way, through pre-described lines of communication, for innovative ideas to emerge. The lengthy process stifles new ideas and suggestions for improvements.

4.3 Current automation efforts

How can this process be improved? How are existing computerization efforts used to overcome the inefficiencies outlined above? Recent developments in commercial CAD programs aim at eliminating the use of paper documents in design, and moving towards the goal of an often heralded "paperless" design. According to this promise, design documents will be produced digitally and sent to participants on diskettes or through modem. The benefits of such a paperless design are clear: the transfer of data will be accelerated, and the costs of multiplying drawings will be almost eliminated. However, as we have posited above, it is unlikely that, without changing the existing process, the mere substitution of paper drawings with CAD drawings, will significantly improve the existing window design process and eliminate the inefficiencies outlined above. This is particular true for window design.

Consider the shopdrawing process one more time. In a sample implementation of a paperless design process we have been involved in, we found that there was almost no time savings achieved by exchanging the submittals digitally as compared to the traditional paper-based shopdrawing process. Although the transfer of files through the modem was faster than photo-copying them and using overnight delivery systems (the typical traditional practice), the total duration of the shopdrawing process didn't decrease. The reason was the following. The real bottleneck in the shopdrawing process was not the time it took to transmit the documents, but the actual time it took for a participant to effectively act on a drawing. Current standards of practice give the professional up to 28 days to review a shopdrawing, and, since there are no incentives for the reviewer to expedite the process, the shopdrawings, whether digital or on paper, almost always end up in the in-basket of the participant, and remain there, inactive, until the time is due (or a reminder is received). Thus, the savings of one day obtained by sending the files electronically are insignificant compared to the 27 days still needed for the participant to process the documents sent in a rush.

This example points once again to the futility of automating an existing process, when the process design itself is inappropriate. It shows that there is a necessity for a deeper understanding of the dependencies between existing processes performed by the different organizations, and to fundamentally reengineer them.

4.4 Process Redesign

Let us thus apply the approach put forward above to accomplish this end. We will perform the following three steps: (i) analyze the deep structure of the *design a window* process, (ii) map its core activities and the dependencies, and (iii) replace the existing coordination processes with alternative ones that use IOS.

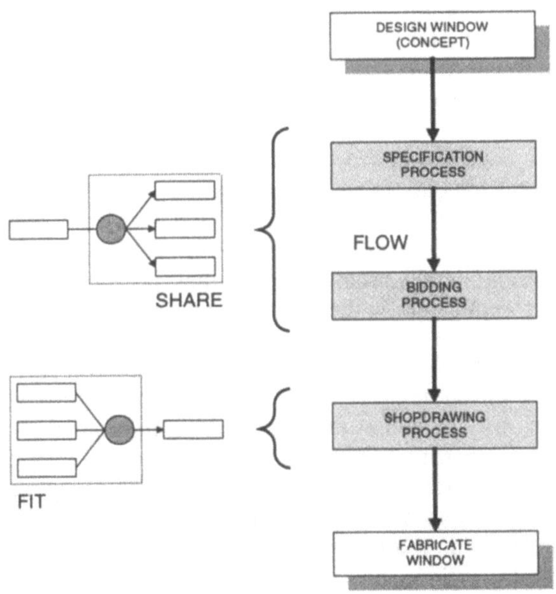

Figure 8. The generalization of the observed process into three coordination process: one possible surface structure of designing a window

4.4.1 Deep structure analysis

What is the deep structure of the window design process? At the very basic level, it is the transformation of an idea (an abstract concept of a window) into a real artifact (the built and installed window). Or differently stated, it is the process of managing the *flow* dependency between the concept design process and the fabrication process. This *flow* dependency, however, is not a simple one. Because multiple autonomous design organizations are involved, it is necessary to decompose the flow dependency into two separate dependencies. (1) a share/flow dependency between select participants/sell service, in which the distributed knowledge required for the process is found and activated, and (2) a fit/flow dependency between develop/approve design, in which the distributed knowledge is leveraged and coordinated.

4.4.2 Existing coordination processes

These share/flow and fit/flow dependencies are currently managed by the three coordination processes: a *specification* process, a *bidding* process, and a *shopdrawing* process.

The first process is the *specification* process which manages the select/sell product dependency. Generally, the designer browses through catalogs, talks to manufacturer's representatives, gets feedback from the client, and specifies the window. Since designers lack the expertise to know exactly what exists and what is possible, advice from window manufacturers and window contractors is solicited at this stage. Manufacturers and contractors provide free information, with the hope of being awarded a contract later in the process. However, they are also aware that the design will ultimately go to a competitive bid, and that the designer will not have the decision right to select them. Accordingly, their unpaid involvement at this stage has its limits.

In the second coordination process, the bidding process, the design and specifications are sent for bid. The bidding process manages the select/sell services dependency. Generally, in this process, a window design and specification is bid by several general contractors. In order to bid, the general contractor solicits price quotations from subcontractors, who in turn get their prices from manufacturers. The bidding process is complicated by the fact that subcontractors don't like to publicize their bids and wait until the last minute to place them. Further, subcontractors often submit bids to multiple contractors bidding on the same project, which allows them to vary their quotes for strategic reasons. Manufacturers, usually fewer in number than subcontractors, provide quotes and availability information to subcontractors.

Finally, the third coordination process is the shopdrawing process in which manufacturers and contractors prepare details of the windows, make suggestions and send these with samples and other submittals, through the general contractor to the designer, back and forth, until all participants have approved the shopdrawings. This usually takes several iterations, because of omissions and errors, ambiguities which need to be clarified, or because the details of the manufacturer do not correspond to the designer's design intentions.

4.4.3 Substitution

Let us now apply alternative coordination mechanisms to redesign this process. Figure 10. shows how the *specification*, the *bidding* and the *shopdrawing* process can be replaced by two coordination mechanisms that use Internet-based interorganizational systems. In this illustration, we have plugged in an *electronic marketplace*, and a *concurrent design platform* for replacing the existing coordination processes.

The *electronic marketplace* manages the share/flow dependency facilitating the finding of the right specialists and products in the window design market [Huang and Pollalis 1997]. The *concurrent design platform* manages the fit/flow dependency, enabling participating designers, engineers, contractors and manufacturers to develop and refine *concurrently* the details of the window.

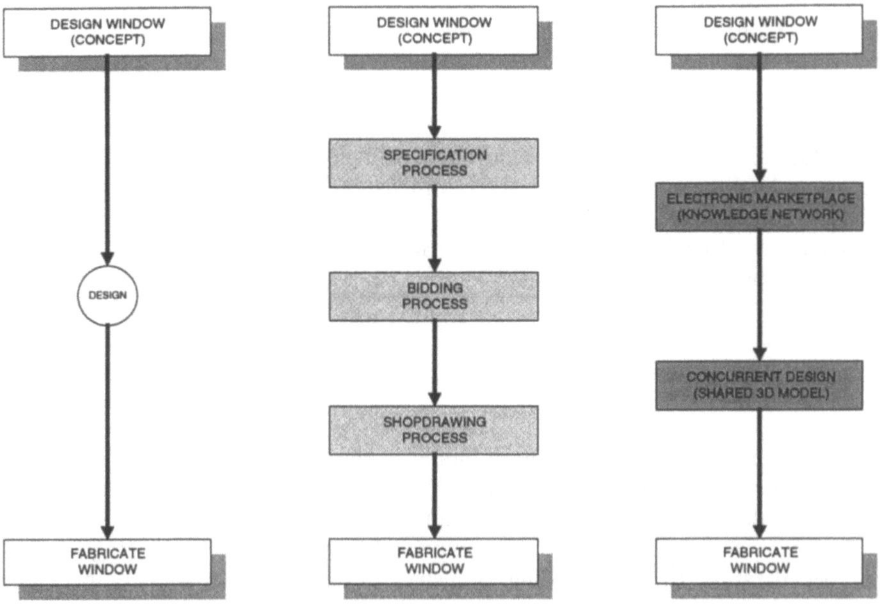

Figure 9. The juxtaposition of the deep structure of "design a window", the existing surface structure, and an alternative surface structure that uses Internet-based coordination mechanisms

4.5 Alternative coordination mechanisms

The new process offers the following coordination advantages of Inter-Organizational Systems, as described in [Huang 1997].

4.5.1 Electronic marketplaces

Electronic Marketplaces reach a wider range of potential candidates, reduce search costs, and enhance comparability in the selection of the appropriate supplier. They augment the likelihood that the right window will be found, and an appropriate fabricator selected, whose capabilities are aligned with the designer's intentions [Bakos 1992]. The contract for design services awarded to the window manufacturer and installer extends the design team to include downstream participants. The architect assigns design responsibilities for detail design to the manufacturer, just as he traditionally assigned responsibilities to the engineers, and to the technical consultants. Active involvement in the design process is beneficial for the manufacturer in many ways. First it gives him an opportunity to earn a good reputation, which, in a transparent marketplace enhances his chances to win the contract. Second, active involvement and interaction with upstream participants brings him closer to the market, and provides him with valuable information about emerging customers' needs, and ideas for new product development. With the existing trend towards mass-customization and tailored products, the manufacturer's involvement in design will become essential for his survival.

4.5.2 Concurrent Design Platforms

The use of the *concurrent design platform* transforms a previously sequential process into a parallel, interactive process. Participating organizations are collocated in a virtual shared space, and are provided with tools for design interaction. The discussion database attached to the three-dimensional model enables the participants to have conversations around particular issues that need design attention. The accountability of each participant's action in a transparent environment anticipates positive competition and true cooperation.

5. CONCLUSION

We have presented a coordination model of the design process and have shown in one particular example how it can enable us to redesign existing coordination processes in distributed inter-domain design settings.

The research was based on the assumption that with the growing fragmentation and specialization of the building industry, design has increasingly become a complicated organizational process, in which specialized knowledge is leveraged from multiple autonomous organizations. Accordingly, design performance depends upon how this distributed knowledge is coordinated and collocated with the decision-making processes of the designer. In other words, a design organization is more likely to produce a good design, if the available knowledge is well leveraged into the design decision making process: more knowledgeable design decisions lead to better built environments.

New information systems promise to improve the coordination of knowledge and decision rights in design organizations. The success stories from other related industries provide clear evidence. Current applications of computers in building design, however, instead of bringing the distributed design organizations together, have so far further aggravated the existing fragmentation by focusing on the mechanization of existing intra-organizational processes and creating islands of automation.

Therefore, in order to challenge and redesign existing ways of design coordination, and improve the collocation of knowledge and decision rights, we need to shift our focus from an organizational view to an inter-organizational perspective, and make explicit the *deep structure* of the dependencies among the activities of the various participating organizations.

This is what we have tried to do. With the explicit mapping of how autonomous design organizations act together, we have put forward a shareable basis for systematically inventing new design processes, and for developing appropriate enabling interorganizational information systems.

6. ACKNOWLEDGMENTS

This paper summarizes the theoretical part of my doctoral dissertation at the Harvard Design School. I wish to thank the members of my thesis committee for their invaluable guidance and support: Professor Spiro N. Pollalis, Professor Gerhard Schmitt, Professor Yannis Bakos, and Professor Carl Sapers. Special thanks also to Professor Thomas Malone, Director of the Center of Coordination Sciences at the MIT Sloan School. The research was supported by a scholarship from the Harvard Design School.

7. REFERENCES:

Bakos, Y. N., "A strategic analysis of electronic marketplaces," *MIS Quarterly*, 15, 3 (1991), pp. 295-310.

Bakos, J.Y., and Pollalis, S.N., *A Framework for the Design Process*, No. LCT-87-2, Harvard University Graduate School of Design, Cambridge, MA, November 1987.

Booch, G., *Object-oriented analysis and design with applications*, Redwood, CA: Benjamin/Cummings Publishing, 1994.

Chen, P.P.S., "The entity-relationship model: Toward a unified view of data," *ACM Trans. Database Syst.*, 1, 1, pp. 9-36, 1976.

DeWitte, P., Pourteau, C., "IDEF enterprise engineering methodologies support simulation," *Manufacturing Systems: Information Technology for Manufacturing Managers* (March 1997), pp. 70-75

Kolarevic Branko, Schmitt G., Hirschberg U., Kurmann D., Johnson B., An Experiment in Design Collaboration, in Proceedings of (CD): *ACADIA'98 conference*, Quebec City, October 22-25, 1998

Huang, J., *Interorganizational Information Systems in Design*, Doctoral Dissertation, University Microfilms International, Ann Arbor, Michigan, 1997.

Huang, J., and S. N. Pollalis, "Knowledge, Agency, and Design Information Systems," in R. Junge (ed.), *CAAD Futures '97*, Kluwer Academic, Dordrecht, Netherlands, 1997.

Jeng, T, and C. M. Eastman, "A Database Architecture for Design Collaboration", *Proceedings of TeamCAD Workshop on Collaborative Design*, Atlanta, May, 1997.

Junge, R. (ed.). CAAD Futures 1997, Dordrecht: Netherlands: Kluwer Academic Publishers, 1997.

Malone, T.W., and Crowston, K., "The interdisciplinary study of coordination," *ACM Computing Surveys*, 1994 (March), 26 (1), 87-119.

Malone, T.W., Crowston, K., Lee, J., and Pentland, B., "Tools for inventing organizations: Towards a handbook of organizational processes," Working Paper, Cambridge, MA: MIT Center for Coordination Science, 1993.

Martin, J., *Object-Oriented Analysis and Design*, Englewood Cliffs: Prentice Hall, 1992.

Pollalis, S.N., "Towards the Automation of Conceptual Design," In Tzonis, A., and White, I. (eds.), *Automation Based Creative Design: Current Issues in Computers and Architecture*, Amsterdam: Elsevier, 1993.

Schmitt G., *Architektur mit dem Computer*, Vieweg Verlag, Wiesbaden, Oktober 1996

Wojtowicz, I. (ed.) *Virtual Design Studio*. Hong Kong: Hong Kong University Press, 1994.

Yourdon, E., *Modern Structured Analysis*, Englewood Cliffs, Prentice-Hall, 1989.

A Performance-inspired Building Representation for Computational Design

Georg Suter, Ardeshir Mahdavi, Ramesh Kirshnamurti
School of Architecture, Carnegie Mellon University, Pittsburgh, PA, USA

Key words: performance simulation, constraint-based design, geometric modeling

Abstract: A building representation for integrated building performance simulation is described, which addresses several important issues. First, it captures informational requirements for detailed performance analysis and maintains geometric integrity. Second, it provides computational support for efficient spatial queries and convenient building model manipulation. Representational elements include partitioning and refinement rules, containment hierarchies and dimension constraints. These features provide for ease of manipulation, geometric variety and spatial queries and are illustrated with examples. The paper concludes with a discussion of the limitations of the representation.

1. INTRODUCTION

A major goal in computer-aided design research has been the development of integrated design for convenient generation and evaluation of evolving designs. Despite progress in the development of integrated performance simulation systems (Augenbroe 1992, Mahdavi 1996), a number of usability issues have not been addressed effectively. These include support for the convenient manipulation of geometric and semantic building information.

Interfaces built on top of existing representations may improve ease of interaction by providing graphical interaction techniques. This is often achieved by limiting building models with regard to geometry (e.g. orthogonal geometry) and materials (e.g. global materials for enclosure types). Moreover, existing representations do not provide significant support for scalable operations and integrity management, both of which are critical for modeling large buildings effectively (Suter and Mahdavi 1998).

Four objectives have guided the development of the representation described in this paper. The representation should capture the informational requirements of detailed, physics-based building simulation; maintain geometric integrity; support efficient spatial queries; and provide scalable operations for design development.

2. REPRESENTATION REQUIREMENTS

2.1 Simulation completeness

Detailed performance analyses demand that representations be space based with access to attributes such as location, dimension, construction and material. Typical domains include energy, lighting, acoustics, and air flow analysis. It has been shown that the definition of generic spaces consisting of voids enclosed by walls, ceilings and floors that form a polyhedron are useful for simulation domains to construct their internal representations (Mahdavi 1996). Such representations can be computed automatically and with relative ease. A representation for energy simulation, for instance, involves the generation of a three-dimensional cell grid that is intersected with spaces. Spaces are bounded by space surfaces. Space surface segments associated with each grid cell need to be distinguished as being air, internal or external, depending on whether neighboring cells belong to the same space, to different spaces, or to the outdoor environment. Space related information is indispensable because it unambiguously distinguishes the indoor from outdoor environment. The concept of spaces has played a central role in the search for generic design representations (Eastman 1979, Khemlani and Kalay 1997). On the other hand, space-based representations may not always be necessary, e.g., for structural analysis or calculation of environmental loads.

2.2 Integrity

Integrity maintenance is an issue that is often ignored in drafting and simulation systems. The term integrity refers to rules that data must satisfy. Lack of integrity in data structures may cause a system failure or meaningless feedback to the user. In the context of simulation, a building model is geometrically valid if none of its elements interfere with each other, and if all spaces are enclosed by surfaces. Gaps between interior spaces, which may unknowingly occur during manipulation, should be excluded. Current simulation tools typically do not check, for instance, whether openings overlap, or whether each space is properly enclosed. A representation with procedures for automated integrity checking could at least partially relieve designers from integrity maintenance tasks.

2.3 Efficient spatial queries

Existing simulation representations currently offer little support for efficient spatial queries (the term efficient is used here in a broad sense). This is because these representations are structured in a manner that makes access to certain information difficult. Lighting and acoustic simulation, for instance, require information about external walls or neighboring spaces. More structured representations could either directly incorporate such frequently accessed relational information, or at least allow for its efficient derivation. This aspect is important for integrated simulation systems, in which common spatial queries could be performed more efficiently based on a rich, shared model.

2.4 Design development

Convenient modification of building models is relevant for design development. Again, existing representations typically do not provide much support. For instance, a designer may be interested in the relationship between the massing of a building and overall energy use. This could involve the modification of a building's overall width, depth, or height. In conventional systems, the designer would have to go through a series of tedious and error-prone changes to change overall dimensions. Additional difficulties with maintaining building models arise when relations across floors and among openings, are to be considered as well. In contrast to existing representations, a richer representation would provide scalable operations, which would allow users to perform discrete and continuous operations in one step, regardless of a building model's size.

3. REPRESENTATION DESCRIPTION

3.1 Partitioning

The above requirements inform the development of the representation for performance based computational design. The representation is organized around building geometry. Each new design session starts with an *initial configuration*, which consists of a default root with a default building on a default site. Designers manipulate spatial information by recursively applying *partitioning rules* to *volumes*, *surfaces*, and *lines*, which are the basic geometrical *types* supported by the representation. The concept of partitioning is analogous to partitioning regions in a subregion representation (Steadman 1973). Volumes are bound and separated by surfaces. Surfaces are bound and separated by lines. Manipulation of surfaces and lines may affect volumes and surfaces, respectively. In other words, a surface cannot exist without volumes, and, similarly, a line cannot exist without surfaces. An important feature of partitioning is the preservation of information with regard to parent entities, which remain accessible for further manipulation.

3.2 Partitioning terminology

Partitioning terminology is introduced through volume partitioning (Figure 1). Partitioning is always performed between opposite surfaces of a particular volume, i.e. between surfaces that do not share edges. Thus, a volume with six faces may be partitioned in three ways. Designers choose a surface pair, which are called *primary bounding elements* (Figure 1b). The remaining surfaces are called *secondary bounding elements*. New or reevaluated *partition* planes are intersected with those surfaces such that they are properly contained within the parent volume, or *entity* (Figure 1c). Each new partition defines a boundary for a new sub-volume, or *sub-entity*. Individual partitions and sub-entities have *labels* that allow for the attachment of semantic information as required by simulation. The *partitioning direction* is stored at the parent level and indicates the order for the insertion of new partitions. Designers may modify the partition direction at any time, even if partitions already exist. A modification of partitioning directions would trigger a reevaluation of the entities affected by that operation. Making partitioning directions available for manipulation facilitates operations such as mirroring entities.

The terminology applies analogously to surface and line partitioning.

Two basic partitioning styles are described in the following: orthogonal and non-orthogonal partitioning. Although non-orthogonal partitioning subsumes orthogonal partitioning, this distinction is made mainly because of simplified user interaction. The distinction between orthogonal and non-orthogonal reflects the pervasiveness of orthogonal geometry in buildings.

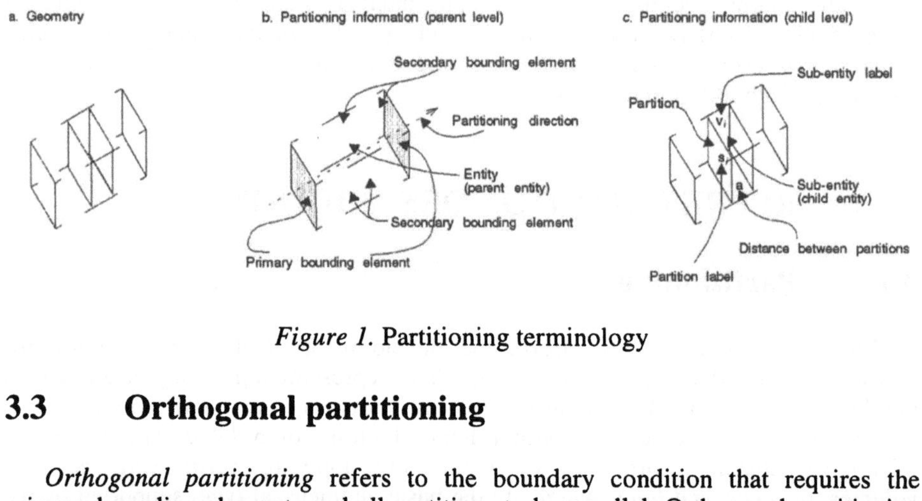

Figure 1. Partitioning terminology

3.3 Orthogonal partitioning

Orthogonal partitioning refers to the boundary condition that requires the primary bounding elements and all partitions to be parallel. Orthogonal partitioning is specified by three rules. Figure 2 illustrates the rules for volumes. Note that the illustration does not provide detailed information about the location of partitions between the primary bounding surfaces. This aspect of rule application is addressed by constraints and constraint satisfaction procedures, and will be discussed in detail later. Rules are presented for configurations with at least two partitions. These are also applicable to configurations with a single partition. In this case, a primary bounding element can be thought of as the second partition.

The first rule allows for the creation of child entities at a new level in the hierarchy tree (Figure 2a). It is only applicable to non-partitioned entities. Once a new hierarchy level has been created, existing entities may be modified, or new ones added. Figure 2b illustrates modification rules for configurations. Whenever a dimension is modified, partitions may adjust their position to accommodate the change. New partitions are parallel to the primary bounding surfaces and sibling partitions. Again, related sibling partitions are adjusted to make room for new partitions and their corresponding sub-entities (Figure 2c).

Examples of building or building component models generated by applying the orthogonal partitioning rule are illustrated in Figure 3. The rule may be applied to building sites, enclosures, openings, and indoor spaces. The examples represent finalized results rather than the sequence of rule applications. Note, again, that no details are provided at this point with regard to surface labeling.

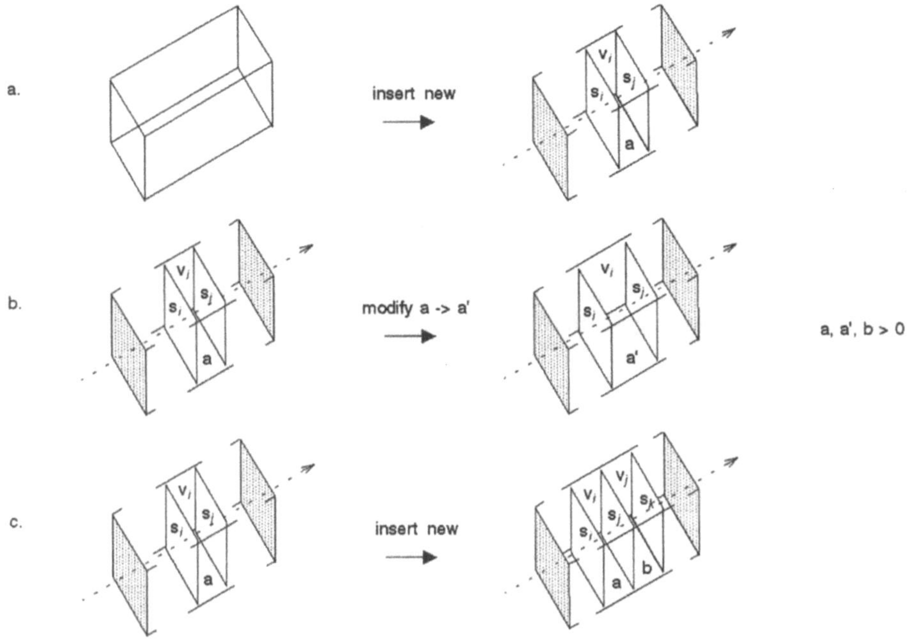

Figure 2. Rules for orthogonal partitioning (shown for volumes)

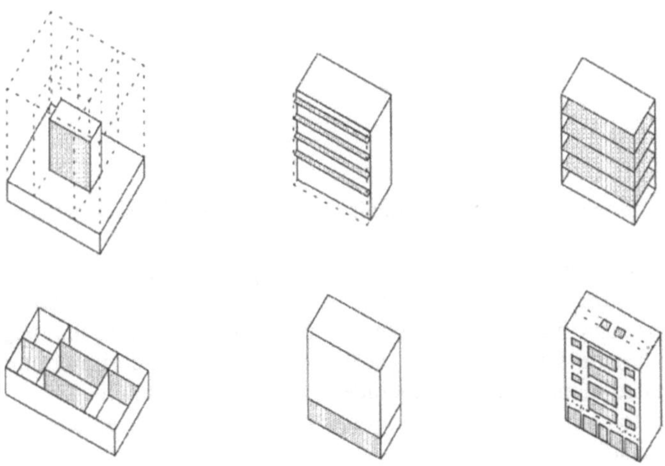

Figure 3. Examples for volumes and surfaces generated by orthogonal partitioning rule

3.4 Non-orthogonal partitioning

Non-orthogonal partitioning is similar to orthogonal partitioning. Again, partitions are inserted between the two primary bounding elements. Conditions are, however, less restrictive, facilitating greater freedom of manipulation Bounding planes or partitions are not required to be parallel. This is achieved by increasing the number of dimensions controlled by the designer from one to two, reflecting the spatial relations between the partitions.

Mapping from orthogonal to non-orthogonal partitioning is initiated by the designer (Figure 4a). According to Figure 4a, at least one orthogonal partitioning is required prior to the application of non-orthogonal partitioning rules. The transition from orthogonal to non-orthogonal partitioning involves the splitting of partitioning directions and distribution of dimensions parallel to the two emerging partitioning directions. As a result, all individual dimensions are available for continuous and discrete manipulation (Figure 4b,c).

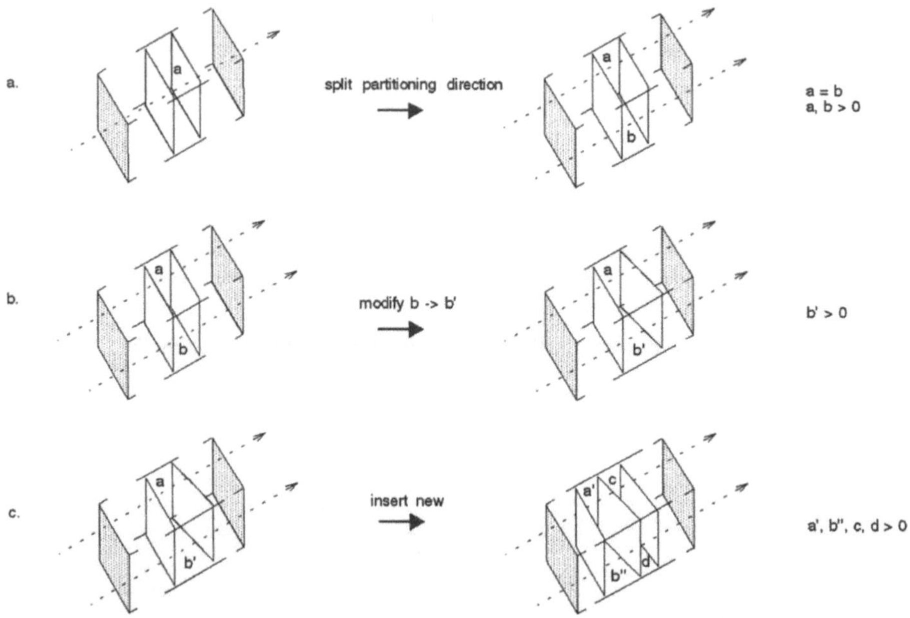

Figure 4. Rules for non-orthogonal partitioning of volumes and surfaces (not applicable to lines)

Mapping from non-orthogonal to orthogonal partitioning is harder, since this may require modification of dimensions in order to satisfy the constraint that partitions and primary bounding elements are parallel. As a result, the mapping may not be unique. It should be noted that, at this time, although desirable, mechanisms for such automated mapping are not included in the representation.

In addition to volumes, non-orthogonal partitioning also applies to surfaces; lines as such are sufficiently covered by orthogonal partitioning rules. Note that the geometric variety of building configurations is expanded by the introduction of non-orthogonal partitioning rules. Examples of sloped sites, walls, roofs, and non-orthogonal space configurations are given in Figure 5.

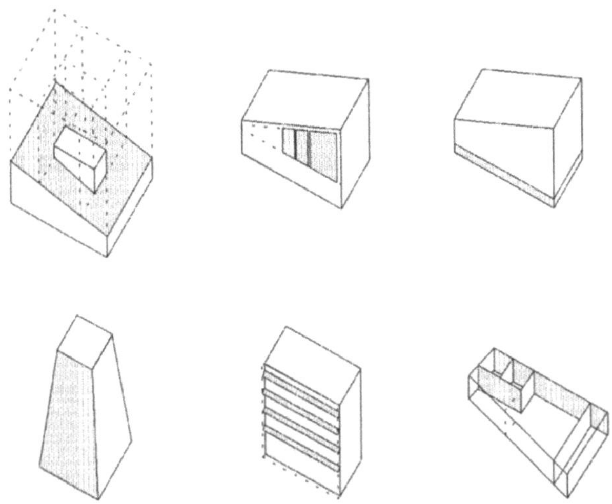

Figure 5. Examples for volumes and surfaces partially generated by non-orthogonal partitioning rules

3.5 Refinement of partitions

While the partitioning rules described above allow for some geometric variety of building configurations, they do not capture certain features that can be found in a large number of buildings. Among those features are gabled roofs, bay windows, dormers, courtyards, and sloped sites. Most of the building elements mentioned can be generated by applying *refinement rules*. They are introduced in order to add resolution to partition-based models. A useful analogy for refinement is the folding of a sheet of paper, which transforms a planar configuration into a spatial one. Refinement rules are applicable to surface and line partitions only; volumes are affected indirectly through refinement of enclosing surfaces. Refinement is always closely linked with and dependent on partitioning.

Surface and line partitions need to fulfill various conditions in order to be refinable. For example, line partitions are only refinable when separating two co-planar surfaces.

Context (of a partition) is important when applying refinement rules. Depending on the context, new surfaces may be required to fill gaps that emerge as a result of refinement operations. We have the following distinctions for refining surfaces:

- if a surface partition intersects with secondary bounding surfaces of the parent volume, no new surfaces emerge as a result of refinement operations;
- if a surface partition does not intersect with secondary bounding surfaces of the parent volume, new surfaces are needed to fill gaps resulting from refinement operations;
- no refinement is allowed if neither conditions are met.

The first case is referred as *refinement without filler surfaces*, the second as *refinement with filler surfaces*. Filler surfaces are inserted automatically by the representation if necessary. Although the shapes of filler surfaces cannot be manipulated directly, they are still available for further partitioning and refinement.

Designers may refine a partitioned surface by pulling a partition away from the plane defined by the original surface. The new location of the refined surface edge is stored in the representation (Figure 6a). It is used to determine the location and dimension of new partitions. In that case, an existing edge shared by two surfaces is split and spread to make room for a new surface (Figure 6b). Examples for configurations that can be partially generated with this refinement rule are included in Figure 7. They include gabled roofs, setbacks, and space layouts with wall setbacks. Analogous rules exist for refinement with fillers, or for refinement in two directions, but are not described here in detail.

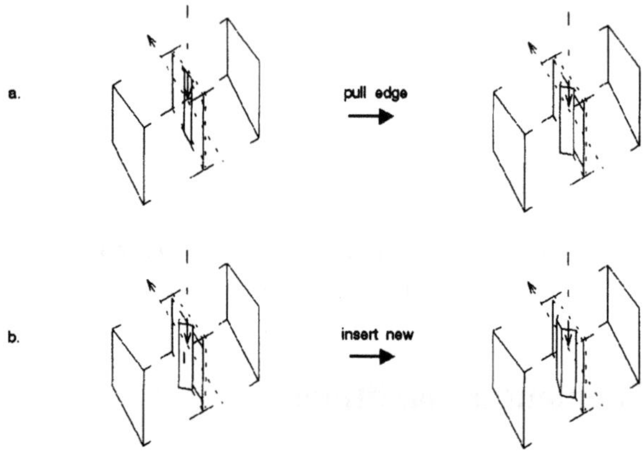

Figure 6. Rules for surface refinement in one direction and without fillers (not applicable to volumes)

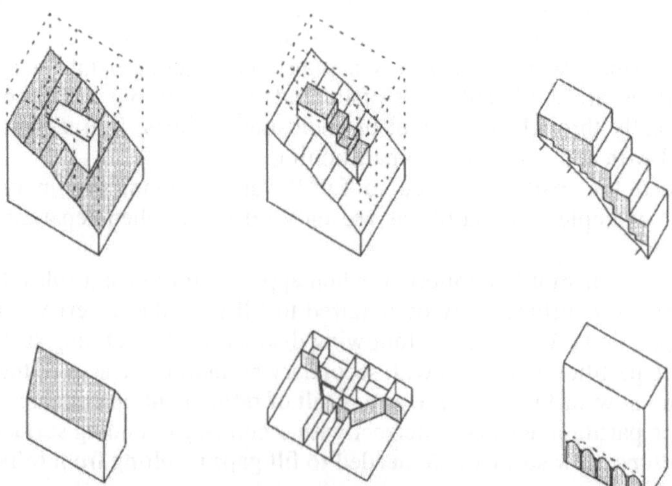

Figure 7. Examples for refined surfaces and lines (without fillers)

3.6 Integrity conditions for refinement

Maintaining integrity of refined surface requires various kinds of geometry checks. In general, no surface is allowed to intersect with other surfaces; only surfaces that adjoin are legal. Refinement rules do not automatically enforce integrity conditions, and additional geometry checks are therefore required. Specific examples of integrity conflicts include: a refined partition intersecting or overlapping with sibling entities; interference of refined surfaces with bounding surfaces or surfaces of neighboring partitions; and interference of a refined partition with secondary bounding elements. Integrity checking for refined partitions may require considerable effort. However, the scope is limited to adjacent elements. No global search of the geometric entities in a model is needed to validate a refined partition. Similar checks are required to exclude intersections with bounding elements.

3.7 Containment hierarchies

Partitioning and refinement rules allow designers to organize spatial information in a hierarchical manner. The main feature of such hierarchies is geometry. The hierarchies envisioned for this representation include strong child-to-parent as well as sibling-to-sibling relationships. The modification of a partition, for instance, may affect lower level as well as same level entities. Hierarchies facilitate constraint satisfaction, labeling, integrity management, and spatial queries.

Figure 8 illustrates the decomposition of a simple, mostly orthogonal building into volumes, surfaces, and lines. The volume hierarchy is the main hierarchy to which additional surface or line hierarchies may be attached. The root volume serves as the initial container for recursive partitioning and refinement, which is performed in the x, y, or z direction. For simplicity, information about bounding elements, partitioning directions, and volume labeling is not included in this illustration. Grayed surfaces indicate labels representing materials and construction types, including air surfaces. All surfaces at the leaves are required to have unique labels. Detailed labeling rules are introduced later in this section.

Surfaces are accessed by traversing the volume decomposition tree. Whenever a surface is further partitioned or refined independent of volume partitioning, a separate surface hierarchy is attached to the original surface, which is included the volume hierarchy. Similarly, line hierarchies may be attached to an original line in a surface hierarchy.

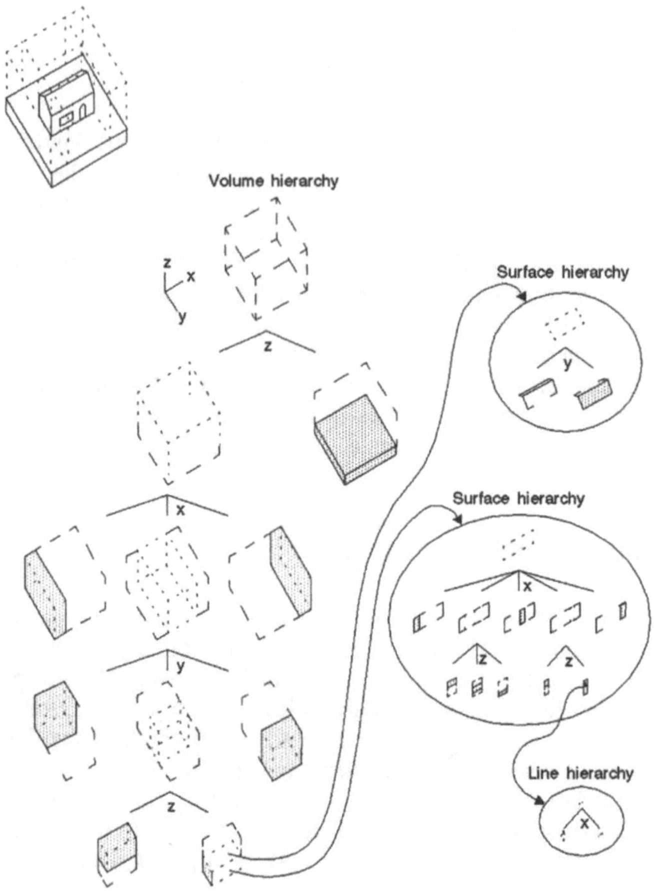

Figure 8. Example for hierarchical decomposition of a house

3.8 Constraints

3.8.1 Dimension constraints

The approach taken in this representation with regard to constraints is deliberately simple. In contrast to space planning systems (see, for instance, Eastman 1970, Mitchell et al. 1976, Flemming and Chien 1996), constraints are used in the representation to facilitate repetitive editing tasks rather than actively support the designer through the generation of design alternatives. Further, only dimension constraints for entities are considered. From a designer perspective, the main advantages of dimension constraints are their universal applicability to volumes, surfaces, and lines alike, minimal constraint maintenance, and relative ease of understanding their consequences. These claims, however, are speculative at this point.

3.8.2 Constraint manipulation

Constraints attached to dimensions of geometric entities are either *free* or *fixed*. Fixed dimensions are never modified. Free dimensions, on the other hand, may be modified as a result of constraint satisfaction. Both fixed and free dimensions of active entities are directly modifiable. By default, dimensions of new entities are free. Although fairly simple, dimension constraints provide designers with powerful means to control the overall behavior of models when local changes are made.

In an ideal scenario, designers would initially be interested in rapid exploration of various design concepts without paying much attention to detail. Designers may fix the few crucial dimensions and accept the others that were determined by constraint satisfaction. More dimensions may become fixed as the design proceeds. According to such a scenario, only minimal constraint maintenance would be required from the user.

3.8.3 Constraint satisfaction

Modification requests are propagated in a building model through a constraint satisfaction mechanism. This may not always be successful and lead to conflicts in the case of contradictory constraints. Constraint satisfaction is therefore performed in two stages. The goal during the first stage is to detect potential conflicts. Modification requests for specific entity dimensions are first passed to sibling dimensions. Once these have been successfully modified for the requested modification, partial modifications are sent to the next lower hierarchy level. This procedure is repeated until the leaf nodes are reached. Dimension conflicts arise in the following situations: a reduction is greater than the dimension of an active entity, an enlargement is greater than the sum of all free sibling dimensions, or all sibling dimensions are fixed. Conflict situations for discrete manipulation are analogous. The second phase of constraint satisfaction consists of finalizing modifications of affected entities. It is only performed when there are no conflicts.

At this point, only top-down or one-way propagation is considered. Bottom-up or two-way propagation is non-trivial; the mechanisms required for non-orthogonal partitioning would involve the resolution of ambiguities.

3.9 Labels

Labels attached to geometric entities provide building element information such as buildings, zones, spaces, openings, or construction types. These are important for integration with simulation, and permit simulation routines to perform selective searches on volume and surface hierarchies for specific semantic information.

Each root entity has a default label that may be modified. All other nodes may either lookup labels at higher levels, or override an inherited label. Overriding labels are not affected by modifications of default labels. *Label inheritance* increases operational efficiency, especially for configurations with a large number of identical labels. For example, label operations enable designers to conveniently explore the impact of various enclosure materials on energy use. In order to model openings, enclosure surfaces are typically decomposed into smaller surfaces (Figure 9a). Label modification without label inheritance could be inefficient because surfaces would need to be accessed on an individual basis. Label modifications at high levels in an enclosure hierarchy, on the other hand, would allow for indirect modification of surface labeling at lower levels.

Another concept is *label instantiation*, which refers to the attachment of properties to entities at the leaves of a hierarchy (Figure 9b). Non-terminal volumes and surfaces do not implement labels. They can be seen as virtual, non-physical entities. In contrast to the label inheritance hierarchy, designers cannot modify label instantiation hierarchies. This ensures that there are neither label definition gaps nor multiple, conflicting label definitions.

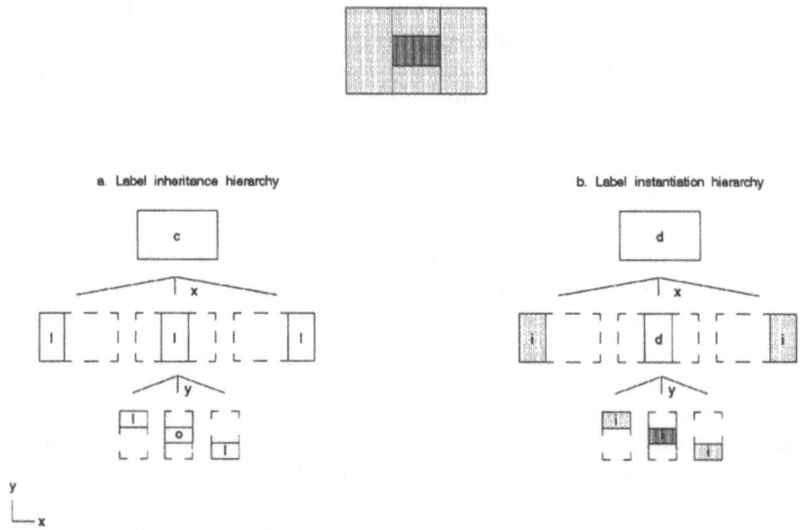

Figure 9. Example for surface label modification and propagation (c = create label type, o = override label type, l = lookup label type; i = instantiate label type, d = decomposed entity, no instantiation)

3.10 Integration with simulation

Labeling of geometric entities is now discussed from a simulation perspective.

Simulation routines look for specific information in the representation. Constitutive building elements include site, building, zones, spaces, and openings. The information exchange between representation and simulation is facilitated by labels. These allow for volumes and surfaces to be marked for simulation queries. Spatial queries provided by the representation are accessible to simulation routines to perform selective searches of a model in order to extract spaces, space boundaries, openings, etc.

The definition of semantic labels is included in higher level operations. These higher level or composite operations consist of sequences of primitive operations, which include adding/modifying/removing new partitions, refining partitions, and assigning labels. The insertion of a space next to an existing space, for instance, is a higher level operation that can be selected by a designer from a menu. It involves three primitive operations. First, a new partition is inserted next to the currently selected space. Second, a default surface label is assigned to the new partition. Last, the new volume is designated as a space by attaching a space label.

As was pointed out earlier, one major advantage of a rectangular subregion representation is the relative ease of identifying adjacencies. Some adjacencies are explicitly stored in the representation, and candidate filtering is efficient. Geometry computation is limited to testing for overlaps at the edges of two regions (Harada 1997). A situation that allows for considerable filtering of adjacency candidates is illustrated in Figure 10. A space configuration is shown in projection on the left-hand side and as a corresponding volume hierarchy on the right hand side. The query consists of finding the neighbors of a large, shaded space. A closer look at the hierarchy reveals that not all spaces need to be visited in order to identify the neighbors. Three out of four spaces at the lowest level do not have to be considered because only one space can be adjacent to the large space in this situation. The other three spaces are therefore grayed out in the volume hierarchy to indicate that these can be filtered out. No geometry checks for overlaps are necessary in this example because the large space does not have any siblings. Neighbor identification for non-orthogonal partitioning is essentially the same as orthogonal partitioning, with only minor differences in geometry checking. In both cases, geometry checking is straightforward. Pruning the search is harder for refined partitioning. It is still possible, but in a more limited fashion and involves further geometry processing.

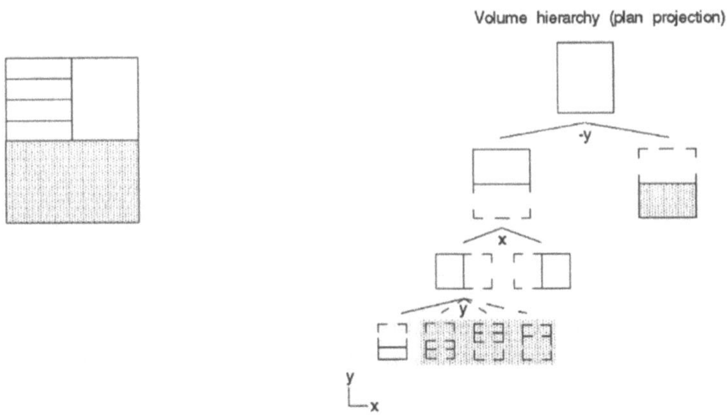

Figure 10. Filtering for orthogonal partitioning (filtered entities are greyed out)

4. IMPLEMENTATION OF A REPRESENTATION PROTOTYPE

A partial implementation of the representation is currently under way. Java has been chosen as an implementation environment, which eventually will allow users to run the prototype over the Internet. It is planned to integrate the prototype with the Semper simulation environment (Mahdavi 1996). Figure 11 includes an illustration of the current status of this development effort. The snapshots of a user interface mockup convey an impression of how a user may eventually interact with an integrated performance simulation system based on the representation. A building is displayed at various levels of abstractions, which are customizable by designers depending on the operations they want to perform. With design and simulation views activated simultaneously, designers would eventually be able to immediately observe the effects of building model modifications on performance. Collapsible views allow designers to switch back and forth between design and evaluation in a straightforward manner. The dimension control region includes an iconic representation of dimensions associated with geometric entities. The designer may modify dimensions by dragging the mouse, and new entities would be inserted by selecting from a popup menu. As noted earlier, these events would trigger a reevaluation of affected entities in the building hierarchy.

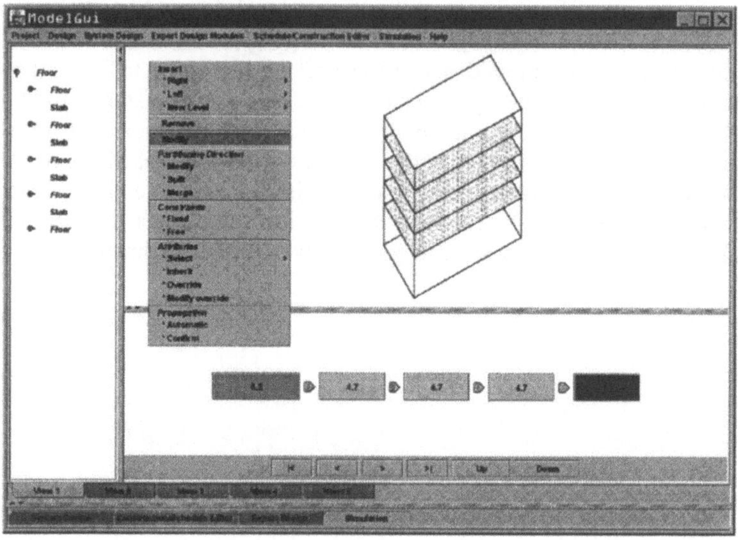

Figure 11. Snapshot of a mocked-up user interface based on the described representation. The interface provides various customizable views of the underlying model.

5. LIMITATIONS OF THE DESCRIBED REPRESENTATION

The description of the representation has focused so far on what kinds of building models can be generated. Like most representations, it has certain limitations.

First, partitioning is based on a bounding polyhedron with six surfaces, quadrilateral plane segments, and line segments. Consequently, model geometries with, for instance, circular or triangular elements may, at best, be approximated. It is conceivable that partitioning and refinement rules can be developed for such geometries that are analogous to the rules for polyhedra with six faces. However, algorithms for spatial queries and hierarchies would be different.

Building hierarchies are probably most useful when represented entities are decomposable in a straightforward manner. A majority of buildings, arguably, can be decomposed into enclosures, floors, zones, and spaces. Some buildings, however, may be harder to break down. While it is relatively easy to recognize a hierarchy of zones, spaces, and circulation in regular layouts, this could be much more difficult in case of irregular layouts.

A further limitation is that the representation currently only allows for a single hierarchy. It may be desirable to allow designers to generate multiple hierarchies. Hierarchies for spaces, structural elements, or HVAC systems could be linked together at a certain level, and otherwise be treated largely independent of each other. According to such a scenario, space partitioning operations would in most situations not affect column locations, and vice versa. Integrity rules would be necessary to establish precedence in case of interferences between space enclosures and structural elements.

6. CONCLUSION

A building representation has been introduced which is rich enough for performance simulation, while at the same time providing support for rapid manipulation and efficient spatial querying of models. The elements of the representation include partitioning and refinement rules for geometric entities, a hierarchical, geometry-centered building description, labels for semantic attachments to geometric entities, dimension constraints, and spatial queries for simulation. We believe that this building representation addresses certain shortcomings of existing implemented representations.

At a conceptual level, it questions the approach typically advocated within the simulation research community with regard to devising effective model generation and modification mechanisms in simulation environments. There is a common view among researchers in the performance simulation community that graphical interfaces integrated with commercial drafting would solve most problems with regard to the communication of building descriptions (Clarke et. al. 1995, Soebarto and Degelman 1995). According to that position, drafting systems are seen as convenient infrastructures providing services that can readily be used to improve the usability of simulation systems. Lack of effective data exchange between drafting and simulation systems is perceived as the main obstacle to integration.

In contrast to that view, it is argued here that such surface level integration, although it constitutes an improvement over purely numerical input, ignores critical issues dealing with manipulation and efficient spatial queries of building models. We consider computational support in these two areas as a necessary requirement for significantly increased usability of building simulation applications.

Augenbroe, G, 1992, "Integrated building performance evaluation in the early design stages", *Building and Environment*, 27 (2), 149 - 161

Clarke, J A, Hand, J W, Mac Randal D F, Strachan D, 1995, "The development of an intelligent, integrated building design system within the European COMBINE project", *Fourth International Conference, International Building Performance Simulation Association*, Madison WI

Eastman, C, 1970, "Representations for space planning", *Communications of the ACM*, 13 (4), 242 - 250

Eastman, C, 1979, "The computer as a design medium", Research report, Institute of Building Sciences and Design Research Center, Carnegie Mellon University, Pittsburgh, PA

Flemming, U, Chien S F, 1995, "Schemantic layout design in SEED environment", *Journal of Architectural Engineering*, 1 (4), 162 - 169

Harada, M, 1997, "Discrete and continuous design exploration by direct manipulation", PhD Thesis, Department of Architecture, Carnegie Mellon University, Pittsburgh, PA

Khemlani, L, Kalay Y, 1997, "An integrated computing environment for collaborative, multi-disciplinary building design", *7th International Conference on Computer-Aided Architectural Design CAAD Futures*, Munich, Germany

Mahdavi, A, 1996, "SEMPER: a new computational environment for simulation-based building design assistance", *International Symposium of CIB W67 (Energy and Mass Flows in the Life Cycle of Buildings)*, Vienna, Austria

Mitchell, W, Steadman J P, Liggett R, 1976, "Synthesis and optimization of small rectangular floor plans", *Environment and Planning B* 3 37 - 70

Soebarto, V, Degelman L, 1995, "An interactive energy design and simulation tool for building designers", *Fourth International Conference*, International Building Performance Simulation Association, Madison WI

Steadman, J P, 1983 *Architectural morphology: an introduction to the geometry of building plans* (Pion, London, England)

Suter, G, Mahdavi A, 1998, "Generation and communication of design information: a building performance simulation perspective", *4th Conference on Design and Decision Support Systems in Architecture and Urban Planning*, Maastricht, Netherlands

About objects and approaches

a conceptual view on building models

Ann Hendricx and Herman Neuckermans
CADLAB, Department of Architecture, Urban Design and Planning, KULeuven, Belgium

Key words: CAAD, integrated design environment, building models, conceptual modelling, MERODE

Abstract: Considering integrated CAAD environments for architectural design, a number of different approaches are possible. This paper presents the policy of the CADLAB at the KULeuven University, where design support right from the first design phases is a basic consideration. After a short introduction on the theoretical framework and additional design tests, we will discuss the core object model that forms the cornerstone for the contemplated design environment. This object model describes all possible data, concepts and operations connected with the architectural design process. For its development, we used the object-oriented analysis method MERODE. The starting-points and main aspects of the model will be discussed, illustrated with examples of implemented prototypes. The architect's point of view and the specific nature of the architectural design process were always kept in mind, thus leading to a model that hopes to make a valuable contribution to the research area of integrated design environments.

1. INTRODUCTION

1.1 The computer in the design process

Architects use a great variety of concepts while designing: geometrical shapes, spaces, building elements, user activities, structure, etc. Within this vocabulary, different design approaches are possible. Some designers follow a top-down procedure; others proceed according to a bottom-up strategy. Whatever the approach, it goes without saying that commitments made in the first stage of the design process have the biggest impact and are the hardest to undo later.

Until now software packages offer a broad range of computational tests to evaluate an architect's creation in the 'post-design' stage: adequate tests capable of handling incomplete data and estimates are rare. Exactly those tests could help the design process in the first stages. Additionally, we should not ignore the fact that in today's practice the computer only comes into play when the design is more or less completed. Computation programmes and the drawing packages themselves should concentrate more on this first stage of sketching and exploration. At least the unproductive translation of a design made by hand into a digital version would be avoided.

To assist the architect throughout the whole design process, an integrated approach is essential. To modelling tools and testing tools alike, they should be part of one and the same design environment, providing the architect with full interactive use of the tools while working on one and the same design model. Both in the field of product modelling - the description of the designed product - and in the area of design and testing tools (e.g. Sequin et al. 1998) a lot of efforts are carried out world-wide. But often the applicability of these efforts is limited: schematic and poor design possibilities, forcing the designer into a specific way of working, concentrating on a particular evaluation test and thus hampering later extension of the system etc. Approaches often underestimate design as a *process*, and neglect the evolutionary nature of design.

An integrated design environment basically consists of the following parts (Kim et al. 1998, Eastman et al. 1997):

1. a core object model or aggregation of models able to describe all actors and processes within the world of architectural design
2. a data management system
3. tools to assist the architect while designing

We like to subjoin the following remarks: First, we enlarge the chance of success by leaning on an underlying theoretical framework. In this way, the integrated approach is secured and developing a complete environment will be easier. Our proposal for such a framework will be discussed further down.

Next, we divide the assisting tools in those tools enriching and helping the design process itself and those tools evaluating and appraising designs. The first category includes both modellers - a fine example is the very attractive Sculptor, a still evolving package that encourages an intuitive way of handling bodies in space (Kurman 1998) - and research topics such as shape grammars, case based reasoning, automatic layout configurations etc. The second category covers tools for predicting thermal behaviour, lighting, insolation, statics etc.

The IT approach is not only beneficial to the designer, all building partners will take advantage of the integrated environment by referring to or using the core product model. A big step forward knowing that communication and passing on of data between professionals in the building industry often happens in a time-consuming and inefficient way. In this respect, research on data transfers and collaboration - also within the design team - and our research join hands.

1.2 A theoretical framework for the architectural design process

Considering these preliminaries, a general conceptual model for computer-aided architectural design has been developed (Neuckermans 1992).

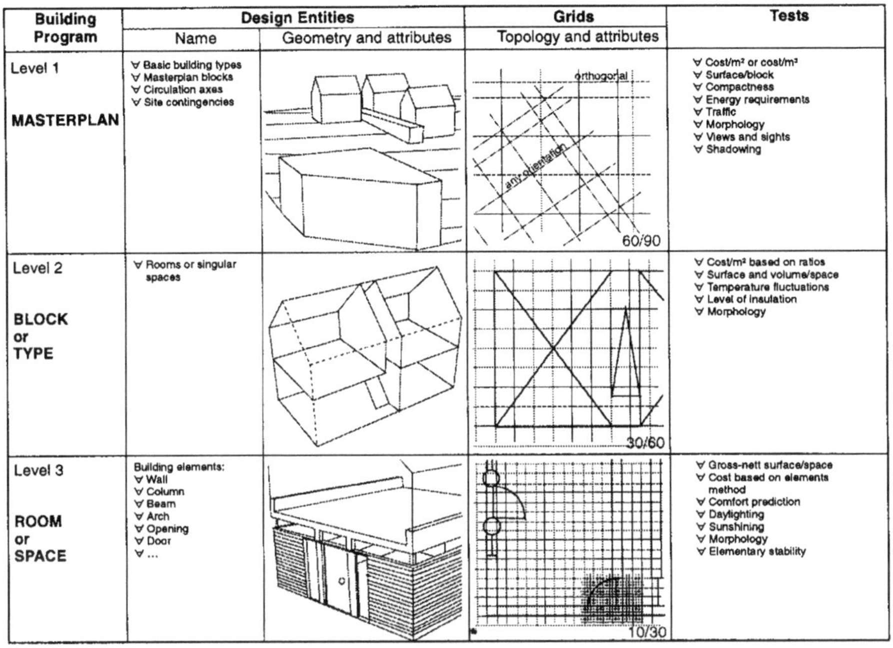

Figure 1. A conceptual framework for the design process

In this model, the building programme is subdivided hierarchically into 3 levels spanning the normal scope of architectural design. Extension of these levels both upward to the urban design scale and downward to the building element is possible, when judged relevant by the designer. Depending on personal preferences or on the design problem at hand, each level can be seen as an entry point for the design process. On any level, the architect can use entities he is familiar with. Building types, spaces and elements like walls, columns, doors and stairs take the place of polylines, ruled surfaces and other "meaningless" CAD objects. Optionally, these elements can be placed on a grid, whose mesh widths vary with the actual level of detail one is working on. In addition to levels, elements and grids, the framework provides a collection of tests allowing the user to analyse building properties. These tests are in tune with the precision of the model at that moment and are relevant to the specificity of the building programme (hospital, school, office, housing etc.). This model presented the notion of a scale dependent representation: the 'intelligent zoom' displays one and the same object with more detail when the scale of the drawing increases (Neuckermans, 1986).

Globally, this framework remains primarily graphical and is an open, not deterministic but interactive aid to the designer. It aims at assisting instead of steering the design process.

2. THE CORE OBJECT MODEL

Based on the logic of this theoretical framework, the first research phase led to the development of a conceptual object model able to describe all elements mentioned in the framework. After a short introduction to MERODE, the analytical method used to develop the model, we discuss the most important aspects of the model. This part is illustrated with samples of code and images of implemented prototypes.

2.1 MERODE, an object oriented analysis method

We have chosen to use MERODE, an object oriented analysis method to produce so-called business models because of its elaborated systematic approach. This approach has proven to be a great help to achieve the contemplated core model. We may as well mention that we were charmed by the fact that the authors specifically refer to architectural design when describing the development of business models (Zachman 1987, Dedene et al. 1996).

MERODE is an acronym for Model driven Existence dependence Relation Object-oriented DEvelopment. For a full description of the method we refer to Verhelst (1996). Since some aspects of our object model are a direct consequence of this method, we tersely mention its main features.

First, MERODE makes a clear distinction between the specification phase and the implementation phase. The specification phase is entirely object-oriented. It makes no assumptions whatsoever about technology that will be used for implementing the system. This means that specification is completely conceptual and therefore independent of technological aspects of a system. Both 'traditional' implementations and object-oriented ones are still possible. Thus, tough choices about relational or object-oriented databases, CAD packages and so on are avoided at this stage of the design. And more important, research is not limited by the chosen hardware and software or left to the arbitrariness of a certain software developer.

During the conceptual specification phase, reliability, ease of correcting and flexibility are the relevant quality measures, whereas during implementation performance, efficiency and user-friendliness are prevailing. In specifying MERODE models you even need to have no knowledge whatsoever about computer-related matters. After specification, a computer specialist can carry out implementation with minimal effort. This supports the idea that implementation is achieved by transforming the specification and not by further elaboration.

Second, a distinction is made between 'business objects' and 'function objects'. Business objects correspond to objects that exist in the real world. Defining methods attached to the business objects enforces business rules. Thus, the business subsystem simulates the real world system (e.g. a car rental company). When real world rules modify, only the business subsystem has to be changed. It can never be influenced by changes in information needs (subject matter of the output subsystem) or by changes in business organisation (subject matter of the input subsystem). In the example of the car rental company, 'car' and 'customer' are examples of business objects. Apart from these business objects, function objects can be of three kinds:

- *Input objects* gather the necessary information related to the actual system and send messages to the business objects.
- *Output objects* gather information from the business objects and present it to the user in an appropriate way, e.g. printing an invoice.
- *Information objects* hold track of an object's history. Because the business model only simulates the real-time situation, this information is necessary for example when you need to know how many times a car has been rented in the past in order to perform a car check.

In addition to an object model, which takes care of the dynamics of the system, MERODE uses a kind of Entity-Relationship (ER) model to express the static aspects of the system. Both models are fully integrated: each object in the object model corresponds to one and only one entity in the ER-model. One of the cornerstones of MERODE is **existence dependency**, with following definition:

> *Object type A is existence dependent from object type B, if and only if each object of type A, during its entire life-cycle, engages in a mandatory relationship with one, only one and always the same object of type B.*

As a matter of fact, all relationships in the ER-model should denote existence dependency. In our car rental company, 'car' and 'customer' can therefore not be linked directly: a car can exist without being rented by a customer and a person does not only exist by means of his renting a car. Thus, the need of existence dependency should involve the creation of a new object 'rental', existence depended both of 'car' and 'customer'. Why this is so important is made clear by Snoeck (Snoeck et al. 1996). Next to simplifying consistency checking in the model, existence dependency is a valuable alternative for the sometimes-vague concept of the Part-Of relation.

Finally, special attention is given to specification of business constraints. A distinction is made between two different types of constraints. Sequence constraints (e.g. a car cannot be rented before it is bought) are expressed by sequence diagrams (as in the Jackson System Development method JSD) or by finite state machine charts. Data constraints (e.g. a client can only rent 3 cars at the same time) are expressed in a declarative way. By modelling constraints, the object model itself will be able to maintain integrity, as advocated for building models by Galle (1995). The different aspects are described in the most appropriate way, keeping in mind that consistency between the different notation techniques is essential. Thus, an object model handles the dynamics of the system, an Existence Dependency graph scheme - the mentioned ER-like scheme - takes care of the static relationships between objects, and finally sequence and data constraints are described by JSD diagrams and Eiffel-like notations. All this leads to the final ADT (Abstract Data Type) description of the objects.

The MERODE methodology was considered a firm base in order to start the research on the construction of the core object for architectural design in the early design phase. Due to the strict distinction between business modelling and the rest of the model, user input (by using a CAD package), output (architectural drawings, data for computational tests, etc.) and creation of information objects (essential for version control) could come in to play in a later research phase. Different views on the model by different members of a collaborative design group do not influence this business model and can be specified later.

2.2 Development of the core object model

The specifications of the object model were deduced from the theoretical framework. Parallel with this framework, a state-of-the-art in product modelling and integrated efforts was made up. Finally, we called upon the design expertise at our department. The following paragraphs present the premises of the object model. They discuss aspects often ignored in current research initiatives. More information on the research project as a whole can be found in Hendricx (Hendricx et al. 1998).

2.2.1 In pursuit of 'good architecture'

The object model should be able to describe architectural designs in a full-fledged way. We do not want to confine ourselves to evident, routine design and 'poor' architecture, but we are aiming at a digital design environment that can lead to a creative and qualitative architecture. The designer cannot disclaim responsibility in this matter, but should at least not be limited by the CAAD environment. In this respect, seminal examples of Belgian architecture - as open to question the definition of good architecture may be - validated our model. Although we know that the realisation of this ambition is closely connected with flexible and user-friendly modelling tools, we limited ourselves in this phase to the - as complete as possible - description of a full-fledged design. Important aspects are:

- the description of boundaries between design elements: between two physical objects, between a physical object and a space, between two spaces
- considering architectural design as a process: one and the same design object can be represented in different ways during this process
- a general object definition capable of describing both physical design objects and abstract spaces
- avoiding restrictions in the use of geometrical shapes to depict building objects

These basic considerations have an important impact, even on the static description of the contemplated object model. As a consequence they were validated in an early stage of the project. It led to the development of a first prototype by means of a relational database, providing easy input and checking of relationships between objects. Images of this first prototype are shown in figures 2 and 3.

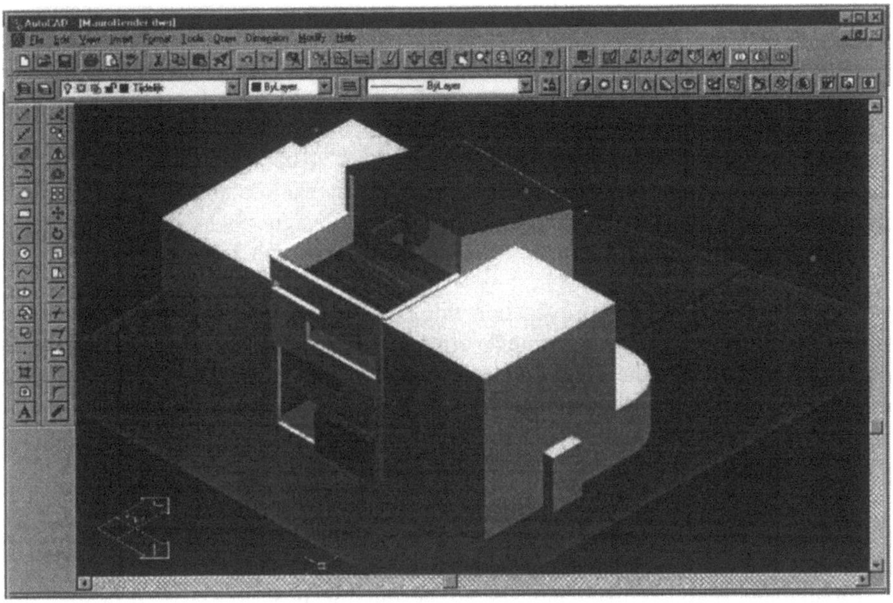

Figure 2. AutoCAD model of the Poponcini dwelling in-between its neighbours

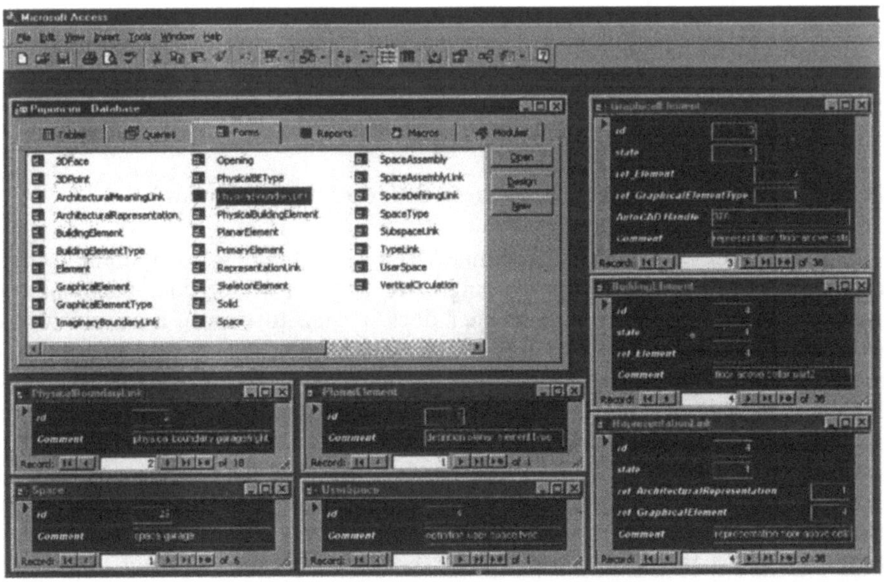

Figure 3. Access database of the Poponcini design

In this example, the general applicability of the model is tested by the design for a single-family row house (architect Mauro Poponcini, Antwerp 1994). The core object structure is implemented in Microsoft Access; AUTOCAD R14 takes care of the graphical representation. The connection between de central object description and the on-screen representation is made by adding an attribute to the geometrical drawing objects holding the AutoCAD entity's Handle or Entity Name.

Also the principle of multiple representations has been tested and its validity was proven. In this product model one and the same staircase for example is depicted by a sloping plane in the early phase and by a solid visualising each step at a later design phase. Both representations are kept in the database. The kind of representation is dependent both of the current design phase and the scale of zooming. This aspect will be further explained in paragraph 2.2.4.

This first prototype has proved two things. On the one hand, the nature of the implementation (relational or object oriented approach) is not dictated by the conceptual model. On the other hand, the model is capable to describe real projects and is not limited to trivial designs.

2.2.2 Supporting the early design phases

The importance of supporting the early design phases has been mentioned earlier in this text and has following consequences:

First, the design environment should permit the description of a design from scratch by offering 'primitive' architectural objects as well as more elaborated ones. Primitive objects can be further specified in the course of the design process. How the model covers this aspect will be shown in paragraph 2.2.4.

Second, the environment should offer appropriate tests for the early phases to turn traditional post-design computation into 'testing-while-designing'. At the moment, our research group is developing a daylight and insolation test for the early design phases. It uses the core object model to extract the appropriate data and leads to both a computational and graphical evaluation of the design at hand. Within the scope of this paper, we will not pursue this matter further.

2.2.3 Different design approaches - different design entries

The design environment is not limited to a given building programme (school, hospital, town planning, office building...) or a given set of computational tests (light, insolation, statics, cost...), neither does it want to impose a certain way or methodology of design upon its user. Both specifications call for the possibility to approach the design process from different point of views. We explicitly model both physical objects (from terrain, wall and floor to door-handle and chalk layer) and the abstract notion of 'space', an approach also mentioned by others (e.g. Eastman et al. 1995). On top of this duality, the designer has the possibility to handle mere geometrical forms, a possibility also defended by Ekholm (1966). This leads us to the notion of 'possible design entries'. This doesn't rule out the mutual connection between the mentioned objects, but comes down to *not* assigning a certain order of importance upon them. To say it the MERODE way: none of the basic elements is existence dependent of another one. These different views on the design process and the corresponding design entries were considered very important. To see how the model copes with it, we first discuss the general structure of the object class named ELEMENT.

To cover all possible aspects of a building object, we create an overall neutral object class ELEMENT (or THING) that can contain the different kinds of data belonging to an object: graphical information, specific architectural information, additional alphanumerical information (figure 4). The last object class can for example handle the data connected with a given activity or function and thus allows the modelling of user activities.

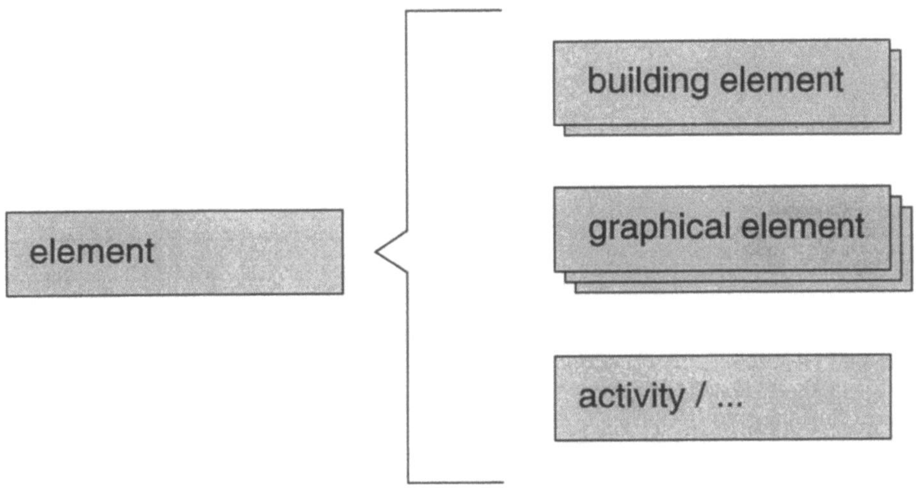

Figure 4. Main construction of a building element

As said before, the different components of an ELEMENT have no predefined order of importance. This makes the contemplated different design entries possible, as depicted in figure 5:

- juggling around graphical drawing objects in space without assigning an architectural meaning to them from the start (upper left arrow)
- positioning physical building elements (lower left arrow)
- positioning and modelling spaces, with a possible 'materialisation' afterwards (lower right arrow)
- organising the layout of user activities, connecting them to specific spaces afterwards (upper right arrow)

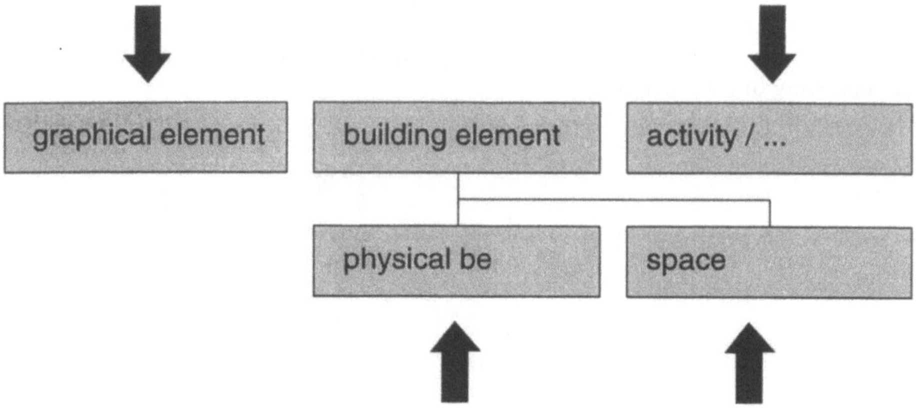

Figure 5. Different entries

2.2.4 Architectural design as a process

Design is a process. Design objects are not defined and fixed once and for all, but evolve during the design process. Closely connected are the notions of design phases and different architectural representations. This results in an additional challenge: providing the possibility of change.

Evolution of objects has to be handled within the core object model, thus making reuse of data possible: when changing/elaborating an existing design element, as much data as possible is reused to build the new version of it, thus avoiding the necessity to reconstruct an element from scratch every time a major change occurs.

Looking at this aspect from the object-oriented point of view, designed objects appear to switch classes. A particular wall can start as a general PLANAR ELEMENT in the early design phase, become a WALL in a later one and result in a specific kind of CAVITY WALL in a final stage.

To meet the requirement of object evolution, we use the following strategy: a building element is divided in an abstract part still named BUILDING ELEMENT and a part containing the specific 'type' information, named BUILDING ELEMENT TYPE. Considering the changing wall above, we will keep the same BUILDING ELEMENT, but its type pointer will switch from PLANAR ELEMENT to CAVITY WALL during the design process. This process is hidden to the user: the object simply seems to change classes. Figure 6 clarifies this structure.

Object classes are represented by rectangles holding the class name. The small white squares suggest contract objects: a construction within MERODE to ensure the existence independence of the object classes it connects. Within the scope of this paper, it is unnecessary to enlarge on this.

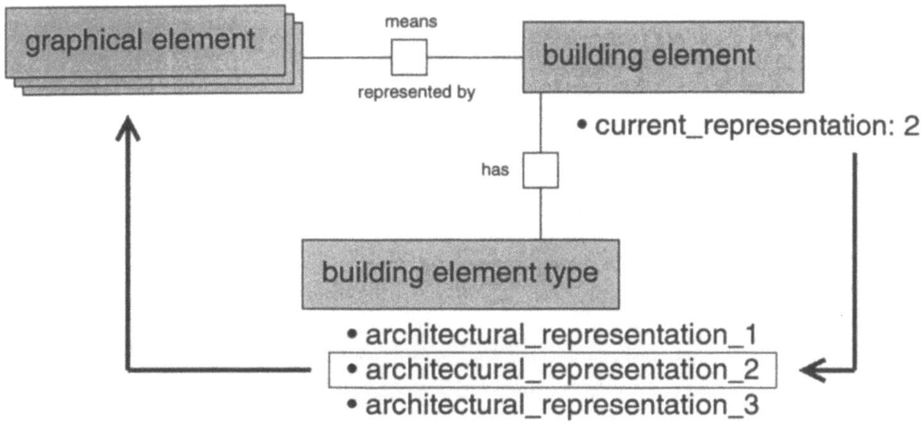

Figure 6. The principle of BUILDING ELEMENT, BUILDING ELEMENT TYPE and GRAPHICAL ELEMENT in the core object model.

The notion of changes in representation is closely connected with this object evolution (see also paragraph 2.2.1). The same figure 6 illustrates our solution.

Several architectural representations are assigned to one and the same BUILDING ELEMENT TYPE. The transition between two design phases (implying a new value for current_representation) leads to the selection of the appropriate representation: the correct member function (architectural_representation_#) is activated and the corresponding GRAPHICAL ELEMENTs will depict the new situation. Doing so, our above-mentioned 'wall' will be displayed as a simple line or surface, as a cuboid in space or as a very detailed representation of a cavity wall. The red arrows in figure 6 show this cycle of changing representations. At this stage, the member function architectural_representation_# uses a parameter description or sequence of drawing commands to obtain the wanted graphical output. A more convenient way to describe the often complex predefined and user-defined representations will be the topic of further research (e.g. description in a suitable script).

The example below shows a possible implementation of the core object model in the C++ programming language. AutoCAD handles input and output. Here, the aspect of different architectural representations is connected with the notion of 'intelligent zooming'. As mentioned in paragraph 2.1, the final conceptual object description in MERODE is the ADT (Abstract Data Type). To illustrate this, we provide the ADT of the core object class BUILDING ELEMENT. Next, the C++ implementation of the same object class as used in the above prototype is shown.

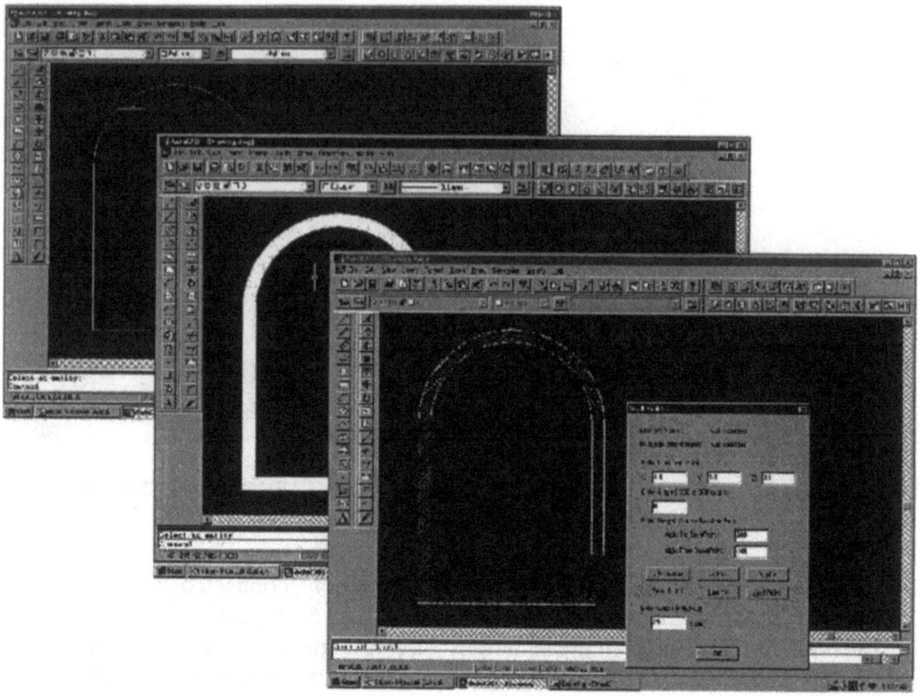

Figure 7. The multiple representation principle (implementation by Bart Anrijs)

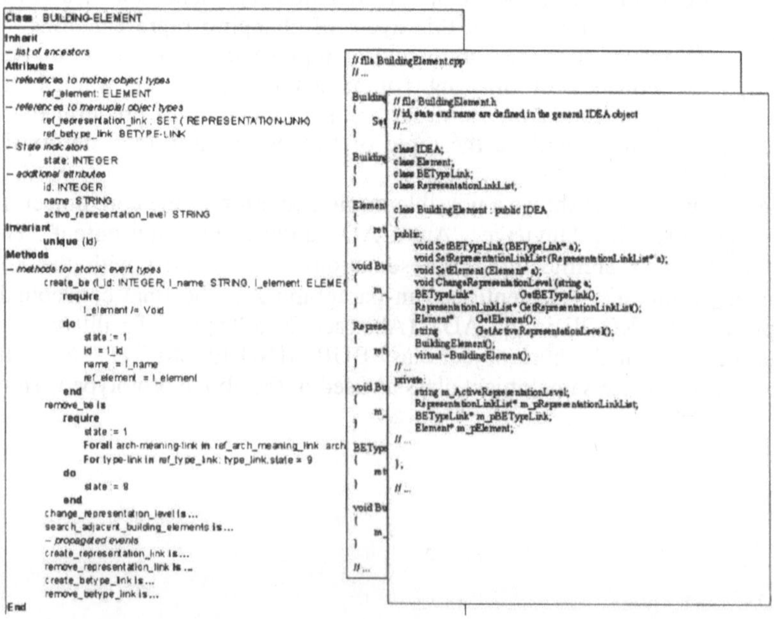

Figure 8. Sample code of an object ADT by MERODE and its C++ implementation

2.2.5 Clear separation between the core model and its implementation

An important starting-point is the neutral description of the core object model. This neutrality has a double meaning:

- The fact that the core object model is developed with no particular design domain or design test in mind
- The fact that the core description of objects is disconnected from its input and output, notably its on-screen realisation or implementation (and thus from the graphical packages that take care of an object's graphical representation)

This paragraph discusses the second kind of neutrality.

To obtain the possibility to work with different graphical applications (e.g. using AutoCAD graphical entities or Open Inventor API), we apply the typical 'bridge' principle (Gamma et al., 1995): a neutral GRAPHICAL ELEMENT is separated from its application dependent implementation by building an intermediate GRAPHICAL REALISATION (see Figure 9).

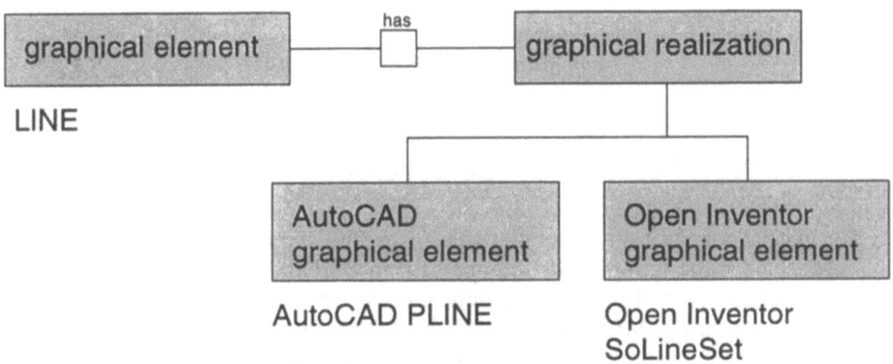

Figure 9. Separation between a graphical element and its on-screen representation

Switching the graphical package to provide input and output implies that the correct GRAPHICAL REALISATION object has to be called. To use a new graphical package you only need to create the appropriate GRAPHICAL REALISATIONs: the core model itself remains intact and does not require changes.

This principle was tested by linking the graphical elements of the core object model to two different graphical output packages: on the one hand AutoCAD by means of ObjectARX (see also the example in 2.2.4), on the other hand Open Inventor. An illustration of a prototype using Open Inventor and steered by a central C++ implementation is shown in figure 10.

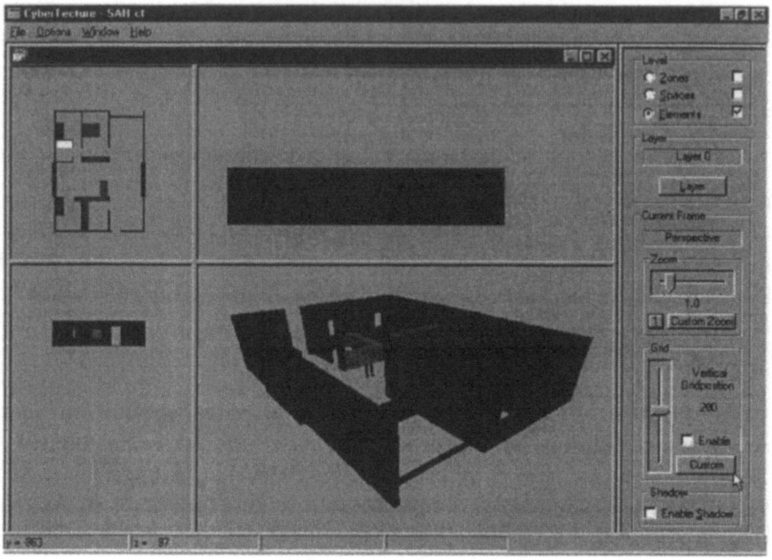

Figure 10. Graphical representation in an Open Inventor environment (implementation by Bart Geeraerts)

2.3 Overall picture of the developed core object model

We like to conclude this section with an overall picture of the developed core object model. In the (simplified) scheme of figure 11 the mentioned principles can be distinguished. Yet, this existence dependency graph is only one of the different facets of the complete MERODE model: here only the core objects and the static relations between them are shown. As mentioned before, contract objects are depicted by a little white square. Only the main scheme is given, the grey object classes stand for an embedded hierarchy of object classes. The system as a whole will be thoroughly covered in a forthcoming PhD thesis on the subject.

The scheme is presented in the MERODE notation method. We considered the use of the more general Express-G notation, but decided to stick with the MERODE method in its entirety. With respect to content, this does not make a difference.

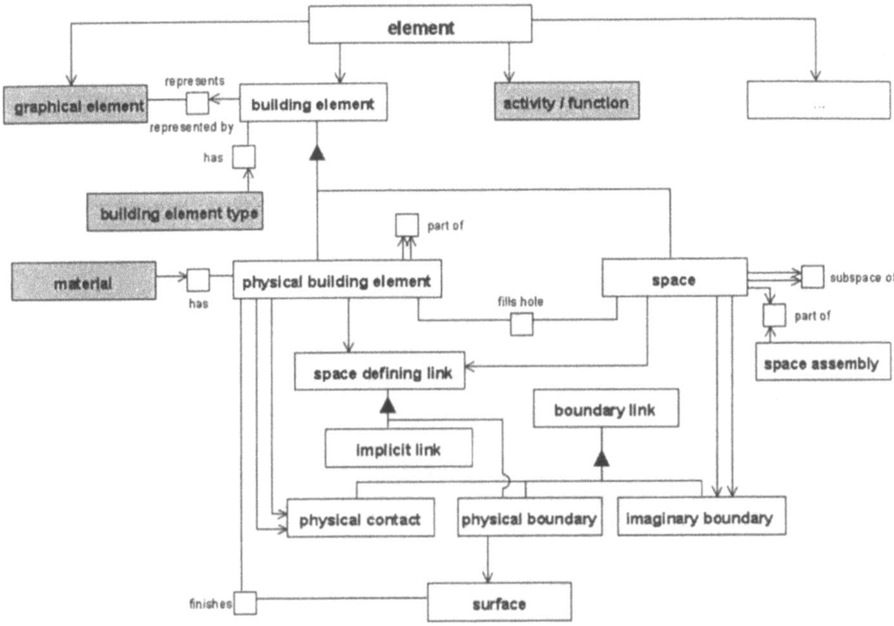

Figure 11. Simplified view on the conceptual object model - existence dependency graph

3. CONCLUSION

This paper discusses the core object model for the architectural design process, developed within the framework of the IDEA+ project (Integrated Design Environment for Architecture) of the CADLAB at the KULeuven. The model was developed with the aid of MERODE, an object-oriented analysis method. After arguing the need of such a model, its strong points were revealed. These points were illustrated by means of examples taken from prototypes developed at the CADLAB. Further research topics are the parameter/script description and representation of building types and the user-friendly interface between the core model and the designer. Although a lot of work still lies ahead of us, the core object model has proven to be a strong base for further developments.

4. REFERENCES

Dedene, G., Maes, R., 1996, "Reframing the Zachman Information System Architecture Framework", *Tinbergen Institute discussion paper* TI 96-32/2, (Rotterdam)

Eastman, C., Jeng, T.S., Chowdbury, R., 1997, "Integration of design applications with building models", *CAAD futures 1997, Proceedings of the 7th International Conference on*

Computer Aided Architectural Design Futures, Munich, Germany, 4-6 August 1997, JUNGE, R. (ed.), 45-49

Eastman, C.M., Siabiris, A., 1995, "A generic building product model incorporating building type", *Automation in Construction*, 3 283-304 (Elsevier)

Ekholm, A., 1996, "A conceptual framework for classification of construction works", *Electronic Journal of Information Technology in Construction* (ITcon), vol 1, Royal Institute of Technology

Gamma, E., Helm, R., 1995, "Design Patterns, Elements of Reusable Object-Oriented Software" (Addison Wesley)

Galle, P., 1995, "Towards integrated, "intelligent", and compliant computer modeling of buildings", *Automation in Construction*, 4 189-211

Hendricx, A., Geebelen, B., Geeraerts, B., Neuckermans, H. (1998) "A Methodological Approach to Object Modelling in the Architectural Design Process", *Design & Decision Support Systems, Proceedings of the 4th Conference on Design and Decision Support Systems in Architecture and Urban Planning*, Maastricht, The Netherlands, 26-29 July 1998 (Eindhoven University of Technology)

Kim, I., Liebich, T., Maver, T., 1998, "Managing design data in an intergated CAAD environment: a product model approach", *Automation in Construction*, 7 35-53 (Elsevier)

Kurmann, D., 1998, " SCULPTOR - How to design space?", *CAADRIA'98, Proceedings of the Third Conference on Computer Aided Architectural Design Research in Asia*, Sasada, T., Yamaguchi, S., Morozumi, M., Kaga, A., Homma, R.(eds.), Osaka, Japan, April 22-24, 317-325

Neuckermans, H., 1992, "A conceptual model for CAAD", *Automation in Construction*, 1 (1) 1-6 (Elsevier)

Neuckermans, H., 1986, "The intelligent pencil, a framework for CAAD in education", *ACADIA Workshop '86 Proceedings, Architectural Education, Research, Practice in the Next Decade Association for Computer-aided design in Architecture*, October 25 1986, TURNER, J.A. (Ed.) (Houston, USA)

Sequin, C., Kalay, Y., 1998, "A suite of prototype CAD tools to support early phases of architectural design", *Automation in Construction*, 7 449-464 (Elsevier)

Snoeck, M., Dedene, G., 1996, " Bestaansafhankelijkheid: conceptueel modelleren "volgens contract" (Existence Dependency: Conceptual modelling by contract)", *BeleidsInformaticaTijdschrift*, 22 (4) 1-36

Verhelst, M., 1996, "Object-oriented application development with M.E.R.O.DE.", course text, Department of Applied Economics, KULeuven

Zachman, J.A., 1987, "A framework for information systems architecture", *IBM SYSTEMS JOURNAL*, 26 (3) 276-292

The BAS•CAAD information system for design – principles, implementation, and a design scenario

Anders Ekholm and Sverker Fridqvist
Computer Aided Architectural Design, Lund University, Lund, Sweden

Keywords: CAD, design, dynamic schema evolution, information systems, object oriented modelling, product modelling, design scenario

Abstract: The objectives of the BAS•CAAD-project are to investigate into theories and methods for computer aided architectural design, with emphasis on requirements of early stages of the design process. Information systems can be characterised as static or dynamic concerning the definition of classes in the model schema, and concerning classification of model objects. The paper presents the BAS•CAAD system, a prototype software that implements the conceptually most important features of a dynamic information system for design. The BAS•CAAD information system is built on a generic ontological framework. The system allows a free combination of attributes, supporting the incremental way that knowledge is built up during design. It provides a generic library structure that allows definition of objects classes in different levels of generalisation that may originate from international standards or the individual designer. For example, in the construction context, it allows modelling of buildings and their parts, as well as user organisations and user activities. The function of the system is illustrated in two scenarios.

1 INTRODUCTION

In a computer integrated construction process, information is generated, used and communicated through computers. In order to enable computer based analyses of the products and processes developed, this information must be model based. This concerns not only building information, but also information about other objects, e.g. the user organisation, the site, the construction process, and the facility management process.

The questions of the structure of building product models and the communication of building product data between different actors and computer systems have been given much attention within construction information research. Analyses of characteristic features of such models in the construction context have been done by Björk (1995) and Galle (1995). Lately, principles for structuring computer based information about the user organisation have been discussed, see e.g. Eastman and Siabiris (1995), Ekholm and Fridqvist (1996 and 1998), Maher, Simoff and Mitchell (1997).

Many approaches to product modelling focus only at modelling, and seem to overlook the process of creating the models. The most outstanding feature of this process is that the information changes and evolves over time, not only in quantity but semantically as well. This would make it hard to use a product modelling system based on a fixed classification schema in the earliest, most dynamic phases of design, since the fixed schema would be at odds with the evolving semantics of design.

This paper presents the implementation of a prototype information system, which has been constructed as part of the BAS•CAAD research project. The project aims to find solutions to both the problem of modelling different products and processes, and the need to reflect and support the evolving nature of the design process.

2 INFORMATION SYSTEMS FOR DESIGN

2.1 Design problems

Design is a problem solving process, it is similar to problem solving both in everyday life and science. Problem solving is a process of exploration, where solutions are alternately hypothesized and tested. In the process, the properties of the hypothesised object are determined in an incremental manner, the designer adds and removes attributes from the conceptual representation of the object.

A problem statement is a description of an object whose state, according to certain presuppositions or objectives, is unknown, or unsatisfactory. The state of an object is its attributes at a certain stage of time. A problem solution, or hypothesis, is information that describes the state of the object, or action that leads to a satisfactory state. Design problems are procedure problems, they deal with questions of human action (Bunge 1998:208). The test of a design hypothesis may be theoretical, relating the solution to existing knowledge, or empirical, involving the construction of an artefact.

A design problem may be characterised as open or closed concerning the determining factors of the designed artefact, e.g. in building design such factors are environmental impact, user requirements and available technology. To a closed problem, the determining factors and their combinations are well known, while to an open problem neither the determining factors nor their combinations are known, but must be explored or invented. Open problems are also called "wicked" (Rittel 1984).

Design can also be categorised as routine or innovative. Routine design is a closed problem solving process, it consists in selecting a prototype solution and determining the values of its attributes. Innovative design is necessary when no such prototype solution can be applied, and new kinds of things or new uses for known things have to be created. Building design is an example of both routine and innovative design, the latter especially during early stages. The approach to building product modelling today, e.g. in object-oriented CAD programs, does not support innovative design and is best suited for the later stages of the design process.

2.2 Dynamic and static information systems

An information system is a computer based system which makes it possible to store and retrieve information of relevance to the information needs of a user. It consists of a conceptual schema, an information base and an information processor (ISO 1985:15). In a traditional implementation, the classes of the conceptual schema refer directly to the objects in the domain of discourse. A specific model of an object is achieved through selecting an appropriate class in the schema and determining the values of the attributes that describe the object in the information base.

In the BAS•CAAD project, we have found that information systems can be characterised as dynamic or static concerning the possibility for the user to 1) define new class concepts in the conceptual schema, and 2) classify model objects. These two characteristics are mutually independent, see Figure 1.

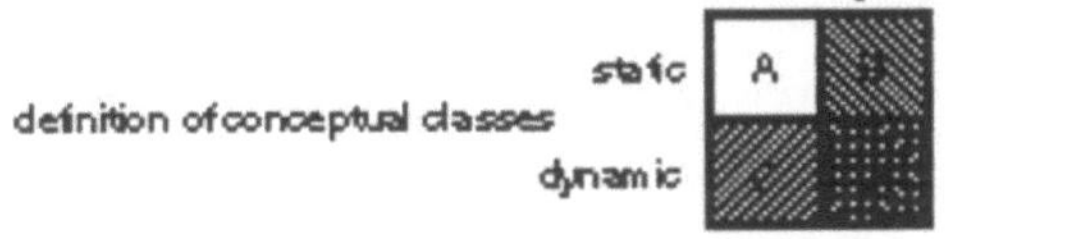

Figure 1: Dynamic and static information systems

The four kinds in Figure 1 are:

a) Static systems: the user is restricted to a predefined set of modelling classes, model objects have to retain their classification once instantiated into the model.

b) Dynamic classification: the user is restricted to a predefined set of modelling classes, but can reclassify model objects between these classes during modelling.

c) Dynamic schemas: the user can create new classes, but cannot reclassify model objects during modelling.

d) Fully dynamic systems: the user is free to create new classes and to reclassify model objects between all classes, predefined and new, during modelling.

The literature on information systems, e.g. (ISO 1985), describe systems that belong to category A, but the terminology and theory can be used for all kinds of system in Figure 1. A static, or closed, classification is often suitable for a routine design process, which presupposes a high degree of detailed knowledge about the domain of discourse; however, it is not suitable for a more search-like innovative design process.

2.3 The BAS•CAAD approach to dynamic modelling

Through developing a design schema (see section 0) that defines and relates classes that only indirectly and in a generic way refer to the domain of discourse, it is possible to create a dynamic modelling system. Instances of these "meta"-classes are used for the development of model schemas that describe and directly refer to the members of the domain of discourse.

The domain of implemented meta-classes is orthogonal to (independent of) the domain of runtime data or instances. The user of an information system has access only to the latter; the former is available to the system developer only. In the static approach, the model schema resides in the domain of implemented classes, and thus is not open for user manipulation. In the BAS•CAAD system, we have made possible both dynamic classification of design objects and dynamic definition of classes by 'sliding down' the model schema from the domain of implemented classes to entirely reside in the domain of instances, see Figure 2.

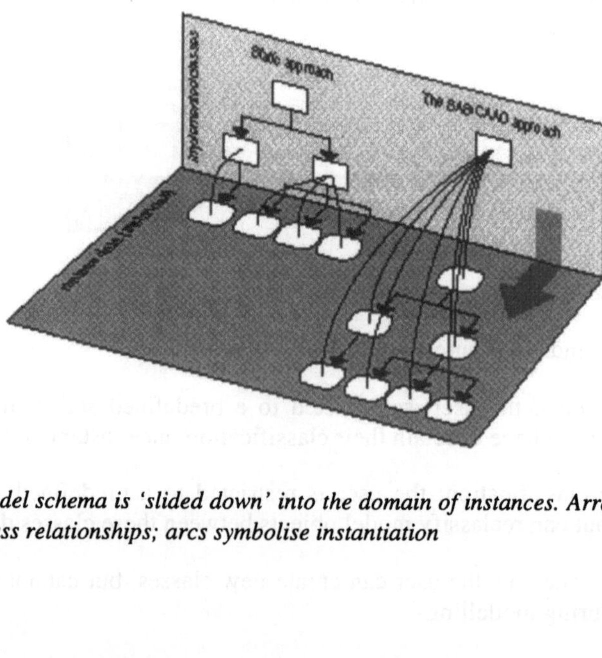

Figure 2: The model schema is 'slided down' into the domain of instances. Arrows symbolise superclass-subclass relationships; arcs symbolise instantiation

The traditional static or hardcoded approach is not in every respect inferior to the dynamic approach; one of its advantages is that it is easier to ensure consistency in a fixed class structure. Dynamic systems, such as the BAS•CAAD system, have to implement mechanisms for dynamic consistency checking. Also, since in the static approach a definite set of classes is implemented, only operations for managing these classes are needed . In contrast, a dynamic system must provide operations on a generic level, which is a much more complex task. The static approach is especially fit in situations where the modelling context is very specific and beforehand well known, such as creating information systems for routine design.

A more thorough discussion of information systems for design can be found in (Ekholm and Fridqvist 1998). There, we defined that information systems for design must: 1) support representing objects in the domain of interest, 2) be able to communicate with other software, 3) support defining the design goal, and 4) have a dynamic object structure.

Most proposals for product modelling software focus on the first two requirements. To be useful not only for describing the results of the design process, but as a tool in this process, an information system for design must have all properties mentioned above. This is also discussed by e.g., Eastman and Fereshetian (1994), Eastman, Assal and Jeng (1995), Galle (1995), Junge, Steinmann and Beetz (1997), Leeuwen and Wagter (1998), and Ekholm and Fridqvist (1998).

3 FOUNDATIONS FOR THE BAS•CAAD INFORMATION SYSTEM

3.1 The ontological framework

The BAS•CAAD information system is intended to cover all levels of generalisation of modelling in the construction context, from international classification standards to specific buildings. It is built on a generic ontological framework including systems and property theory, with the object classes *thing class*, *relation* and *unary attribute* see Figure 3. It allows models to be multi-contextual; that is, several contextual views can be co-ordinated in one model. For a detailed account of this framework, see Ekholm and Fridqvist (1998). The BAS•CAAD information system supports generic design operations, like generalising and specialising, aggregating and decomposing, and adding and removing attributes; for an account of design operations, see (Fridqvist and Ekholm 1996).

In Figure 3, the ontological framework is depicted as an EXPRESS-G diagram. The figure differs from the one in (Ekholm and Fridqvist 1998) in that the entity Attribute is now named UnaryAttribute. In addition, the superclass BAS_CAAD object is removed.

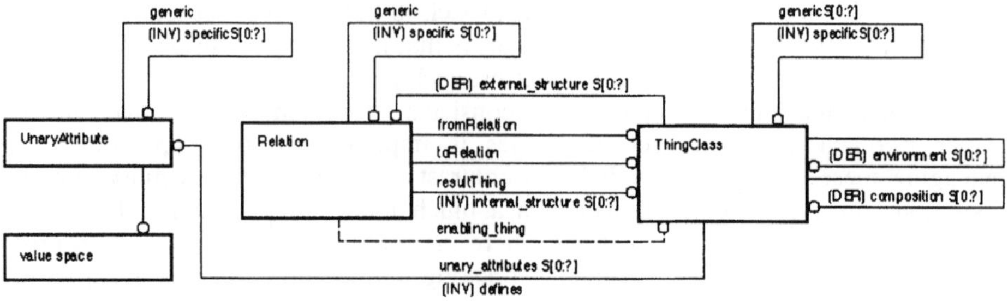

Figure 3: Ontological framework of the BASoCAAD information system

3.2 The design framework

Artefact design can be characterised as making statements about the designed thing and a desired 'satisfactory situation' that the artefact is to support. Statements, or predicates, consist of attributes of different kind; in case of design, these represent properties of the designed artefact. The classes in conceptual schemas are such statements or predicates, made up of one or many attributes. For a more detailed account of this, see (Ekholm and Fridqvist 1998). An information system for design should accordingly at least be able to record such statements. The statements, of course, need not be verbal. Also 2D drawings or 3D models or parts thereof can be treated as statements. The BAS•CAAD information system is designed to collect such statements and to support analysis and communication of them.

A model is defined through a consistent collection of correct statements involving the three basic classes of the ontological framework in Figure 3. The schema in Figure 3 is not suitable for a direct implementation as a design tool, since it implies that all connections between objects must be established. The trouble with this is that some connections might imply statements that the designer would not intend. For example, some things are used in numerous different environments; it would not be desirable to have to include all these environments in the definitions of thing classes like 'bolt' or 'nut'. Instead, we have found that all information necessary for the creation of such a more suitable design schema is contained in the set of design statements made possible through the schema in Figure 3.

The attributes of the entity ThingClass in Figure 3 reflects that it can be defined as a 6-tuple of sets of attributes, T = (T_G, T_C, R_I, T_E, R_E, A_U). Given the BAS•CAAD ontological framework, only seven different types of statements are possible. The first is the statement "there is a kind of things, called T"; this corresponds to creating ThingClasses. The other six statements implies to add attributes to the six sets mentioned above:

- T_G: A T-thing is a kind of Y-thing (= kind T is a subkind or specialisation of Y).
- T_C,: A T-thing is composed by a P-thing part.

- R_I,: A T-thing is internally structured, so that any part of kind P_1 is related to any part of kind P_2 by an R_I -relation.
- T_E,: A T-thing has an environment which includes an E-thing.
- R_E: A T-thing is related by an R_E-relation to any E-thing in its environment.
- A_U: A T-thing has the unary property Q.

The BAS•CAAD design schema in Figure 4 has been developed for the data structure to be implemented. Design data are stored in instances of *ThingClassDefinition*, which collects instances from the six subclasses of *Attribute* that correspond to the six different kinds of attributes. The classes *RelationDefinition* and *UnaryAttributeDefinition* define relations and unary attributes, and ensures that attributes referred to in several places are identically defined.

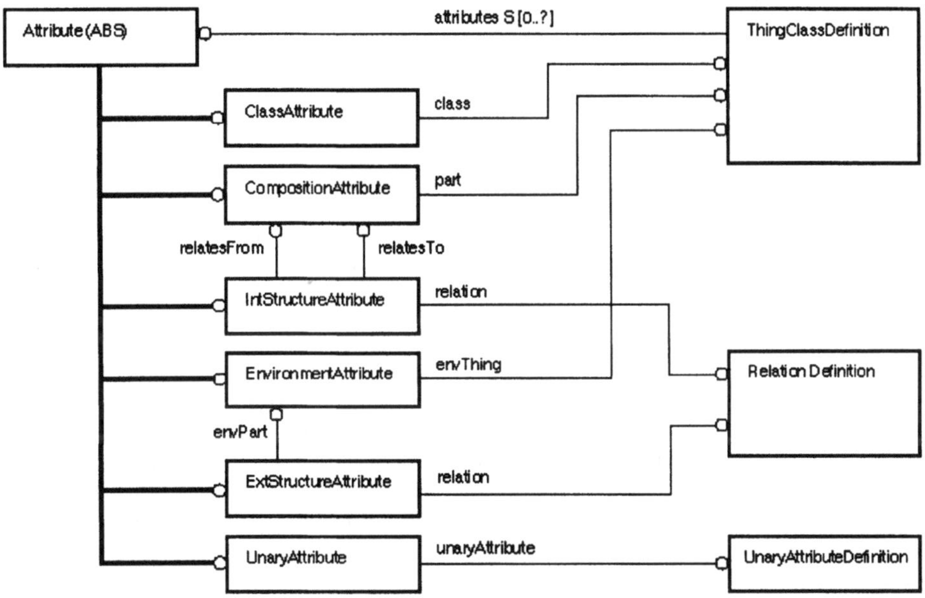

Figure 4: Object structure of the BASoCAAD design schema

3.3 Other features of the object structure

Besides generating and maintaining the object structure, the prototype currently supports a simple schema for object identification, inheritance and specialisation of attributes, and class libraries. Additionally, class subsumption is planned to be implemented.

3.3.1 Object identification

The BAS•CAAD information system for design allows thing classes to be defined by references to predefined libraries (see section 4.3). Such a mechanism requires that libraries and classes can be unambiguously identified, in order to ensure that the correct libraries are used. The current version can identify objects within a library, but there is no secure identification of libraries. The development of such a mechanism is a task for database specialists and international standardisation organisations, and outside the scope of this phase of the BAS•CAAD research project.

3.3.2 Attribute inheritance

Class attributes indicate a more generic kind, also called a superclass. All statements defined for the superclass are also valid for the subclass; the attributes of the superclass are inherited attributes of the subclass. An example: the class House has the attributes 'man-made' and 'provides dwelling'. If Palace is defined by the

	House	Palace
Defined attributes	man-made provides dwelling	kind of House
Inherited attributes		man-made provides dwelling

Figure 5: Defined and inherited attributes

attribute 'kind of Building', Palace inherits the attributes 'man-made' and 'provides dwelling' (Figure 5) A mechanism for attribute inheritance has been implemented.

3.3.3 Attribute specialisation

However, with the above set of attributes, the Palace class doesn't reflect that a palace is a specific kind of house. This type of difference can be defined in to ways; either by adding a new attribute to the specific class, or by specialising an inherited attribute. Which way to choose is a question of clarity in modelling and usefulness of the model; se Figure 6 for a illustration of the differences of the two methods in terms of sets.

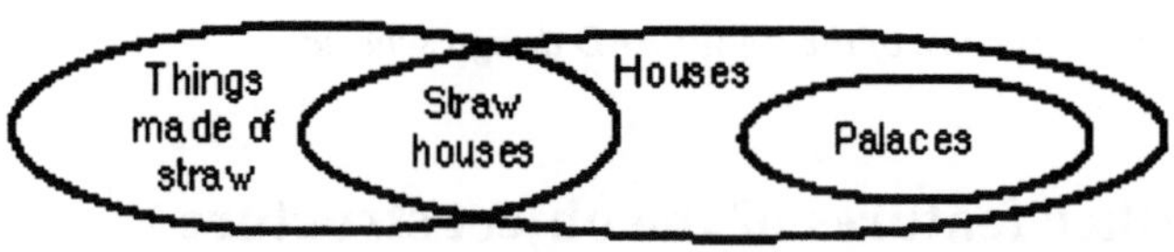

Figure 6: Classes and subclasses

In this case, we have chosen to specialise the attribute 'provides dwelling' into the more specific 'provides luxurious dwelling', and to define the class Palace with the specialised attribute. A mechanism for this has been implemented. The new attribute is defined to substitute the inherited attribute 'provides dwelling' in the Palace class, Figure 7.

	House	Palace
Defined attributes	man-made provides dwelling	kind of House *provides luxurious dwelling*
Inherited attributes		man-made ~~provides dwelling~~

Figure 7: Attribute specialisation

3.3.4 Class subsumption

Description logics (see next paragraph) provides a mechanism called subsumption. This could be described as the inverse of inheritance, since it answers what superclasses a given class has according to its set of attributes. For instance, if a thing Bungalow has the attributes 'man-made' and 'provides dwelling', then it is subsumed to be a kind of House, i.e. the class Bungalow is a subclass of the class House. Subsumption is not currently implemented in the BAS•CAAD prototype system, since it would make implementing inheritance more complicated. However, since subsumption could arguably provide a powerful tool for case retrieval, it will probably be implemented in a future version of the BAS•CAAD system.

Description logics (DL) is a field within artificial intelligence research, and is based on first order predicate logic. It aims to develop mechanisms for describing concepts, and to automatically classify concepts. . DL system implementations are usually similar to programming languages (Lambrix 1996). Although DL is similar to the BAS•CAAD approach in many ways, it seems that it cannot serve as the base for a design information system.

4 THE BAS•CAAD PROTOTYPE INFORMATION SYSTEM

An objective of the BAS•CAAD project is to support expression of objects in a computerised database during design. The prototype is intended to study the feasibility of organising an information system based on the BAS•CAAD schema.

The prototype displays some features we consider important in an information system for design, but it is not intended for productive design work. The current version is centred on building symbolic schemas for concepts referring to things, and lacks most of the abilities to specify measurable values necessary for a production tool.

4.1 Smalltalk

The BAS•CAAD prototype information system is currently implemented in Smalltalk under the Macintosh operating system. The reason for choosing the

Smalltalk computer language was that it is object oriented, and that it supports explorative program development.

In product modelling, products are structured as objects that are assembled of objects in several levels. This makes it natural to choose an object oriented programming language for implementation. In explorative program development, the software needs only to be partly defined before it is executed and tested. Thus, experimental solutions can be tried and kept if successful; otherwise, they are discarded. This way the final software solution is obtained through exploration.

In Smalltalk, code segments can be run and tried directly, without any time-consuming compilation or linking procedures. Actually, Smalltalk allows the programmer to change the code while the software is running, thereby relieving the programmer from repeatedly punching in lots of test data. This is not possible in C++, a popular programming language that also supports object oriented programming. Smalltalk is profoundly object oriented; every concept is treated as an object class. Thus, in Smalltalk it requires an effort to not be object oriented, as opposed to C++. A drawback with Smalltalk is that the resulting programs do not run as fast as those created with e.g. C++.

4.2 Implementing the system

The foundation for the BAS•CAAD prototype is the three concepts thing class, relation and unary attribute that are implemented through the corresponding object classes *ThingClassDefinition*, *RelationDefinition* and *UnaryAttributeDefinition.*

To create and manipulate these objects, there are, in principle, two choices: a command line user interface or a graphical user interface (a GUI). In a command line user interface, the user types commands, and then usually hits a key to get the command executed. A command line interface requires developing a command line interpreter, a piece of software that can translate the command lines into software actions and input data. A problem in software design is to create software that correctly interprets the user's intentions. A GUI features a set of standardised tools, each having a specific function. The user controls the software via the buttons, menus, text input fields etc. Thus, the software developer can restrict the user to input only such instructions or data that are valid.

Regarding these concerns, the choice fell upon a GUI as the method to interact with the software. The Smalltalk dialect used for the BAS•CAAD prototype (Smalltalk Agents, STA), provides generic classes of GUI components, such as windows, menus, buttons etc. These can be subclassed and modified to suite the particular needs of the software being developed.

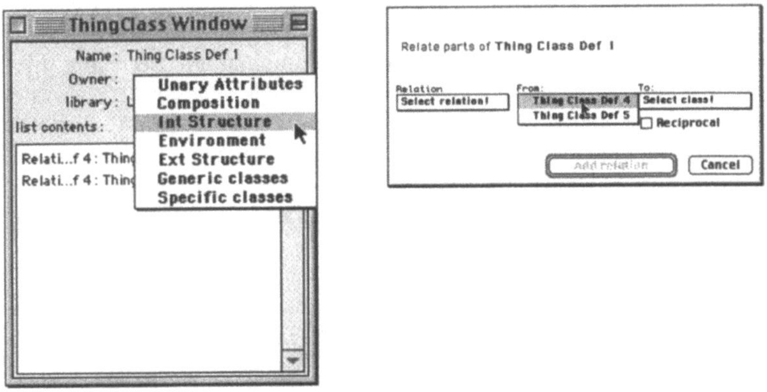

Figure 8: Examples of user interface windows

Windows communicate with the corresponding model objects through messages. Each user interaction with a window causes at least one of the window's methods to be executed. This results in messages being sent to the model object, which responds with the appropriate actions. Finally, some messages are sent from the model object back to the window, to ensure that the window reflects the new state of the model object.

In the current BAS•CAAD implementation, there are windows for the three main classes, and in addition to this, there are windows for specific tasks, such as adding different kinds of attributes to thing class definitions, see Figure 8. Most of the work with the development of this prototype system has been, and continues to be, to decide how to obtain the desired functionality. To perceive a user action in the terms of objects exchanging messages is a complex task. One particularly difficult question is to decide what object is the agent, and what objects are acted upon. The reason for this is that many objects may be involved in the execution of one user action, with many messages exchanged. If the resulting web of such message interchanges is too entangled, the programmer is likely to get lost. Additionally, there is a need to make the software code as intuitive as possible to any future programmer (which, incidentally, often is the same person, but a couple of days later).

4.3 Libraries

The purpose of the BAS•CAAD system is to record information about concrete things during the design process. Acknowledgeably, defining a coherent conceptual understanding of the concrete world from scratch is a very demanding task. To allow the BAS•CAAD system to be used in normal design situations, pre-defined well-considered schemas of Thing Classes have to be provided in libraries, from where the designer can fetch definitions to build up his design database.

In order for the BAS•CAAD system to function in the construction context, it is necessary that its libraries contain attributes and classes belonging to established building classification systems. Today, these systems are not structured for use in the early design stage. The classes are mostly combinatory, i.e. built up from combinations of several attributes. One example of such a classification system is the BSAB system, a de facto Swedish construction industry standard. In order to implement a classification system into a system for design, including the earliest stages, it is necessary to make the constituent simple attributes availlable. The BSAB system would be the natural choise for analysis for an application to the Swedish context.

Principles for structuring product libraries are developed as part of the STEP standardisation activity, the so called Parts Lib work (ISO 1997). An example of the development of an operational class library is POSC/Caesar with its Reference Data Library for documentation of facilities for the oil and gas industries (POSC/Caesar 1999).

The current BAS•CAAD prototype provides an ontology for and supports class libraries. Objects in libraries can refer to objects in other libraries, so that hierarchies of libraries can be created. This feature is implemented to explore a mechanism that would allow international and national standardisation organisations to create generic libraries. Clients, building material providers, and others can then create intermediate level libraries that can be used as references in specific building projects.

The concept of class library has been generalised, so that library is the sole format for databases and files. Thus, the work of any level of specificity can be the foundation for further specialisation. For instance, it would be possible to use the project file from a window designer as a library file in a building design project without any reformatting or reclassification.

5 TWO DESIGN SCENARIOS

5.1 Coordinated design of user-organisation and building

This section describes how the BAS•CAAD system can support actual design through two scenarios. The first scenario describes the task of designing a building for Mr and Mrs Smith's small business. The object of design is on one hand the building, and on the other hand the activity to be carried out inside the building. To document this knowledge, we create two Thing Classes, *Business-building* and *Business-activity*. We also add attributes that define them to be each other's environments.

THINGCLASS *Business-building*
    ENVIRONMENTATTRIBUTE *Business-activity*
    EXTERNALSTRUCTUREATTRIBUTE *provides-environment-to Business-activity*

THINGCLASS *Business-activity*
ENVIRONMENTATTRIBUTE *Business-building*

Our task is now to develop the design, and to document the knowledge we may collect during the process. In the BAS•CAAD system this is accomplished by creating additional Thing Classes and by adding attributes to them.

When consulting the Smiths, we learn that the main activity of their business is decomposed into two parts, workshop and office work. We document this with two Thing Classes, *Workshop-activity* and *Office-activity*. The new Thing Classes are attributed as parts of *Business-activity* through composition attributes. The Smiths also tell us about the shape and size of the two activities. This information is added to *Workshop-activity* and *Office-activity*, respectively, through Unary Attributes.

THINGCLASS *Workshop-activity*
UNARYATTRIBUTE *shape* VALUESPACE *(rectangle 8 x 15)*

THINGCLASS *Office-activity*
ENVIRONMENTATTRIBUTE *Workshop-activity*
EXTERNALSTRUCTUREATTRIBUTE *administrates Workshop-activity*
EXTERNALSTRUCTUREATTRIBUTE *is-adjacent-to Workshop-activity*
UNARYATTRIBUTE *shape* VALUESPACE *(rectangle 5 x 8)*

THINGCLASS *Business-activity*
ENVIRONMENTATTRIBUTE *Business-building*
COMPOSITIONATTRIBUTE *Office-activity*
COMPOSITIONATTRIBUTE *Workshop-activity*
INTERNALSTRUCTUREATTRIBUTE *Office-activity administrates Workshop-activity*

Since we were told that the two activities don't co-exist very well, we conclude that each activity needs a space of its own. Therefore the *Business-building* is divided into two spaces, *Workshop-space* and *Office-space*, which become parts of the building. The primary function of the spaces is to provide environments to the activities. This is reflected by the attributes in the definitions, which relate the activities to the spaces.

THINGCLASS *Workshop-space*
ENVIRONMENTATTRIBUTE *Workshop-activity*
EXTERNALSTRUCTUREATTRIBUTE *provides-environment-to Workshop-activity*

THINGCLASS *Office-space*
ENVIRONMENTATTRIBUTE *Office-activity*
EXTERNALSTRUCTUREATTRIBUTE *provides-environment-to Office-activity*

THINGCLASS *Business-building*
ENVIRONMENTATTRIBUTE *Business-activity*

EXTERNALSTRUCTUREATTRIBUTE *provides-environment-to Business-activity*
COMPOSITIONATTRIBUTE *Office-space*
COMPOSITIONATTRIBUTE *Workshop-space*

In the example above, we have seen how a design can be represented by thing classes from the most generic notion, and through consecutive steps of refinement to a more developed model. Using the BAS•CAAD system, the design process would progress in a similar manner until all things in the building were described to a sufficient detail and uniquely identified.

5.2 Detailed building design

The second scenario describes how the BAS•CAAD system supports detailed building design including aspects like spatial, functional, and compositional views. Our task is to design a building, and we will use pre-defined library Thing Classes, representing common things and systems in building.

To begin with, we create the Thing Class *building* to represent our building, see Figure 9 a. Since we are interested in the spatial aspect of buildings, i.e. the ability to define a space, we decide our new Thing Class to be a subclass of the pre-defined Thing Class *space*. Here, the Thing Class *space* represents both the spatial and enclosing aspects of things. Being composed of *enclosures* is an attribute of the *space* Thing Class. The *enclosure* attribute is inherited by the *space* subclasses, e.g. our new *building* Thing Class. An *enclosure*, in this example, represents both the shape and enclosing aspects of things, e.g. prohibiting pass-through.

In the next step of the scenario we want to subdivide our building into subspaces. This is done by creating new subclasses of *space*, and by adding composition attributes to the class definition of *building*, that point to these new subspace Thing Classes, see Figure 9 b. In this schema only one of the building's subspaces is shown.

When the spaces are defined, we need to specialise the composition attributes of the spaces and the building, so that they are defined to be composed of things that specificly enclose them. The specialised *enclosures* are subclasses of the generic *enclosure* thing class, see Figure 9 c.

The enclosures of the building represent the enclosing aspect of the building's exterior walls, e.g. shape and climate enclosing. Since the *space 1* faces the exterior of the building, some of its enclosures, too, are aspects of exterior walls. A Thing Class that represents an exterior wall would need to include both aspects. This is done through multiple inheritance, where the wall *Thing Class* inherits all attributes from both enclosures, see Figure 9 d.

The design process will continue by consecutively adding attributes to the model objects until a satisfactory design has been achieved.

In order to have the building specified for cost calculation, tendering and construction, we need to define ***building elements*** and *work results.* Building elements are pre-defined classes of functional parts of a building, and work results are pre-defined classes of compositional parts of buildings, i.e. things that results from constuction work. In the Swedish classification system BSAB 96, we can find such classes. If these classes are expressed as thing classes in the BAS•CAAD system, we can transfer the attributes that define the BSAB classes directly through mutiple inheritance, thus including these aspectual views into the building model, see Figure 9 d.

The tests in these relatively simple scenarios indicate that the BAS•CAAD system provides a suitable generic structure for product documentation, also during the design stage.

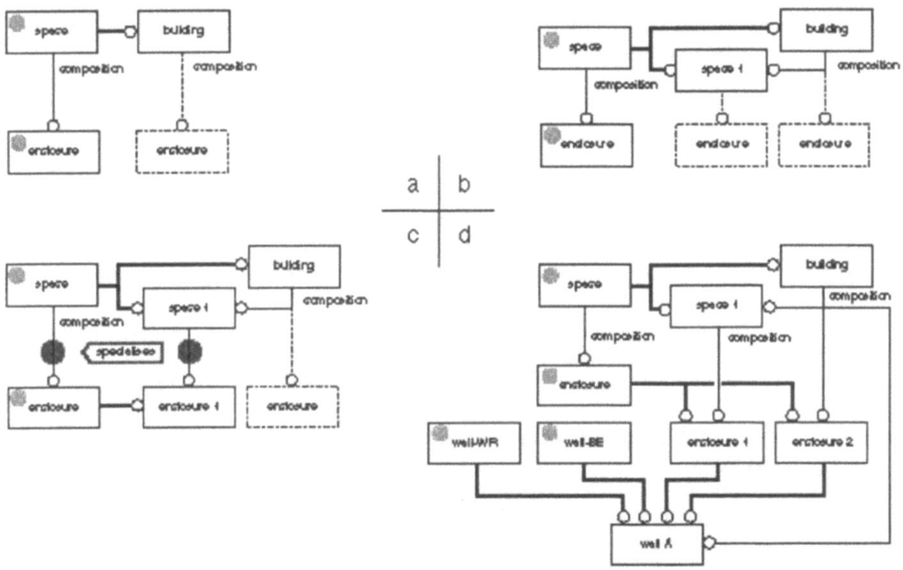

Figure 9: Four stages of the design of a building (in modified EXPRESS-G). Grey dots indicate library classes. Thick connecting lines indicate subclass relationships, thin connecting lines indicate attributes, dotted enitities indicate inherited attributes.

6 REFERENCES

Björk B C, 1995, Requirements and information structures for building product data models VTT Publications 245 (Technical research centre of Finland, Espoo)

Bunge M, 1998, *Philosophy of Science Vol 1 From Problem to Theory* (Transaction Publishers, New Brunswick, New Jersey)

Eastman C M, Fereshetian N, 1994, "Information models for use in product design: a comparison", *Computer-Aided Design* 7 (26) 551-572

Eastman C M, Assal H, and Jeng T, 1995, "Structure of a database supporting model evolution" *Modelling of buildings through their life-cycle.* CIB Proceedings Publication 180, (Eds. Fisher M, Law K, and Luiten B) (Stanford University, Stanford, Ca, USA)

Eastman C M, Siabiris A, 1995, "A generic building product model incorporating building type information" *Automation in Construction*, 4 (3) 283-304

Ekholm A, Fridqvist S, 1996, "Modelling of user organisations, buildings and spaces for the design process" *Construction on the Information Highway* CIB Proceedings Publication 198, (Ed. Ziga Turk) (University of Ljubljana, Slovenia)

Ekholm A, Fridqvist S, 1998, "A Dynamic Information System for Design Applied to the Construction Context" *The Life-cycle of Construction IT Innovations* (Eds. Björk, B-C. and Jägbeck, A.) Proceedings from the CIB W78 workshop, 3-5 June 1998, Stockholm, Sweden

Fridqvist S, and Ekholm A, 1996 "Basic ObjectStructure for Computer Aided Modelling in Building Design" *Construction on the Information Highway* CIB Proceedings Publication 198, (Ed. Ziga Turk) (University of Ljubljana, Slovenia)

Galle P, 1995, "Towards integrated, "intelligent", and compliant computer modeling of buildings" *Automation in Construction* 3(4) 189-211

ISO, 1985, *Concepts and terminology for the conceptual schema and the information base* ISO/DTR 9007 (TC97), also SIS teknisk rapport 311 (SIS Stockholm)

Junge R, Steinmann R, Beetz K, 1997, "A dynamic product model" *CAAD futures 1997* (Ed.Junge R) (Dordrecht: Kluwer Academic Publishers)

Lambrix P, 1996, *Part-Whole Reasoning in Description Logics* (Diss. Linköping University, Linköping, Sweden)

Leeuwen J P, Wagter H, 1998, "A Features Framework for Architectural Information", *roceedings of the Artificial Intelligence in Design Conference 1998* (Ed. Gero, J. And Sudweeks, F.) (Dordrecht: Kluwer Academic Publishers)

Maher M L, Simoff S J, Mitchell J, 1997, "Formalising building requirements using an Activity/Space Model" *Automation in Construction* 1(6) 77-95.

POSC/Caesar, 1999, URL: http://www.posccaesar.com/snapshote/snaptoc.htm, POSC/Caesar Reference Data Library Accessed January 1999.

Rittel H, 1984, In Cross N, *Developments in Design Methodology* (London: John Wiley and Sons)

Information Sharing in Building Design

James Snyder and Ulrich Flemming
- Lockheed Martin Advanced Technology Laboratories, Camden, NJ 08102
- School of Architecture and Institute for Complex Engineered Systems (ICES), Carnegie Mellon University, Pittsburgh, PA 15213

Key words: building models, conceptual modeling, product modeling, information modeling, design tool integration

Abstract: We address information modeling and exchange in asynchronous, distributed collaboration between software applications or design agents that are heterogeneous, that is, developed independently based on application-specific data models. We identify the requirements an integration environment must satisfy if it is to support the semantically correct exchange of selected, locally generated information between the agents. These requirements are distilled from both the literature and our own experiments with the Object Modeling Language OML. The resulting requirements were then formalized into an information modeling and exchange environment constructed around the modeling language called SPROUT (supported by a compiler) and an associated software architecture that can be targeted toward many different hardware and software platforms. A unique capability supported in this environment is formal support for integrating existing applications: Given a schematic description in SPROUT, a formal specification can be used to generate computer programs that provably map data to and from the applications.

1. INTRODUCTION

In the present paper, we address conceptual and technical issues that arise in a specific type of collaborative design scenario. We assume that the firms or design teams collaborating on a building project use local software to the degree that is appropriate to them. We assume furthermore that the local software uses algorithms and internal representations/data models tailored to the particular design task it supports. Each of these local models captures a particular universe of discourse that differs - to some degree at least - from those of the other teams. Each team uses local persistent data storage facilities, i. e. a local database, and it does this again to the degree it deems necessary and appropriate. It is likely that the individual pieces of software were developed independently of each other. It is furthermore to be expected that different teams enter the project at different times, which are, in addition, unpredictable at the outset.

On the other hand, many of the decisions made by one team must be communicated to some of the other teams because they have ramifications for the work done by those teams. Each team intends to take advantage of the fact that it has a computable design representation and to use that representation to automate information exchange to the largest possible degree.

Clearly, our scenario does not deal with synchronous collaboration in which teams work on the same data model or 'document' over some network connection; that is, we do not address situations handled by networked 'groupware' or shared 'electronic whiteboards'. It has been observed that the approaches suggested for a more asynchronous information exchange typically fall into two categories [Rolland and Cauvet 1992] pages 39 and 40: (1) The fully integrated approach, which relies on a unique, shared representational schema to which all subsystems have to conform; and (2) the federated approach, in which each local system has its own schema and is responsible for controlling its interactions with other systems by deciding which information should be imported and exported. With respect to building design, the integrated approach has been described as follows [Augenbroe and Winkelmann 1990]:

> A building project requires generating, updating, and communicating an enormous amount of data... Traditionally this description is stored and displayed in analog, causing numerous problems due to ambiguity, incompleteness and inconsistency of information. There is a strong consensus in the computer industry that the key to integration will be the definition of complete data models for each product type that will satisfy all...information needs. This complete data model is generally called a product model.

Attempts to develop such an approach are too numerous to be listed here. To the best of our knowledge, the federated approach has been tried in the context of building design only in the Agent Communication Language (ACL) project, in which both authors participated [Khedro et al. 1995].

We argue below that neither approach is appropriate in the context of the scenario sketched above, and we introduce subsequently work that attempts to overcome the shortcoming of each approach in this scenario. That is to say, our paper should not be misunderstood as a general critique of fully integrated or strictly federated architectures. Each may well work in different scenarios.

To start with the fully integrated approach in the present scenario: It is unrealistic to expect that the ever-changing cast of teams with their heterogeneous software support can be made to subscribe to a single, comprehensive data model that is able to satisfy the information needs of every participant. In fact, it may not even be desirable to have such a model because individual teams may use data that are of no interest to other teams or even proprietary. The following example, taken from a different industry, illustrates this and shows - at the same time - that this situation is not unique to the building industry.

Figure 1 shows a high-level, simplified decomposition for an airliner that identifies five major subcomponents, each of which is designed by a different team/firm specializing in a specific domain/discipline using internal software, which we call an application below. Information exchange occurs in this situation within and between domain specializations. An example of the former is wing design, which is proprietary, and proprietary data are exchanged freely within the team. An example of the latter is the design of the wing connection to the fuselage, which is contracted out, and information must be exchanged selectively between the different teams responsible for these respective tasks.

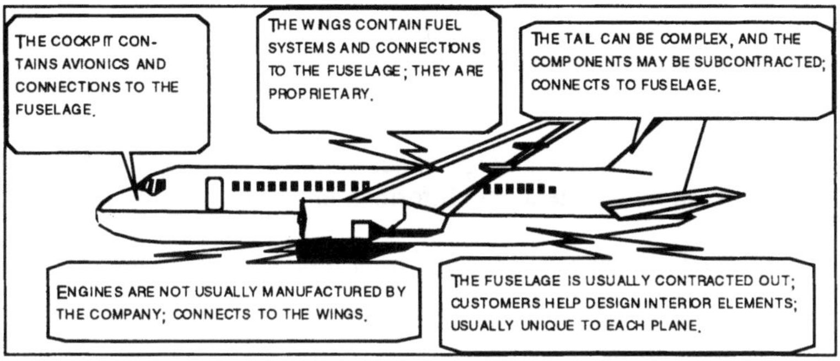

Figure 1. Information Exchange in Airplane Design

We cannot expect that the internal models used by the different applications are conceptually equivalent. An example that arose in the context of the ACL project described in greater detail below should make this clear. Two teams, one representing architectural design, located at Carnegie Mellon University (CMU), and one representing structural design, located at the Massachusetts Institute of Technology (MIT), had to exchange information about spatial configurations and structural systems aimed at supporting these configurations. The MIT team, after receiving an initial spatial configuration from CMU, had to develop a preliminary structural design. It used a very efficient and fully recursive internal representation that made use of the fact that structural members in different locations often are based on the *same* design that satisfies some maximum load condition [Chiou and Logcher 1996]; Figure 2 shows a portion of this representation. The representation therefore recorded the shared design only once and associated it with the different locations where it occurred. The architects at CMU, on the other hand, were not interested in structural details; they were only interested in the geometry of the structural members, which needed to be completely enumerated because they had to be displayed graphically and checked visually. The CMU application therefore represented structural members in a flat list, from where it exported data to a CAD package used only for purposes of 3D display. But it had to remain possible to re-synthesize from this enumerated list particular locations and designs, for example, if the architects did not like the shape or location of a specific element. Note that this type of communication goes beyond the kind of type conversion called transduction in [Eastman 1994]. It requires mappings between the different conceptual schemas involved and may include synthesis and re-synthesis of concepts.

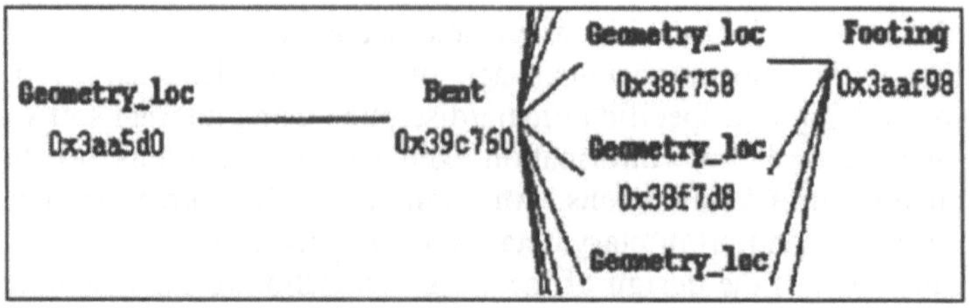

Figure 2. Parsimonious Representation of a Structural System [Chiou and Logcher 1996]

Semantical correctness must be maintained in both directions. More generally, the exchange of (partial) information must satisfy the following criteria under the present scenario:

1. *Robustness of information exchange*: Since applications do not export every piece of information they generate. Information loss will occur during the exchange.
2. *Semantically correct concept mapping*. It must be possible to map bi-directionally information defined in terms of one conceptual schema to information defined in terms of another conceptual schema, and the mapping may require synthesis and re-synthesis of concepts.
3. *Variable rate of information changes*. Some information is only relevant if it is delivered in a timely manner, while other information is relatively static. The information exchange has to be able to take such varying rates into account.

We use the term *communication fidelity* to refer to the three requirements in combination. Note that the example under point 2 above illustrates why a federation architecture as used in the ACL project is inadequate to guarantee communication fidelity as defined here. The facilitator used in this project had no internal memory: it 'flushed out' all information it received immediately to the interested applications; it was thus unable to support the type of synthesis and re-synthesis suggested by the example, which in turn requires that the exchange agent have some form of intermediate representation and internal memory. Indeed, the approach that suggests itself for the present scenario is one that relies on some form of mediated architecture as described in [Wiederhold 1995] in the context of client/server architectures.

We can now state more succinctly the issues addressed in the present paper. We deal with information exchange that maintains communication fidelity for the information exchange in asynchronous, distributed, computer-supported design collaboration as captured in our scenario. Especially, we explore the potential of a mediated architecture that provides not only communication fidelity, but also a large degree of openness and flexibility: agents can enter or leave the collaboration at any time; they are not restricted to a specific universe of discourse or data model, nor to a specific programming paradigm. This scenario raises a host of complex issues that we cannot fully present, even mention, here. We must restrict ourselves to a highly selective presentation; a full account can be found in [Snyder 1998].

The remainder of this paper is structured as follows: We review briefly in the next section research or industrial efforts that influenced our own approach or address similar issues in contrasting manners. The ACL project mentioned above has been especially significant because it provided us with a first test ground for experiments with a schema specification language, OML. In section 3, we briefly describe the OML experiments and the lessons we learned from them. Both the ACL project and our literature survey convinced us that communication fidelity can best be achieved in an integration environment constructed around a schema specification language; we outline in Section 4 general requirements for such an environment. In section 5, we introduce the schema specification language SPROUT, the successor to OML. In section 6 we outline briefly the kind of integration architecture that can be built with a language like SPROUT and outline a first prototype implementation.

2. BACKGROUND

Computer-Aided Design (CAD) software aimed initially at replacing manual drafting - to some degree - with computer-based tools and was not integrated with any other computer-based design tool. Since then, at least two generations of systems and projects have explored integration issues at increasing levels of generality and sophistication.

2.1 First Generation Integration

A first generation of more integrated systems emerged during the mid-1980's and early 90's. This work concentrated on research prototypes or software-based laboratories and pursued the explicit goal to discover and investigate the issues involved in systems integration in building design. Development on these prototypes has been completed. The following projects were of particular importance to us because of our active participation (see [Snyder 1998] for a more detailed account):

- The Intelligent Computer Aided Design System (ICADS), a project in the Computer Aided Design (CAD) Research Unit, Cal Poly, San Luis Obispo [Pohl et al. 1992] [Myers et al. 1992]
- The Integrated Building Design Environment (IBDE) project in the former Engineering Design Research Center (EDRC) at Carnegie Mellon University [Fenves et al. 1994]

2.2 Second Generation Integration

Second generation integration systems make use of results from first generation systems and benefit generally from a significantly larger suite of commercial or research tools that could be used as subsystems of the large whole (e.g. object-oriented databases). The projects presented below represent solutions to the information exchange problem that are - within their own premises - persuasive and successful.

2.2.1 STEP and STEP-based Extensions

The Standard for Exchange of Product Model Data (STEP) (ISO 10303) is an ISO committee with the purpose of developing an infrastructure for product data exchange across disciplines. The Product Data Exchange Using STEP (PDES) project is a U.S. project that provides industrial input to the ISO committee [Nell 1996]. In order to facilitate the definition of product data models, the EXPRESS language continues to evolve [ISO 10303 Part 11]. Both STEP and EXPRESS are quite large, and we cannot provide a comprehensive review in the context of this paper.

EXPRESS was initially designed to facilitate file-based data exchange using a neutral file format and provides several syntactic mechanisms to support this type of exchange [Eastman and Fereshetian 1994]. EXPRESS's usefulness has been demonstrated in several application areas. It is at its core an instantiation specification language, not a schema specification language (see [Snyder 1998]).

Among the STEP-based extensions, the COMBINE project is of particular interest to us because it reinforces insights from the ACL project, which revealed severe problems associated with schema mappings and translations that must be coded manually [Dubois et al. 1995].

2.2.2 DICE

The Distributed and Integrated Environment for Computer Aided Engineering (DICE) project is a broad research effort of the Intelligent Engineering Systems Laboratory at MIT [Sriram and Logcher 1993]. DICE is a network of collaborating design agents, where communication and coordination are achieved through a global database and control mechanism. The product and processing modeling component of DICE is SHARED [Wong and Sriram 1994a]. Again, DICE is too extensive to be reviewed here in full; we must restrict ourselves to a few comments about shared, the most interesting component of DICE in the present context.

SHARED can be described as a rich object-oriented modeling environment. It also provides database capabilities such as version, alternative and transaction management. [Sriram et al. 1994] view the SHARED approach "... as a first cut in developing a semantic vocabulary for design, where the constraints (through behavior) implicitly define the grammar for checking valid designs." In its current form, SHARED should be viewed an as implementation framework for shared workspaces and not a conceptual modeling environment because concepts are defined in the implementation language, not in a programming language-independent way.

2.2.3 EDM

The Engineering Data Model (EDM) is a modeling environment [Eastman and Siabiris 1995]. EDM-2 [Eastman et al. 1995] added object structuring techniques in terms of a specialization lattice and composition hierarchy. Both constructs are general and have direct analogies to other object paradigms, in particular object/relational database systems, that support schema evolution. EDM-2 can be viewed as a database schema definition language with the added capability of constraint definition and management of the objects. Among 2nd generation integration projects, EDM is in spirit closest to our own work because it explicitly addresses the scenario outlined in Section 1.

2.3 Context

Starting in 1992, groups at the School of Architecture and Department of Civil and Environmental Engineering at Carnegie Mellon University (CMU) have been developing the Software Environment to Support Early Phases in Building Design (SEED) with participation of a group at the University of Adelaide, Australia [Flemming and Woodbury 1995]. SEED builds upon the experience gained with the IBDE project and several projects dealing with 'generative' design systems in the EDRC.

The SEED project intends to develop the prototype of a software environment to support the early phases in building design. The goal is to provide support, in principle, for the preliminary design of buildings in all aspects that can gain from computer support. This includes using the computer not only for visualization, analysis and evaluation (i.e. *design tools*), but also more actively for the generation of designs, or more accurately, for the rapid generation of computable design representations describing conceptual design alternatives and variants of such alternatives at an appropriate level of abstraction, but with sufficient detail that enables sophisticated evaluation tools to receive all of the needed input data from the representation.

The SEED team specifically develops generative software components, called modules, each as it relates to a specific discipline or application domain. Design tools are to be provided to the largest possible degree by third-party packages. SEED is to be used is exactly as described in Section 1's senario, except that the individual modules rely on a shared object server for persistent data storage and version management and share basic design concepts and are not required to use a specific programming language. The environment has to remain heterogeneous and open simply because the SEED developers cannot predict which software has to be integrated in the future and when.

3. OML EXPERIMENTS

We describe in this section experiments with a high-level specification language, called the Object Modeling Language (OML), that were conducted in the context of the ACL project mentioned in Section 1

3.1 ACL Project

The multi-institutional ACL project aimed at investigating concurrent engineering (CE) technologies to support collaboration between participants in the facility design and delivery process [Khedro et al. 1995]. Teams of researchers representing four major universities participated along with a team at the U. S. Army Construction Engineering Research Lab (USACERL), the project sponsor: MIT, Stanford University, the University of Illinois at Urbana-Champaign, and CMU. The collaboration between the teams was to make use of technologies and applications developed independently by the four universities and USACERL. The CMU team was to use SEED-Layout, a module of SEED, to deal with the architectural aspects of early building design [Flemming and Chien 1995].

The sponsor of the ACL project had established at the outset that the cooperation of the cooperating systems was to be based on the agent-based Federation Architecture developed by Stanford, which intends to allow independent software agents to exchange information by communicating with facilitators through appropriately designed APIs [Khedro et al. 1995]. The ACL project it is the only project known to us that addresses the type of asynchronous, distributed collaboration in building design underlying our scenario through a strictly federated architecture that had to be implemented in order to give researchers the opportunity to investigate the more general problems posed by this scenario through realistic experiments, that is, experiments using applications that were truly distributed as implied by the scenario. More specifically, this project lead to the OML experiments that proved so crucial for the present work.

3.2 Object Modeling Language

During the course of the project, the team members realized that because each team had its own internal agent architecture, some kind of unifying syntax and semantics of object descriptions was needed to allow message exchanges to occur. The CMU team developed OML as one of its contributions to the ACL project. OML provides an object description schema with execution semantics, a messaging protocol, and an associated language binding complier that generates data structure mapping code from an OML schema, that is, formal specification [Snyder and Flemming 1994a], [Snyder and Flemming 1994b]. Based on an agent's internal representation, an OML schema builds an OML model external to the agent, which describes the agent's model to the outside world in terms of four basic language constructs: domains, relationship types, and classes (a complete description of OML is given in [Snyder 1993]). OML was designed to provide semantic consistency for inter-agent information exchange.

Given an OML schema, each agent developer must provide a (concise) language binding specification that can be used to map the local representation to and from the OML schema. The mapping between different local representations involves a large amount of monotonous binding assignments, which, in addition, must be verified. This verification is error-prone and time-consuming, which is especially true when the mapping routines need several programming language linkages (e.g. C, C++, Fortran). [Dubois et al. 1995] make the same observation (see also [Lockley et al. 1995]); *IDM* stands for Integrated Data Model and *DT* stands for Design Tool:

> For the time being, too much is still dependent on individual skills and inspiration. Proofs that a resulting model is somewhere near a desired optimum or even semantically correct are still outside of our reach. A particular aspect of the latter is the way that teams now have to check whether their DT model can be mapped to and from (a particular subschema of) the IDM. This in fact is the way that the IDM was checked and refined in the latter stages of the project. As it happens there is no other way available than to try to perform the mapping for a range of examples given the IDM and the DT's aspect model. This is a time consuming task which moreover can lead to a false conclusion depending on the completeness of the chosen examples.

The problem is exacerbated when the schema itself is not entirely static, for example, because the schemas underlying the individual applications are still evolving or new applications are added to the environment. Reconfiguring the schema in these cases may be practically impossible.

These problems are largely circumvented by the type of language binding compiler provided in the OML standard distribution: The developers can use the OML schema and the language binding specification to generate automatically language binding code in a host language like C++. This code can be compiled into a language binding library and linked to an agent's domain code along with OML run-time libraries.

3.3 Observations

OML was tested in two contexts, the ACL project for which it was designed, and the Modular Design System (MDS) project, another USACERL-funded project (not presented here). The language binding compiler was tested both internally in the communication between SEED-Layout and the other agents communicating with SEED-Layout to form the architectural agent in the ACL project (an object database and CAD package used strictly for 3D displays). It was also tested in information exchange between the CMU- and MIT- based agents. But these experiments were conducted in parallel to the main project focus.

The system as a whole was never complete or tested. The Federation Architecture was not able to support the integration of applications of the given type over the Internet for several reasons, among which the following are most significant in the present context: (1) We never knew if information was exchanged correctly because we could not verify concept translation rules or data structure mappings in an application; translations must be based on a formalisms and automated code generation. (2) The Federation Architecture did not provide any tools for object-centered information modeling. Some researchers claim that it should not have such tools, but became clear to us that state-of-the-art object-based modeling techniques, such tools are essential. [Flemming et al. 1996] provide a full report of the research results of the ACL project from the SEED perspective; see [Chiou and Logcher 1996] for a complementary assessment.

Combined with the OML experiments, the results of this project strongly suggest that a robust, programming language-independent conceptual modeling language is needed for information exchange techniques to be scalable. Specifically, the language binding compiler demonstrates a promising approach for generating mapping code from a formal specification rather than by the current techniques that rely on code generated by hand.

4. REQUIREMENTS FOR AN INFORMATION MODELING AND EXCHANGE ENVIRONMENT

The reviewed literature and OML experiments provided us with a wealth of insights and suggestions about promising ways to deal with the information exchange issues raised in our scenario. In the present section, we distill from these a set of requirements an information modeling and exchange environment should meet in this context. The requirements are somewhat motivated by the SEED perspective. They remain however general because SEED is conceived in general terms.

4.1 Modeling Language

The cornerstone of the environment is a modeling language. Before we elucidate the requirements such a language should meet, we provide a brief overview of modeling languages as an emerging research area.

The complexities associated with developing large-scale information systems have pushed existing programing languages beyond their capabilities. This has led to the introduction of high-level modeling languages that can be used to model functional application requirements and information system components at a conceptual level [Loucopoulos 1992]. Given the different areas from which modeling languages historically evolved (knowledge representation in AI, programming languages, data bases), it is difficult provide a concise definition that would cover all instances because different languages tend to retain some of the flavor of the field in which they originated. But certain common traits do emerge. The most unifying force is an object-centered approach to representation.

The most practical way to distinguish modeling from programming languages is to describe the features incorporated in the language as opposed to what constructs can be represented in it (which may have the same names). Several such distinguishing features have emerged in the literature:

- **Semantic Relationships**. Object-oriented programming languages have no explicit constructs for representing the semantics of object relationships; that is, complex object compositions must be maintained manually by the programmer. However, some modeling languages have the ability to manage object compositions using relationships between objects as an explicit construct in the language. Therefore, a more consistent and uniform mechanism can be employed to address relationship management. In fact, several authors suggest that relationships can be generalized with their own properties and definitions ([Rumbaugh 1987], [Parent and Spaccapeitra 1992]).
- **Behavioral Properties**. In addition to the more traditional encapsulation capabilities found in programming languages, modeling languages allow for the specification of reactions to events that occur on an object. To borrow from database terminology, an event can *trigger* an action. Other forms of behavioral properties include constraints on the structural or data properties of an object.
- **Dynamic and Temporal Capabilities**. Modeling languages handle changes to objects over time while maintaining some type of historical record of these changes; in database terminology, this is usually version and configuration management. In some cases, semantic relationships and configuration management are highly interconnected (see below).
- **Representational Homogeneity**. Object-centered representations have the desirable property of using similar representations at different levels of conceptual abstraction. Traditionally, product models were specified and represented independently from the process models that utilized them. However, a homogeneous representation should support both simultaneously.

4.2 Overall Schema Requirements

The modeling language supporting the application integration environment envisaged is used to formally specify a conceptual schema serving the following functions:

1. Provide a 'lingua franca' used by participating applications for mappings between concepts of mutual interest. The OML experiments clearly demonstrate the power and flexibility gained from basing information communication on such an exchange medium. The modeling language is used to specify a shared schema to and from which the schemas underlying the participating applications are mapped through language bindings.
2. Support the semi-automatic generation of language binding code. Again, the OML experiments, supported by the literature, demonstrate the feasibility, indeed, the necessity of this approach. In the SEED context, this includes also the automatic generation of the database schema from a high-level specification.
3. Provide basis for negotiations between application developers about the shared concepts.

The last point needs some elaboration because it is universally neglected in the literature. We can assume that at the outset, only the developers of the individual applications understand the semantics their underlying the data models. In order to develop a shared schema, the individual developers have to engage in rounds of negotiations in which they 'hammer out' both the content and form of that schema. The OML experience as well SEED-internal negotiations suggest to us that the modeling language itself can provide crucial support for this process by establishing from the outset a common communication medium. Indeed, the language can provide a firm structure for the negotiations.

For example, the following process proved successful for negotiations between the SEED teams. Before the first meeting, every team has to do its 'homework', that is, express the shareable aspects its own internal data model in terms of the modeling language (as a side effect, this puts the expressive power of the language to the test). These specifications serve to explain to each team the content of the other teams' models. Next, each team indicates which pieces of information it is interested in receiving from the other teams thereby determining both the content of the shared schema and the information that can safely be lost. The group as a whole is then ready to design the shared schema. This process can take advantage of the fact that the information of interest is already expressed in the target language; that is, it consists largely of remixing what is already specified. Central to this last, all-important step is to decide where information loss occurs. The mediated architecture envisaged here gives developers flexibility in this process.

If a new application enters the cooperation, this same process is repeated at a smaller scale. In all likelihood, the shared schema has to be updated as a result and the language bindings recompiled, a process that is largely automated if the schema is conceived along the lines envisioned here.

We conclude this section with some general schema requirements:

1. **Implementation Independence**. One of the more challenging requirements information modeling languages try to meet is the elimination of hardware and software platform dependencies. While the success in this effort must be

achieved at various levels of abstraction and with various resources, particular attention must be paid to schematic features that may diminish platform independence. For example, something as simple as upper and lower case names within a model can make implementation difficult or impossible in some environments. Additionally, the modeling environment should be independent of a programming language.

2. **Application Integration** must be supported by formal methods that allow for the (semi-)automatic generation of language binding code so that any changes to the schema require only recompilation. Specifically, given a shared schema, an application binding specification should be used to generate code to perform the mapping.

4.3 Schema Elements

During the OML experiments, we discovered that the representations used to specify the mentioned above mappings should reflect an object-oriented structure if some applications use object-based representations. Both the OML experiments and the SEED experience suggest that an object-centered modeling language needs at least the following constructs to gain the needed expressiveness:

- **Domains**: An important distinction is the differentiation between well-formed data and representations of design objects of interest. For example, when a module retrieves the altitude of a sun ray at a specific day and latitude, it should not have to check if the angle is in the expected range (between 0 and 90 degrees). Consider, on the other hand, a structural engineer who starts with the design of the roof structure for a building and abandons work for the day without designing any of the support elements. Clearly, the structural system is physically ill-formed, but the engineer should still be able to communicate this incomplete design to others. The sun ray information represents data, whereas the structural system is a design object of interest. A schema should be able to enforce well-formedness of data in the above sense; it has no business examining design data beyond that.

  The most appropriate formal representation for well-formed data is abstract data types (ADTs) as defined by [Meyer 1988]. But modeling language ADT implementations are different from programming languages. Preconditions and postconditions can be offered in both kinds of languages, but the modeling languages require additional support that is almost never provided by programming languages. For example, it should be impossible to describe nonsensical operations on dimensions measured in a unit system, for example, the multiplication of two volumes. To make the distinction between ADTs in the different languages clear, we call the representation of a data type a *domain* in modeling languages. Domains may support inheritance, but domain-based inheritance is essentially different from class-based inheritance (discussed in [Snyder 1998]).

- **Relationship Types**: A relationship type represents a kind of association that can exist between design objects; if it has an inverse, it has to be managed by the schema. They may need associated behavior of the association under the significant data base events described as described below. We decided not to model relationship types as classes for both practical and conceptual reasons.

- **Classes**: Design objects of interest, which differ from well-formed data, are instances of classes within an information modeling language. A class can be defined as an aggregation of attributes of different types that incorporate domains and relationship types. Classes must support inheritance in the more traditional sense, but should support multiple-inheritance for reasons given below.

- **Multiple Classifications**: Clearly, objects of interest need to be multiply classified, for example, to support various queries both during evaluations (e.g. is this space public so that a pathfinding algorithm can treat is as node). Multiple inheritance is many times inappropriately used to classify objects having nothing or very little to do with the behavioral aspects of a class (see [Kiloccote 1996]). A classification scheme should support operations such as subsumption and the detection of conflicts. The work most relevant to SEED is the CLASSIC knowledge representation system, not the least because it incorporates algorithms that are known to be tractable [Brachman et al. 1991].

4.4 Database Support

To overcome the limitations of relational databases, researchers have proposed various approaches able to support an information modeling environment that supports object-centered representations, which we cannot review in detail (see [Snyder 1998]). We point out a specific aspect required by a system like SEED, which is capable of generating large numbers of alternative designs that can undergo version changes, where the designs are modeled as complex object configurations, some of which need to be stored persistently.

Relationships between objects in a configuration behave differently under different database events and must be managed appropriately. Take again the structural system from Figure 2 as an example. An alternative version may use the same designs, which are already stored in the database, but in different locations. When storing the new version, the database has to know that the locations have to be created, but not the associated design, and be able to link correctly, all objects, new or existing. Conversely, when a design is deleted, all associated locations have to be deleted, too. We call this configuration management.

In general, the SEED database must be able to manage properly the following database events:

1. **Delete**: When an object in a configuration is deleted, the database must know which of the associated objects should also be automatically deleted.
2. **Copy**: Copying a configuration means the construction of a complete, 'deep' copy. Usually all objects in the configuration are copied and the copies linked in a network that is isomorphic to the original but exceptions must be specifiable.
3. **Anchor**: Anchoring a configuration means creating a different version of a configuration that can be retrieved based on a current time stamp. In this case, configuration management depends entirely on how a given schema is defined.
4. **Overwrite**: Overwrite replaces a configuration in the database with a configuration that has the same identifier, but a new time stamp. Overwrite then proceeds similarly to anchor.

If a schema is able to distinguish relationship types by the way in which they behave under these events, configuration can be automated based on these specifications.

5. THE SPROUT INFORMATION MODELING LANGUAGE

We briefly describe in this section the SPROUT (SEED representation of Processes, Rules, and Objects Utilizing Technologies) modeling language. The overall objective of SPROUT is to provide a syntactic and semantic framework that embodies the requirements defined in the preceding section. This section introduces the schematic language elements in SPROUT depicted in Figure 3. The issues surrounding domain and class schematic elements were introduced in Section 4. The notion of a unit of measure is a relatively unique modeling language concept that we cannot further discuss here. SPROUT specifications are detailed in [Snyder 1998].

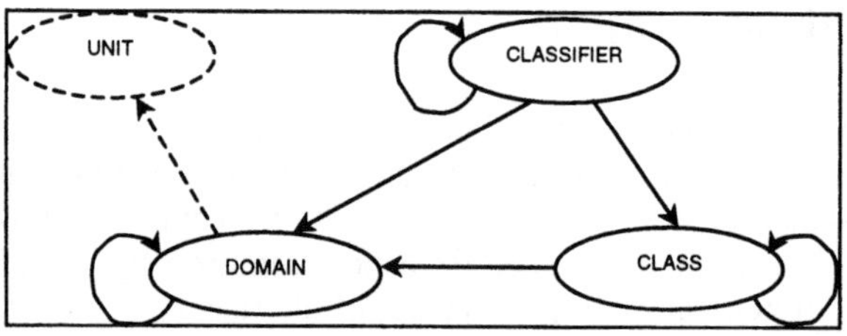

Figure 3. Relationships Between SPROUT Schematic Elements

The SPROUT syntax is heavily influenced by the Eiffel object-oriented language [Meyer 1988]. Eiffel embodies good language design principles and has the added benefit of having abstract data types at the core of its language design including support for exceptions due to invariant violations. It was therefore a natural and important reference work. It is important to note however, that the overall software architecture and semantic assumptions of SPROUT differ from Eiffel because Eiffel is a programming, not a modeling language.

Unlike Eiffel, SPROUT supports qualified schematic element naming via a package naming, thereby defining a package scope in a manner similar to, but not identical with the Java programming language [Gosling et al. 1996]. Package naming is an essential feature in the definition of shared schemas. Without this construct, it becomes virtually impossible to manage element naming within a SPROUT schema (or any schema language for that matter). Something as simple as a two-dimensional point can be defined in several different contexts, and without package names, there would almost certainly be collisions between the names.

5.1.1 Domains

Domains support the notions described in Section 4.3. To give readers a flavor of the SPROUT syntax, Figure 4 shows the SPROUT specification of a simple domain, Meter (*sprout.lang* is the default package name for built-in elements) that is based on built-in domains. The elements of a complex domain must refer to an existing domain to make up structured or recursive domain definitions. The complex domains found in SPROUT are *sets*, *sequences*, *ordered sets*, *ordered sequences*, *tuples*, and *constraint variables*.

```
forward domain SquareFeet, SquareMeter specializes sprout.lang.Real
  implements Area end

domain Meter specializes sprout.lang.Real implements basecase Length
  operator Meter ^ Meter: error is default end
  operator Meter * Meter: SquareMeter is default end
  operator Meter / Meter: sprout.lang.UnitlessReal is default end
  operator Meter < Meter: sprout.lang.Boolean is default end
end
```

Figure 4. Example Simple Domains

SPROUT also supports domain transformation capabilities called *function*, *operator*, and *conversion*. These transforms, in particular function, resemble methods found in object-oriented programming with some important difference in behavior. In particular, inheritance plays a significantly different role.

5.1.2 Relationship Types

Relationship types establish a behavior for a relationship and have, as a minimum, a name and an associated container (i.e. *set*, *vector* or *bag*, and *link*). Additional information can be specified in a relationship type, for example, what should be done with objects in the container under the database events introduced in the preceding section.

5.1.3 Classes

Classes are defined as collections of attributes. Due to space limitations, a complete definition is not provided here (see [Snyder 1998), rather the kinds of attributres are simply stated. In SPROUT, attributes have the following types: value, relationship, method, trigger, derived, and constraint. An example SPROUT class is shown in Figure 5.

```
class Point specializes sprout.lang.SproutObject
    value x sprout.lang.UnitlessReal;
    value y sprout.lang.UnitlessReal;
end

class Rectangle specializes sprout.lang.SproutObject
  value length Meter default 10.0;
  value width Meter;
  relationship lowCorner hasOne(jim.Point);
  relationship highCorner hasOne(jim.Point);
  method m: Meter is
    local
      p: jim.Point;
    do
      width:= 10.0;
      p.Create;
      p.x:= 10.0;
      p.y:= 20.0;
      lowCorner.assign(p);
    end
  derived area     SquareMeter is length * width;
end
```

Figure 5. Example Class Specifications

5.1.4 Classification

During the course of this research, it became clear that, for implementation reasons, support for classification was better developed in the context of the case-based components of the SEED system. As a result, plans to include a formal classification specification in SPROUT were abandoned. However, it is important to include clear and specific definitions of what classification means in the context of an information modeling language.

5.2 Implementation Prototype

At the time of this writing, we are still working on a first implementation of the modeling and exchange environment envisioned here. We describe in the present section briefly the main components of this first implementation.

In order to assure platform independence, the initial middleware server environment is implemented in the JavaServer toolkit. Figure 6 shows the overall process of compiling a SPROUT schema into this type of environment. Note the differences between the runtime and compile-time relationships.

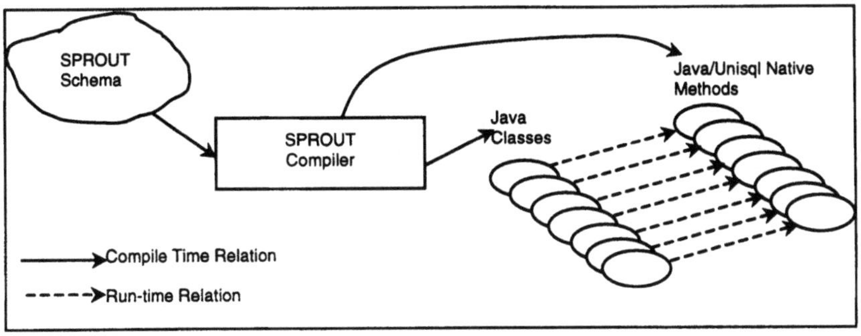

Figure 6. SPROUT Schema Compliation

The specific outputs generated using this target environment consist of a collection of public Java class files and an associated native method implementation file. These Java classes are specializations of Java classes developed as part of the SPROUT runtime environment. Using this technique, many behaviors can be properly implemented without any intervention on the part of the compiler back-end developers or the schema designers. Additionally, these classes are installed as part of a middleware server environment not described here.

ACKNOWLEDGEMENTS

The work reported here has been supported by the US Army Corps of Engineers Construction Engineering Research Lab and, until 1996, by the Engineering Design Research Center at CMU.

REFERENCES

Augenbroe, G. and F. Winkelmann (1990). *Integration of Simulation Into the Building Design Process.* TU Delft Research Report, TU Delft, Holland.

Brachman, R. J., McGuiness, D. L., Patel-Schneider, P. F., and L. A. Resnick. (1991) "Living With CLASSIC: When and How to Use a KL-ONE-Like Language" in *Principles of Semantic Networks: Explorations in the Representation of Knowledge* J. Sowa (ed.) San Mateo: Morgan Kaufmann.

Chiou, J.-D. and Logcher, R. D. (1996) *Final report. Testing a Federation Architecture in Collaborative Design Process.* Research Report R96-01. Department of Civil and Environmental Engineering. Massachusetts Institute of Technology. Cambridge, MA.

Dubois, A. M., Flynn, J., Verheof, M. H. G., and G. L. M. Augenbroe (1995). "Conceptual Modelling Approaches in the combine Project" in *Proceedings of ECPPM '94 - The First European Conference on Product and Process Modelling in the Building Industry* R. J. Scherer (ed.), Dresden, Germany, October 5-7, 1994, Rotterdam: A. A. Balkema.

Eastman, C.M. (1994), "A Data Model for Design Knowledge" in *Knowledge-Based Computer-Aided Architectural Design* G. Carrera and Y. E. Kalay (eds.), Elsevier Science B.V., Amsterdam, The Netherlands. pp.95-122.

Eastman, C. M. and N. Fereshetian (1994). "Information models for use in product design: a comparison" *Computer-Aided Design* 26, pp. 551-572

Eastman, C. M. and A. Siabiris (1995). "A Generic Building Product Model Incorporating Building Type Information" *Automation in Construction*, 4(4), pp. 283-304.

Eastman, C., Cho, M. S., Jeng, T.S., and H. H. Assal (1995). "A Data Model and Database Supporting Integrity Management" in *Computing in Civil Engineering* J. P. Mohsen (ed.), volume 1 of 2, pp. 517-524, Proceedings of the Second Congress held in conjunction with the A/E/C Systems '95, Atlanta, GA, June 5-8, New York: American Society of Civil Engineers.

Fenves, S., Flemming, U., Hendrickson, C.,Maher, M. L.,Quadrel, R.,Terk, M., and R. Woodbury (1994). *Concurrent Computer-Aided Building Design*, Englewood Cliffs, NJ: Prentice-Hall.

Flemming, U. and Chien, S. F. (1995). "Schematic Layout Design in SEED Environment" *Journal of Architectural Engineering*, 121(4), New York: American Society of Civil Engineers. pp. 162-169.

Flemming, U. and R. Woodbury (1995). "Software Environment to Support Early Phases in Building Design (SEED): Overview" *Journal of Architectural Engineering*, 121(4), New York: American Society of Civil Engineers. pp. 147-152.

Flemming, U., Aygen, Z., Snyder, J. and J. Tsai (1996). *A2: An Architectural Agent in a Collaborative Engineering Environment.* Technical Report EDRC-48-38-96, Engineering Design Research Center, Carnegie Mellon University, Pittsburgh, PA.

Gosling, J. Joy, B., and G. Steele (1996). *The Java Language Specification*, Reading: MA

ISO 10303 Committee. Part 11 — *The EXPRESS Language and Reference Manual.*

[Khedro, T., Case, M. P., Flemming,U., Genesereth, M. R., Logcher, R., Pedersen, C., Snyder, J.,Sriram, R. D. and P. M. Teicholz (1995). "Development of a Multi-Institutional Testbed for Collaborative Facility Engineering Infrastructure" in J. P. Mohsen (Ed.) *Computing in Civil Engineering: volume 2: Proceedings of the Second Congress held in conjunction with the A/E/C Systems '95*, Atlanta, GA, June 5-8, New York: American Society of Civil Engineers. pp. 1308-1315.

Kiliccote, H. (1996). *A Standards Processing Framework* Ph.D. Thesis. Dept.of Civil and Environmental Engineering. Carnegie Mellon University, Pittsburgh, PA.

Lockley, S.R., W. Rombouts, W. Plokker (1995). "The COMBINE Data Exchange System" in *Proceedings of ECPPM '94 - The First European Conference on Product and Process Modelling in the Building Industry* R. J. Scherer (ed.), Dresden, Germany, October 5-7, 1994, Rotterdam: A. A. Balkema.

Loucopoulos, P. (1992). "Introduction: Section One: Conceptual Modeling" in *Conceptual Modeling, Databases, and CASE: An Integrated View of Information System Development* P. Loucopoulos and R. Zicari (eds). New York: John Wiley & Sons, Inc. pp. 1-26.

Meyer, B. (1988). *Object Oriented Software Construction*. Prentice-Hall, Englewood-Cliffs, NJ.

Myers, L., Snyder, J. and L. Chirica (1992). "Database Usage in a Knowledgebase Environment for Building Design" *Building and Environment*, 27(2), pp.231-241.

Nell, J. (1996). "Step on a Page" Technical Report. National Institute of Standards and Technology. Gaithersburg, MD. (see STEP Web site at http://www.nist.gov/sc4).

Parent, C. and S. Spaccapierta (1992). "ERC+: An Object-Based Entity Relationship Approach" in *Conceptual Modeling, Databases, and CASE: An Integrated View of Information System Development* P. Loucopoulos and R. Zicari (eds). New York: John Wiley & Sons, Inc. pp. 69-86.

Pohl J., L. Myers, J. Cotton, A. Chapman, J. Snyder, H. Chauvet, K. Pohl, J. La Porta (1992). *A Computer-Based Design Environment: Implemented and Planned Extensions of the*

ICADS Model. Technical Report, CADRU-06-92, CAD Research Center, Design Institute, Cal Poly, San Luis Obispo, CA, 93407.

Rolland, C. and C. Cauvet (1992). "Trends and Perspectives in Conceptual Modeling" in *Conceptual Modeling, Databases, and CASE: An Integrated View of Information System Development* P. Loucopoulos and R. Zicari (eds). New York: John Wiley & Sons, Inc. pp. 27-48.

Rumbaugh, J. (1987). "Relations as Semantic Constructs in an Object-Oriented Language" OOPSLA '87 as *ACM SIGPLAN* 22(12). pp. 466-481.

Snyder, J. (1993). A Semantic Modeling System for CAD, M.S. Thesis, Department of Architecture, Cal Poly, San Luis Obispo, CA.

Snyder, J. D. (1998) *Conceptual Modeling and Application Integration in CAD: The Essential Elements*. Ph. D. thesis, School of Architecture, Carnegie Mellon University., Pittsburgh, PA.

Snyder, J. and U. Flemming (1994a) *The Object Modeling Language (OML) Specification*. Internal Technical Report (URL: http://seed.edrc.cmu.edu/ACL).

Snyder, J. and U. Flemming (1994b). The Object Modeling Language (OML) Language Binding Specification. Internal Technical Report (URL: http://seed.edrc.cmu.edu/ACL).

Sriram, D. and R. Logcher (1993) "The MIT Dice Project" *IEEE Computer* (Jan), pp. 64-65.

Sriram, D., Logcher, R., and A. Wong (1994). "Computer Supported Collaborative Engineering: Research Issues in Product Modeling" in *Bridging the Generations: International Workshop on the Future Directions of Computer-Aided Engineering* (position papers submitted to the workshop) D. Rehak (ed.), July 18-19, Carnegie Mellon University, Pittsburgh, PA.

Wiederhold, G. (1995). "Mediation in Information Systems" *ACM Computing Surveys*, 27(2), New York: ACM Press, pp. 265-267.

Wong, A. and D. Sriram (1994). *Shared Workspaces for Computer-Aided Collaborative Engineering*. Technical Report. Intelligent Engineering Systems Laboratory, Dept. of Civil and Environmental Engineering, Massachusetts Institute of Technology, Cambridge, MA.

Aspects of S2

Ardeshir Mahdavi, Mustafa Emre Ilal, Paul Mathew, Robert Ries, Georg Suter
Carnegie Mellon University, Pittsburgh, PA 15213-3890, USA

Key words: building performance simulation, distributed system, internet

Abstract: We present in this paper the essential aspects of the S2 system. This is the internet realization of SEMPER, an active, multi-domain, space-based, object oriented design environment for integrated building performance modeling. The key features of the S2 environment are as follows: A user can access the system regardless of the computer hardware, operating system or the location on a network; geographically distributed users can asynchronously generate a building model through the user interface; this building model can then be simultaneously evaluated with multiple simulation applications running on remote simulation servers; persistent storage is provided for project data and evaluation results; designers using the system have access to multiple libraries that contain building information such as material data, construction types, schedules, and weather data.

1. BACKGROUND

1.1 SEMPER

SEMPER is a multi-aspect prototype design environment that incorporates an object-oriented, space-based building model, with *dynamic* links to different building performance evaluation applications (Mahdavi 1996). It is thereby intended to computationally support the evaluation of buildings across *multiple* performance mandates concurrently, with a view toward achieving total building performance and systems integration.

SEMPER incorporates the unique synthesis of four principles:

1. *A methodologically consistent performance modeling approach through the entire building design process*
   The performance simulation modules incorporated within SEMPER rely on a consistent (first-principles based) modeling of the physical processes relevant to the building's performance. SEMPER currently includes modules for energy, air flow, HVAC, thermal comfort, lighting, acoustics, and life-cycle analysis.
2. *Seamless and dynamic communication between the simulation model and the general building representation*
   This provides "on-line" building performance feed-back to the user while eliminating the need for explicit definition and updating of the underlying simulation model. SEMPER exemplifies this by dynamically linking the building's general representation (essentially a shared object model) with structurally homologous representations of the simulation modules (essentially the domain object models) (Mahdavi and Wong 1998, Mahdavi et al. 1997). *Figure 1* illustrates, for example, the homology-based mapping from the shared building representation to the thermal and air flow simulation domains. The shared building representation embodies topological information on the architectural spaces and their relationships. The energy simulation module utilizes a spatial representation consisting of spatial units (cells) with nodes that define finite control volumes for heat-balance calculations (Mathew and Mahdavi 1998). This nodal representation of the building is configurationally homologous to and is directly derived from the space-based representation of the building in the shared object model. This homology-based mapping mechanism allows for rapid feedback and is therefore a powerful design instrument, since evolving building designs can be made subject to comprehensive parametric studies in multiple domains without having to input the building model separately for each domain.
3. *Comprehensive performance modeling*
   In order to analyze the implications of design decisions within and across multiple performance domains, SEMPER incorporates modules to deal with thermal quality, indoor air quality, visual quality, acoustic quality as well as environmental impact (Mahdavi et al. 1996).
4. *Active design support using a "preference-based" performance-to-design mapping technology*
   Active design support enables a designer to interactively and dynamically explore the relationships between design decisions and performance variables (Mahdavi 1993). The active design support in SEMPER would provide two functionalities: *a)* the user makes a change in a performance variable in order to see the corresponding changes in building design variables; and *b)* the user makes a change in a building design variable and observes resulting changes in other building design variables when one or more relevant performance variables are constrained.

The proof-of-concept SEMPER prototype 1 was implemented on UNIX with a Tk-TCL user interface, which uses AutoCAD for geometry input and display. This prototype incorporates two simulation modules for energy and thermal comfort analysis.

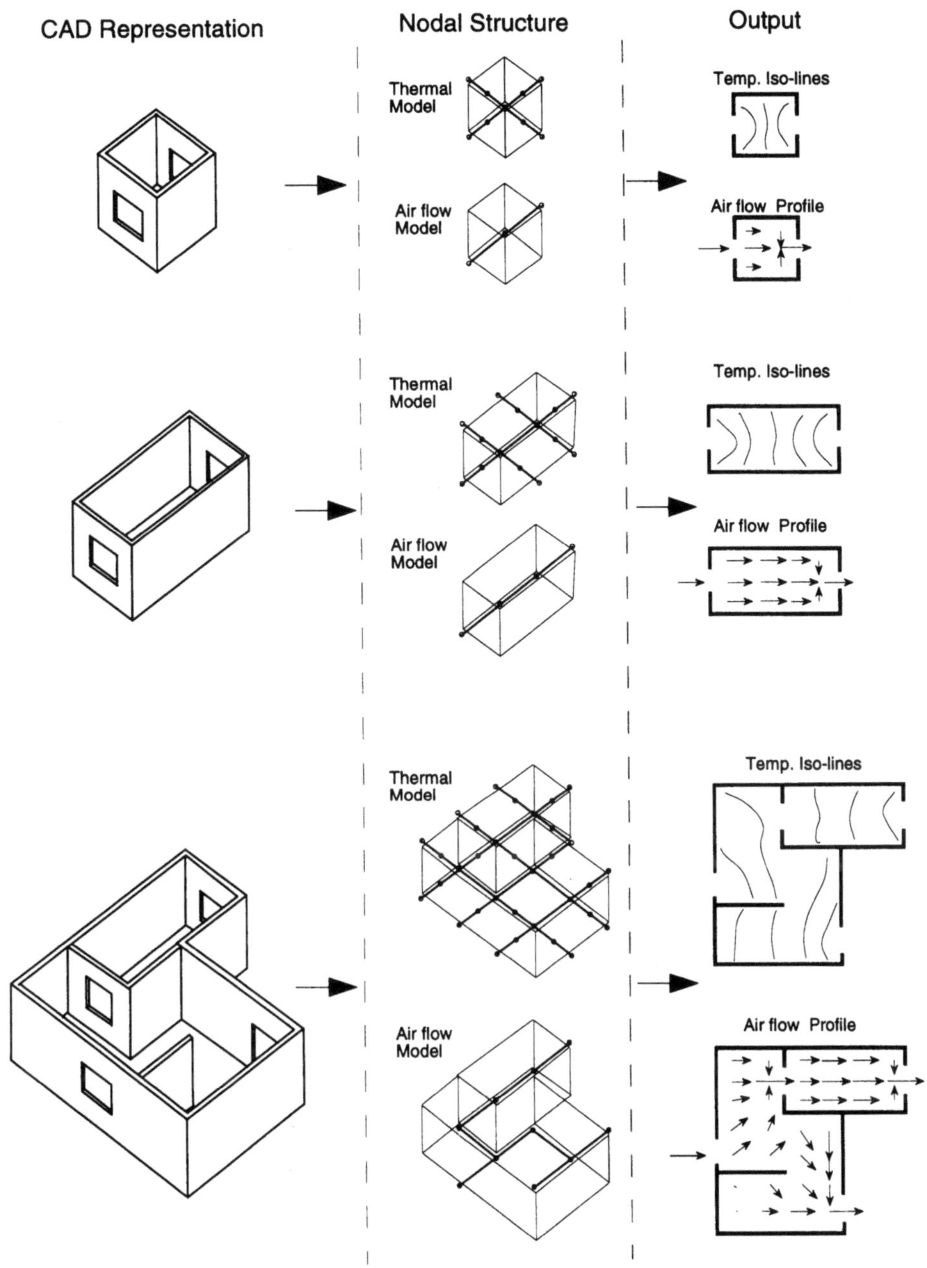

Figure 1. Homology-based mapping of the space-based architectural representation (left) to the nodal building representations in SEMPER's thermal and air flow simulation modules (middle) with their corresponding simulation results (right).

1.2 S2: SEMPER on the Internet

Building on the proof-of-concept prototype 1, SEMPER Prototype 2 (S2) seeks to realize SEMPER on the internet, toward supporting collaborative performance-based building design between geographically distributed users. The various components of S2 (graphical user interface, S2-Kernel, simulation modules, etc.) are distributed over a network, using Common Object Request Broker Architecture (CORBA) as their communication framework. A preliminary implementation of S2 on the internet will involve users at Carnegie Mellon, and two institutions in Singapore - National University of Singapore and Temasek Polytechnic (*Figure 2*).

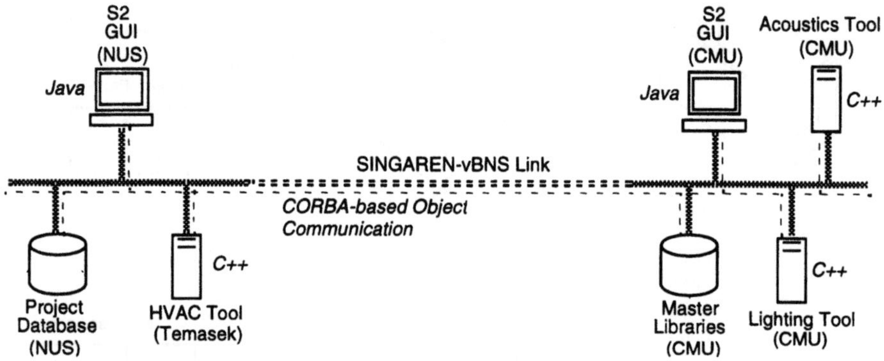

Figure 2. Schematic illustration of components of S2 in the case of two distributed users in Singapore and the U.S. collaborating on a building design/engineering project

2. S2 IMPLEMENTATION

2.1 S2 System Architecture

The S2 system is made up of the following components:

- Graphical User Interface (GUI): Allows users to create and edit building descriptions. It is implemented in Java and includes a simple geometry editor.
- Gateway: The entry point for any GUI. It holds administrative functions and coordinates component interactions.
- S2-Kernel: The application used to instantiate and edit the Shared Object Model (SOM). The SOM reflects the building representational requirements of seven simulation modules. It is implemented in Java.
- Geometric Modeling Engine: A common utility for pure geometric processing.
- Database: Persistent storage for project information as well as shared resources.
- Simulation modules: Applications that can derive their own building representations (DOM's) from the SOM for domain specific processing. The first release of S2 will include modules for thermal analysis, HVAC system simulation, thermal comfort analysis, lighting, and environmental impact analysis.

The component based system architecture of S2 is shown in *Figure 3*.

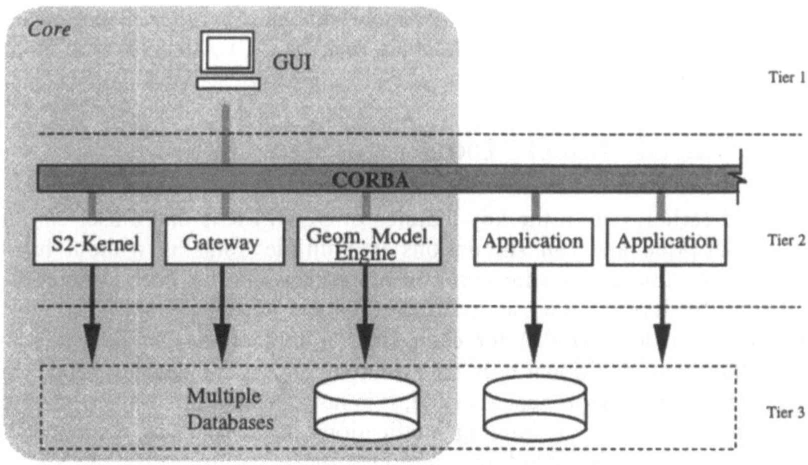

Figure 3. S2 system architecture - components and communication

CORBA was chosen as the communication framework between components. CORBA provides a connection mechanism that is independent of implementation language. This mechanism enables objects implemented in different languages to communicate without the need for any language mapping code. This is an important feature considering S2's implementation which is partially in C++ (Simulation

Modules) and partially in Java (S2-Kernel, GUI). CORBA development also supports a modular structure that simplifies the addition of future components to the system, which is essential for S2.

The GUI is the mediator between the user and S2-Kernel. Whenever a user wants to modify the existing design, the GUI forwards the request to the S2-Kernel, which conducts the requested operation. The result is sent back to the GUI, which displays it.

The Gateway is a connection manager whose URL is published and known by all users. On start-up, a GUI, first contacts the Gateway for authentication. Once the user is authenticated, the Gateway releases an Attendant and assigns it to the GUI. The Attendant acts as a higher level liaison for the GUI. Attendants function as process coordinators and file managers. For example, when the GUI requests the available databases, the Attendant dynamically creates a list of databases which the user has access to and returns the list to the GUI. Similarly, when a user wants to browse the projects inside a selected database, the Attendant conducts the queries and creates the appropriate list. When the user wants to start working with a given project, the GUI again sends this request to its Attendant, which in turn, selects an available server, spawns an S2-Kernel process (which initializes the database connection for the selected project) on it, and connects the GUI with this S2-Kernel.

S2-Kernel is the functional core of the system. It provides access to the SOM which is kept in the database. It also provides all necessary editing mechanisms for it. It populates the SOM and maintains its consistency.

The Geometric Modeling Engine is a library of functions for pure geometric queries. It is a utility available for S2-Kernel and other applications.

When the GUI requests the execution of simulations, its Attendant finds (through CORBA services) the appropriate, available simulation modules and initiates the simulation processes. At the end of such a process, the results are attached to the SOM directly in the database and the GUI-Attendant pair is notified of this event through an event service.

2.2 Shared Object Model

As stated earlier, one of the key features of SEMPER is that a user can input a building description and run simulations in multiple domains without having to manually create domain-specific building representations for each simulation tool. S2 facilitates this through the use of a Shared Object Model (SOM), from which the domain object models (DOM) for each simulation module can be automatically derived without any additional user intervention. The SOM was developed through an iterative bottom-up approach. As such, it is not intended to be a general building model for a broad array of building applications. Rather, it has been specifically developed to reflect the building representational requirements of S2's seven simulation modules. The SOM can, of course, evolve to support additional applications, and additionally, can inform the integration of representations that support simulation into more general models.

The following key features of the SOM bear mentioning:

- The SOM uses a space-based representation i.e. the building is composed of non-intersecting spaces aggregated into sections. This feature was deemed necessary in order to generate the domain representations needed for detailed simulations.
- The primary elements of the building representation each have a geometry (described via an associated geometry class), and a set of non-geometric attributes (described via a "type" class, referenced by type name). For example, a "SOMWall" object has an associated "SOMPolygon" geometry object and a "SOTConstruction" type object that contains layer and material specifications. The type classes, which essentially represent specification objects, are organized into libraries (e.g. wall construction types, space occupancy types, window shade types, etc.).
- The SOM has a number of "visitor" classes (Gamma et al., 1994) that encapsulate the knowledge for an operation that requires coordination among multiple objects in the SOM. For example, adding a space to the SOM involves a complex series of queries and operations on many objects, requiring a complex traversal of the SOM hierarchy. Visitor classes in the SOM extend from a common superclass so that operation specific visitor subclasses can be added without change to the primary SOM elements.

Table 1 describes the package structure for the SOM. *Figure 4* shows the SOM class hierarchy for the primary physical elements.

Table 1. Package structure for the Shared Object Model

Package	Description
Semper	Root package - does not contain any classes
Semper.som	Classes that describe the primary elements of the building model - spaces, walls, etc.
Semper.som.visitor	Visitor classes that encapsulate the knowledge for operations that require coordination among multiple objects in the SOM
Semper.geometry	Classes that describe gemetric entities used in the SOM (e.g. points, lines, polyhedra, etc.)
Semper.sot	"Type" classes that encapsulate specification information for the primary physical elements of the building model
Semper.util	Utility classes

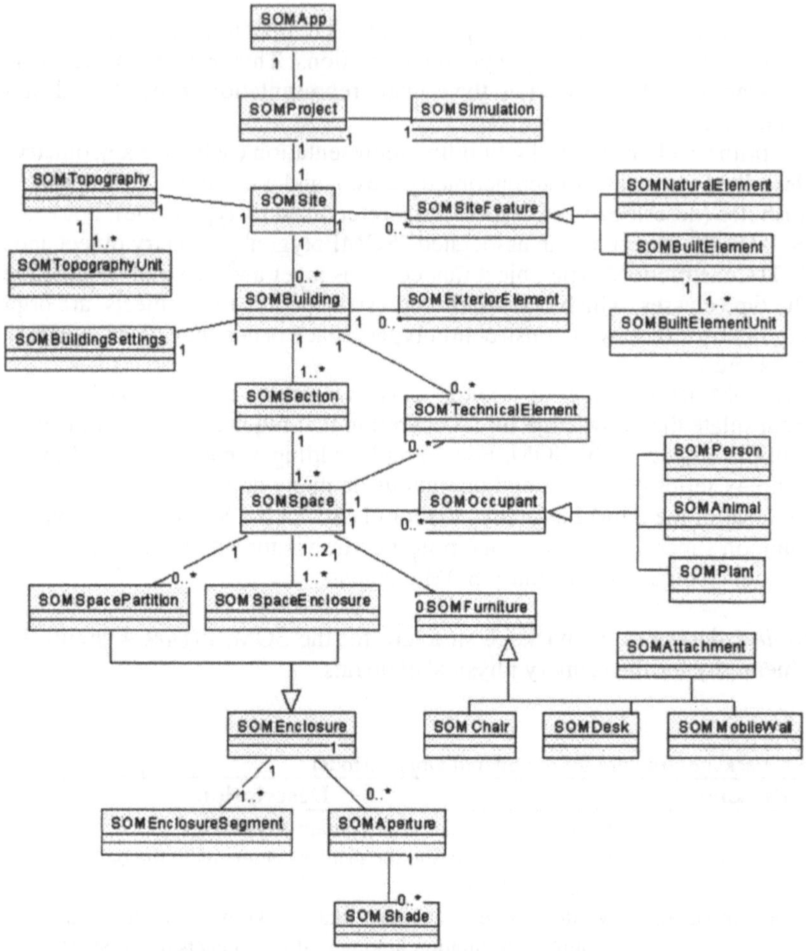

Figure 4. The class hierarchy for primary elements of the SOM

2.3 Graphical User Interface

2.3.1 User interface overview

The system architecture of S2 allows for any number of different interfaces to connect and work with building descriptions that are stored in the database. While any CAD package will be able to connect and update design descriptions, the S2's current GUI is aimed at demonstrating core functionality of the overall system. It is modest yet is designed to reflect the main principles of S2.

Building simulation tools typically require detailed input models and often generate enormous amounts of output data. Both the specification of building models and the interpretation of simulation output are tasks that should be supported in an efficient manner by user interfaces. An important aspect of the user interface in S2 is modularization, which is consistent with the overall system architecture. In the S2 environment, designers specify building geometries by drawing spaces in a graphical editor. This editor serves as a starting point for navigating and manipulating non-geometric building attributes such as constructions, materials, and schedules. Similar editors are used in expert design modules to facilitate the communication of more domain-specific information. A simulation setup window allows designers to define space-time boundary conditions for individual as well as groups of simulation modules. After running a simulation, designers browse simulation results in a visualization module, which allows for the navigation of various kinds of simulation data as well as customization of data filters and presentation types.

2.3.2 Input

Designers input building geometries by drawing space-constituting polygons on a canvas. These polygons are communicated to S2-Kernel. In addition to drawing tools, the editor also includes navigation aids for cycling through building sections. Semantic information associated with particular spaces can be accessed by directly selecting a space on the canvas.

2.3.3 Expert design modules

The objective of S2 to facilitate design evaluation at various levels of abstraction is reflected in the interface by the separation of general and more domain specific input functionality. Expert design modules are implemented as plug-ins to the main user interface and are activated only at the user's explicit request. This organization is aimed at reducing the information load on the designer and also allows for increased operational flexibility. Expert design modules may be run as stand-alone as well as part of S2.

2.3.4 Construction and schedule editor

In S2, constructions and schedules are assigned in two ways. Semantic properties may be assigned at the space level, or global building settings for constructions and schedules may be applied to all the spaces in a particular building. This feature is intended to support rapid modification of space aggregates. Enclosure constructions or shading schedules may be modified in one operation at the building level rather than multiple operations at the space level. The only drawback of such an operation is the loss of individual space information. However, the global settings, once they have been applied to a specific space, may be subsequently overridden by designers at the space level without affecting other spaces.

2.3.5 Simulation setup

Simulation setup in an integrated simulation environment such as S2 requires the definition of multiple space-time parameters for each simulation module involved. Individual parameters include grid resolution, time steps, time period, and spatial range. Users would typically select predefined simulation settings for each module and specify only whether modules are run separately or together.

2.4 Domain Object Models

It has been noted earlier that the building representations for specific simulation modules i.e. the domain object model (DOM), is derived from the Shared Object Model (SOM) without any user intervention. This approach has been successfully implemented for domains such as thermal analysis, air flow analysis and environmental impact analysis using homology-based mapping.

However, domains that involve detailed design of physical components, such as HVAC, present a complex set of issues vis-à-vis the relationship between the SOM and the DOM. Indeed, the role of the HVAC analysis within the overall SEMPER environment exemplifies a number of challenging questions in terms of the shared object model, domain object model, and GUI:

- Building energy analysis can be performed at various levels of depth and resolution. From the software engineering point of view, this represents a number of problems. A building model that is too restricted, may allow only for a limited and ultimately less useful set of analysis options. On the other hand, a model that would capture all the requirements of disciplinary analysis may become too large, leading to the classic problems of massive product models. The question is, to what extent should the object class hierarchy of the primary shared building model reflect the requirements of various detailed applications?
- Certain levels of building energy analysis are more relevant to the types of questions that need to be asked and answered by primary building designers (particularly architects). For example, decisions pertaining to the effects of enclosure and glazing, massing, orientation, and natural ventilation, should be covered by the core analysis capabilities of a building design tool. However, analysis of detailed building support systems and controls may be more relevant to the activities of the domain specialist (e.g. HVAC designer). The question is, how will users with different interest and expertise communicate with a multi-aspect software systems that should allow for different level of analysis?

Our recent research findings have lead us to the following position in the case of the HVAC module's relationship to the overall S2 environment:

- A distinction is made between general user (interacting with the S2 SOM) and the specialist (interacting with DOM).
- The type of HVAC analysis that can be initiated at the S2 level by the general user would be limited to default system types. The user selects the spatial domain to be supplied with HVAC, the type of the system, and optionally the thermal zone boundaries. This information is (apart from the simulation results) the only HVAC-relevant specification included in SOM.
- The HVAC domain derives the DOM based on SOM. Beyond SOM's objects, DOM will include domain- specific objects such as generators (e.g. chillers and boilers), distribution components (e.g. ducts and fittings), and terminals (e.g. diffusers and radiators). The basic configuration of these elements is established by HVAC domain's expert system, based on the previously mentioned, limited, HVAC-relevant information in the SOM.
- Once simulation is performed, the structured set of results are sent back to SOM as simulation results. This is the only exception to the otherwise one-way (SOM-to-DOM) information flow in S2. As a general rule, objects that are not defined and generated in SOM are not "visible" to it and cannot be manipulated by it.
- A domain specialist (or a general user interested in other-than-default HVAC analysis) can directly interact with DOM objects thus circumventing the HVAC domain expert. This interaction does not occur via S2's general GUI, but rather via an independent user interface (or a plug-in to the general GUI). Objects and system designs defined by the specialist in DOM are not transparent to SOM, nor are they saved as part of the SOM. They must be saved as part of the DOM and maintained by the specialist. However, simulation results due to such designs can be properly tagged and sent back to SOM as simulation results.

In summary:

- It is not appropriate to include the detailed classes of all applications in a single SOM. The resulting size and complexity is likely to hamper modularity, make interacting with legacy applications difficult, reduce the ease of system modification, and dilute the structural transparency of the software architecture.
- A distinction needs to be made as to the level of analysis appropriate for the general system user. SOM must include all necessary classes for such analysis.
- Detailed designs by specialists need their respective class hierarchies and user-interaction venues. They must maintain their own designs, alternatives, and analyses. A representation of domain-specific classes in SOM, even in informationally reduced form, is not considered appropriate, given considerations of integrity management of mappings between class hierarchies with different levels of abstraction/resolution. As a general rule, design elements that cannot be manipulated via SOM's GUI should not be represented in SOM.
- While various DOMs will not be integrated within a single unified SOM, applications are conceivable that would allow for concurrent viewing of all object information (e.g. their geometric properties), regardless if they are SOM or DOM objects. This integrated display utility would allow for detection of system integration problems (i.e. conflicts between elements of various domains). Active manipulation of domain elements, however, can be only realized via each domain's user interaction mechanisms.

3. CONCLUDING REMARK

The S2 system has been designed, pilot-tested, and as is currently under implementation. S2 is designed to support collaborative performance-based building design between geographically distributed users, over the internet. This is done by modularizing and distributing the components of S2, using CORBA as a communication framework. The S2 system will be further tested and refined as part of a collaborative projects between Carnegie Mellon University and two academic institutions in Singapore.

ACKNOWLEDGEMENTS

The authors would like to extend their thanks to the S2 development team which, besides the authors, also includes: Rohini Brahme, Seongju Chang, Beran Gurtekin, Youngsoo Kwon, Prechaya Mahattanatawe, Zhengchun Mo, Vineeta Pal, Nyuk-Hien Wong.

REFERENCES

Mahdavi, A., 1996, "Computational Support for Performance-based Reasoning in Building Design", *Proceedings of the CIB-ASTM-ISO-RILEM International Symposium "Applications of the Performance Concept in Building"*, Tel Aviv, Israel. Vol. 1, pp. 4.23 - 4.32.

Mahdavi, A., 1993, "'Open' Simulation Environments: A 'Preference-Based' Approach", *Proceedings of CAAD Futures '93*, CMU, Pittsburgh, Pennsylvania, USA. pp. 195 - 214.

Mahdavi, A. - Wong, N. H., 1998, "From Building Design Representations to Simulation Domain Representations: An Automated Mapping Solution for Complex Geometries", *Computing in Civil Engineering; Proceedings of the International Computing Congress, 1998 ASCE Annual Convention*, pp. 1-10.

Mahdavi, A. - Mathew, P. - Wong, N. H., 1997, "A Homology-based Mapping Approach to Concurrent Multi- domain Performance Evaluation", *Proceedings of the The Second Conference on Computer Aided Architectural Design Research in Asia: CAADRIA '97* (Ed.: Yu-Tung Liu, Jin-Yen Tsou, June-Hao Hou). Hsinchu, Taiwan. pp. 237 - 246.

Mahdavi, A. - Brahme, R. - Kumar, S. - Liu, G. - Mathew, P. - Ries, R. - Wong, N. H., 1996, "On the Structure and Elements of SEMPER", *Design Computation; Collaboration, Reasoning, Pedagogy: Proceedings of the 1996 ACADIA (The Association for Computer Aided Design in Architecture) conference* (Editors: P. McIntosh and F. Ozel). Tucson, Arizona. PP. 71 - 84.

Mathew, P. - Mahdavi, A., 1998, "High-Resolution Thermal Modeling for Computational Building Design Assistance", *Computing in Civil Engineering; Proceedings of the International Computing Congress, 1998 ASCE Annual Convention*, pp. 522-533.

Gamma, E. - Helm, R. - Johnson, R. - Vlissides, J., 1994, *Design Patterns: Elements of Reusable Object-Oriented Software* (Addison Wesley).

Interactive 2D Constraint-Based Geometric Construction System

Benachir Medjdoub
The Martin Centre, University of Cambridge, UK

Key words: Sketching, Geometric Constraints, Interactivity, Geometric construction, Dynamic modifications.

Abstract: This paper presents a 2D Constraint-Based Geometric Construction System where positioning and manipulating geometry is very precise. An unusually simple interface makes this system particularly interactive and easy to use.
In our approach, the geometry types supported are: points, lines, circles, ellipses, circular arcs and b-spline curves. All the fundamental topologic constraints, i.e. tangent, parallel, perpendicular, coincident and concentric, are provided. Metric constraints, i.e. dividing the shapes into equal parts or fixing the geometric parameters, are also provided. These constraints are automatically applied by the application in response to the implied intentions of the end-user. Dynamic modifications of partially dimensioned models are supported, whereby the design is modified while enforcing the constraints. A graph-constructive approach is used to solve the model. As we are dealing with partial modifications, this resolution technique is quite sufficient, and makes our system stable and flexible. Our approach focuses highly on interactivity. Positioning a shape constrained to another is made directly through the graphic interface. Constraint relaxation is also done by direct manipulations. Modifications are made by dragging the geometry, or by typing into a numerical panel displaying the free shape parameters. Again, existing constraints are maintained as those numbers are applied. Well-constrained and under-constrained problems are discussed. This approach was developed in Java, JDK 3.0.1 of SGI's Java software.

1. INTRODUCTION

Geometry is a fundamental tool in the hand of any architect. Associative geometry encourages the plastic interaction with design form, provides geometric accuracy and allows for the exploration of variation in architectural design [Ais92]. Present day architectural CAD systems are not good at handling geometric constraints. How many would allow you to construct the ellipses tangent to two circles and one line, interactively choose one, then divide the elliptical perimeter into equal parts? Having done this, how many allow you to drag one of the circles, and see the ellipse and its subdivisions change, while maintaining all the constraints?

This problem has been considered by several communities using different approaches often in mechanical engineering and manufacturing [Ald88] [Kon90][Kra92] than architecture. Two major issues have arisen from previous research: the constraint structure and the method of resolution.

From consideration of architectural applications, our purposes and goals are as follows:

- We enlarge the geometric object structure to ellipses and splines with constraint model enrichment.
- We treat both well and under-constrained problems. These two classes of problem are driven by the user interface.
- We present a new user interface where constraining, modifying and relaxing constraints is done interactively and precisely.

We present related work in section 2. The representation of geometric constraints and the user interface are presented in sections 3 and 4. The constraint solver is described in section 5. In section 6 some examples are presented. Before concluding in section 8, the implementation is presented.

2. RELATED WORK

In previous work the geometric elements are generally restricted to four types, namely point, line circle and circular arc [Bou95][FH93][Fud93][GC98][LG82] [Ala93]. Systems which handle ellipses and splines are rare.

The repertoire of constraints differs from system to system. Although there is a common kernel of constraints that consists of distances and angles, the parallel, perpendicular and *on* constraints are usually treated separately. We could have a hierarchy of constraints (like the one presented in [BM89]), where the explicit constraints override the implicit constraints. Usually the geometric constraints are classified in two categories, the topologic or structural constraints (tangency, perpendicular, parallel, on and concentric) and the metric constraints [Ald88] (which fix coordinates, distances, lengths and angles).

Four main approaches have been developed to solve geometric constraints: the numerical approach, the symbolic approach, the constructive approach and the propagation approach.

In the numerical approach the constraints are translated into a system of equations and solved using iterative methods such as Newton iteration [SB93][LG82]. The exponential number of solutions and the large number of parameters make this method inappropriate when the initial sketch is just topologically defined. Sketchpad, described in [Sut63] was the first system to use the method of relaxation as an alternative. Others system such as ThingLab [Bor81] kept the relaxation as an alternative to other methods. The Newton-Raphson method has proved to be faster than relaxation [HN94].

The symbolic approach is quite similar to the numerical approach. The geometric constraints are also transformed into algebraic equations. However instead of using numeric methods to solve the algebraic equation, the system is solved with symbolic algebraic methods, such as Grobner's bases [Buc88], or the Wu-Ritt method [GC98]. Both methods can solve general non-linear systems of algebraic equations, but may require exponential running times.

The constructive approach is based on the fact that most configurations in an engineering drawing can be solved by ruler, compass and protractor. In these methods, the constraints are satisfied constructively. The first constructive approach was based on rules. It used rewrite rules to discover and execute the construction steps [Ald88][Bru86][Sun88]. It was shown that this method is correct and solves all problems that can be constructed using ruler and compass. The other constructive approach is based on graphs, and has two phases. First, a graph representing the constraints is analysed and a sequence of construction steps is derived. Second, the construction steps are carried out to derive the solution. This approach is fast and more methodical than the rule-constructive approach. However, as the repertoire of possible constraints increases so the graph-analysis algorithm has to be modified. Requicha [Rec77] uses dimensioned trees that allow only horizontal and vertical distance. Todd [Tod89] generalises the dimension trees and gives a characterisation of the expressive power of the solver. Kramer [Kra92] describes a 3D constraint solver that deals with constraints from kinematics. Owen [Owe91] presents an extension of this principle to include circularly dimensioned sketches, and DCM is a commercial constraint solver [DCM93] using this method.

The propagation approach based on the constraint programming technique does not guarantee to derive a solution or to have a reasonable worst case running time [Doh95].

3. REPRESENTATION OF GEOMETRIC CONSTRAINTS

We have defined our constraint model and the 2D geometric elements according to the most common constraints used in architectural drawing. In [Bou72] a parallel was made between the architectural design process and the geometry classification proposed by the mathematician Felix Klein. Five levels of geometry are defined, among them topologic and Euclidian geometry. Topologic geometry defines continuity and neighbourhood in a building; it corresponds to the sketch step. Euclidian geometry defines distances and angles, and corresponds to the last step of design. From these two levels we can deduce the two categories of constraints which are known as the topologic and metric constraints.

3.1 Geometric entities

The description of 2D geometry in this paper is based on six types of geometric entities, namely point, line, circle, circular arc, ellipse and b-spline curve. Each geometric entity is characterised by a set of attributes which can be described in three categories:

- Geometric attributes: a circle defined by its radius and its centre. An ellipse is defined by its two radii, its centre and its orientation which generate the coefficients of the corresponding conic equation. A spline is defined by its knots.

- Topologic attributes: list of points which define the type of the constraints and the other geometric entities with which the current geometry is constrained.

- Degrees of freedom: boolean variables which correspond to the geometric parameters. In the case of an under-constrained problem these variables allow us to fix some geometric parameter. A priori some parameters are fixed (i.e. degrees of freedom corresponding to line's slope are fixed).

Points can be generated in different ways; explicitly by the user from the menu bar, by intersection points between different geometry, or implicitly by applying a constraint between two elements. In all the cases the point has a label referring to the kind of the constraint, i.e. if the point is a tangent point the label *tangent* is activated. This label is represented by a boolean variable. The points and their label play a central role in the constraint solving method.

3.2 Geometric constraints

We have considered the two constraint categories: topologic constraints and metric constraints (some examples are shown in Figure 1).

Topologic constraints comprise:

- *Tangent* constraint between lines, circles, arcs and ellipses, in all combinations.

- *Perpendicular* constraint between line and line, line and circle, line and arc and line and ellipse.

- *On* constraint: we can constrain a point to be on any other shape, by the way the shape can be constrained to pass over a set of points (i.e. circle by one, two or three points...). These points can be intersection points.

- *Parallel* constraint between lines.

- *Concentric* constraint between circles and arcs.

Metric constraints: are needed more in case of the under-constrained problem. The user can fix according to his need some parameter such as the radius of a circle, or the slope of a line. An example is shown in section 5.2. Among the metric constraints are dividing shapes into equal parts, which is very useful in architectural drawing.

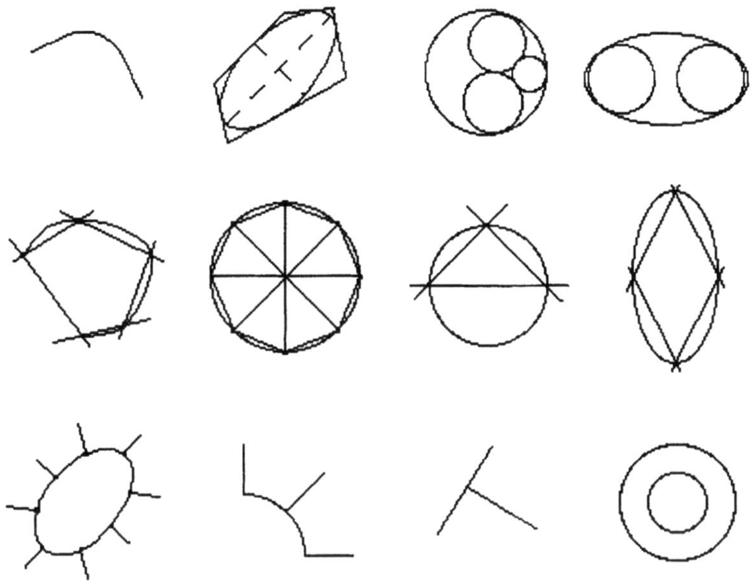

Figure 1. Some geometric constraints supported by our system

The mathematics of a constraint depends on its type. Some constraints are reduced to a quadratic equation. Other constraints such as tangents between ellipses are more complex and need iterative methods to be solved.

4. USER INTERFACE AND INTERACTIVITY

Architectural drawing is a process of adding and modifying. Drawings are built up gradually by adding to what has already been drawn [And92]. Van Sommers refers to this as *accretion* [VSo84]. This build-up tends to take place by locating new elements in some relation to what is already there – *anchoring*. Modifying is done in relation to what is already drawn. Two kinds of modifications can be made: topological and geometrical. The user interface of our approach is based on these two principles of adding and modifying. As an example we are going to construct and modify a geometric model, and demonstrate how easy it is to use this system.

All the constraints defined are done so explicitly by the user. Constraints are located using the right mouse button. Drawing without constraint is done using the left mouse button. In the first step of our example (Figure 2a), we draw an irregular polygon and an ellipse. After that we position five points constrained to be on the ellipse (Figure 2b). In order to do this we simply have to select the button *"point"* on the left menu bar and, with the right mouse button, click five times on the ellipse. The system recognises the approximate position on the ellipse. The *on* constraint is then applied and solved exactly. These actions release a flow of information. The points take the label *on*. The topologic parameter of the ellipse includes the fives points, and the fives points include the ellipse in their topologic parameters. The points may not be on more than two shapes, which avoids conflicts in solving the constraint.

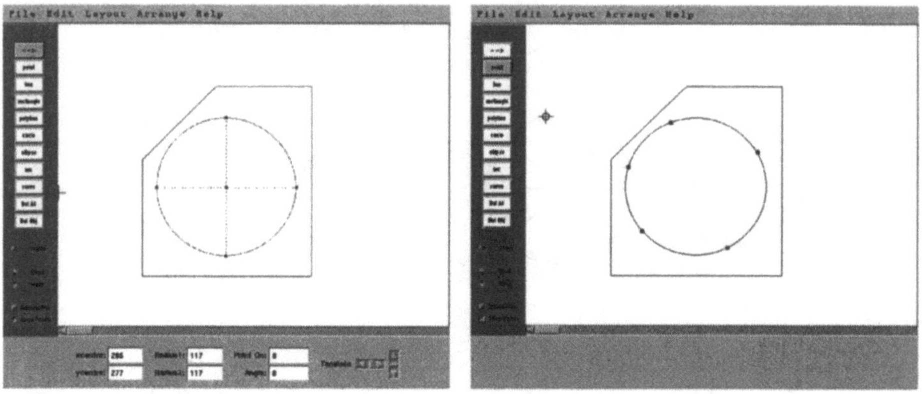

Figure 2(a,b). Drawing an irregular polygon and five points constrained to be on an ellipse

As it is shown in Figure 3, positioning the ellipse tangent to the line is done directly by dragging the point on the line. When an approximate tangency is recognised by the dragging routine, a tangent constraint is applied, and solved exactly. Dragging a second point to touch a second line, a new tangent constraint is added, while maintaining the first. As soon as the tangent constraint is applied the point becomes a tangent point, in this case its label which was *on* becomes *tangent*. The point is added to the line's topologic parameter, and a line into the point's.

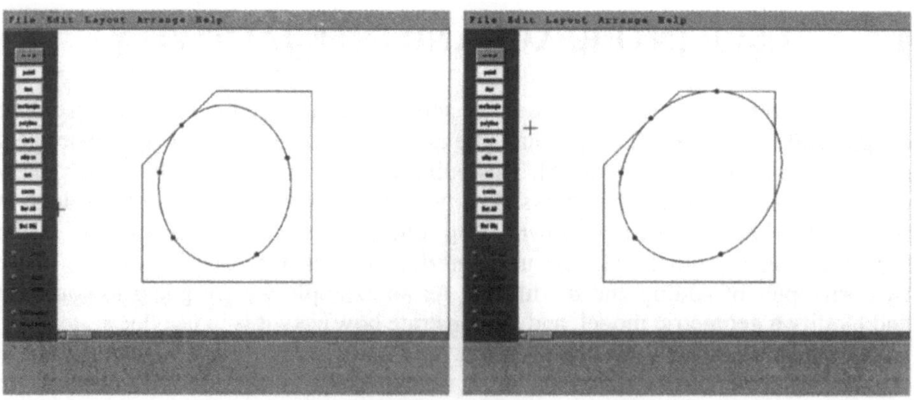

Figure 3(a,b). Drag a point on the ellipse to the line; a tangent constraint is applied

We continue until the ellipse becomes tangent to the fifth line (Figure 4a). In Figure 4b we show eight points constrained *on* the ellipse and dividing the elliptical perimeter into equal parts. This constraint is applied after selecting the ellipse and validating the number of the points on the appropriate numerical panel.

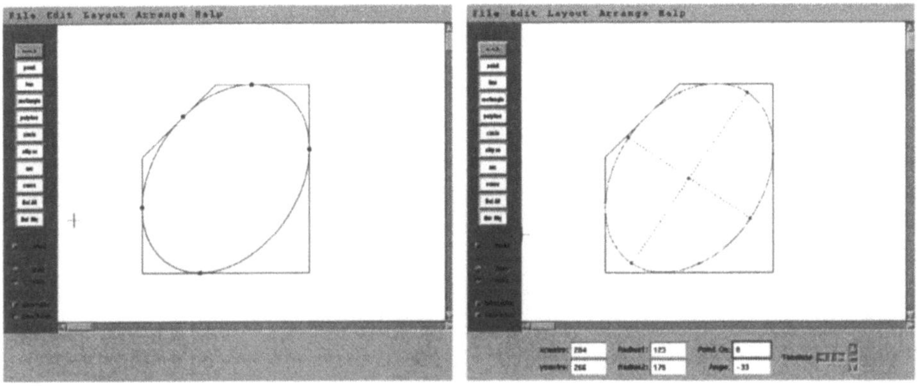

Figure 4(a,b). Continue until ellipse is tangent to fives lines (On the left), constraint 8 points to be on the ellipse dividing the elliptical perimeter into equal parts

For the next step we select three construction lines defined by three points on the ellipse and draw a circle (Figure 5).

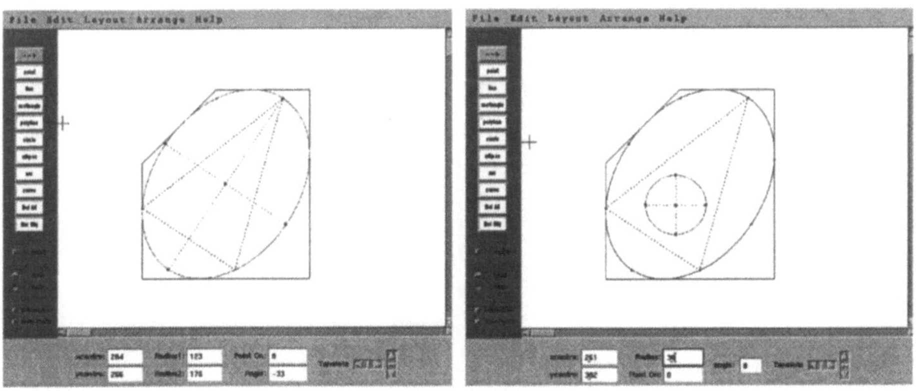

Figure 5(a,b). Three construction lines defining a triangle between three points on the ellipse are selected

We drag the circle from any point on the circle to the construction lines. When the approximate tangent is recognised by the dragging routine, the tangent constraint is solved exactly. We drag a second point to the second construction line. A new tangent constraint is added, while maintaining the first. We continue until the circle becomes tangent to the three construction lines (Figure 6 and Figure 7). The three tangent points take the label *tangent* and their topologic parameters are updated. In the case of the circle, the points are generated implicitly by the system as soon as the tangent constraint is recognised.

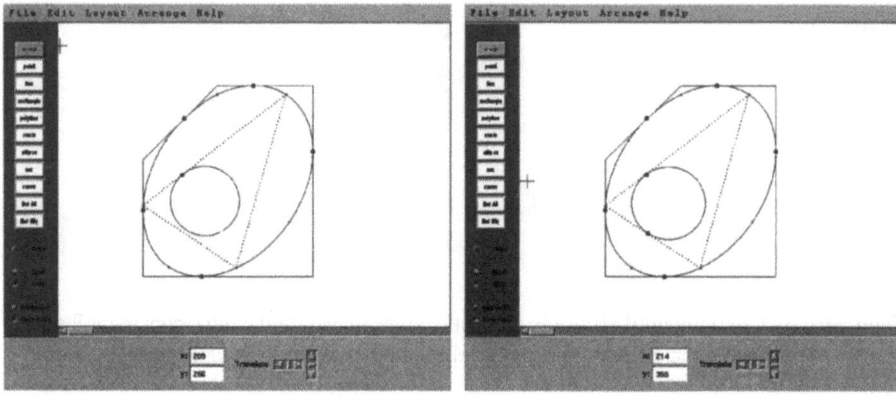

Figure 6(a,b). Drag a point on the circle to the construction line; a tangent constraint is applied

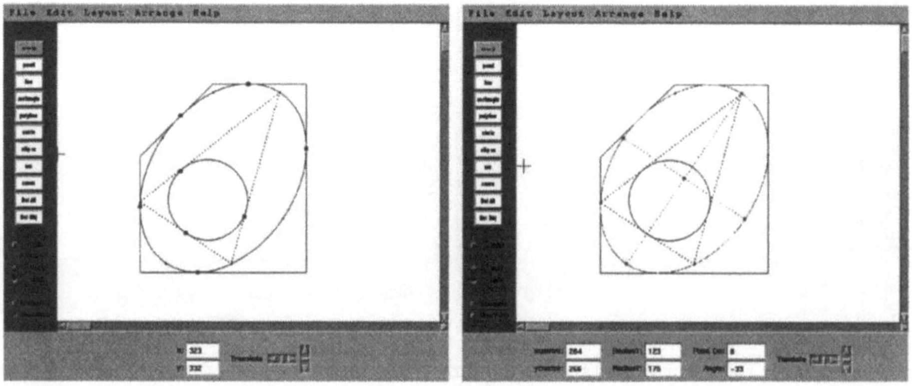

Figure 7(a,b). Continue until the circle becomes tangent to the three construction lines

Modifying the geometric model can be done in two ways; by the numerical panel or by direct manipulation. In the Figure 8 we move the line by fixing its slope parameter (angles between the lines). The geometric model change as the constraints are maintained.

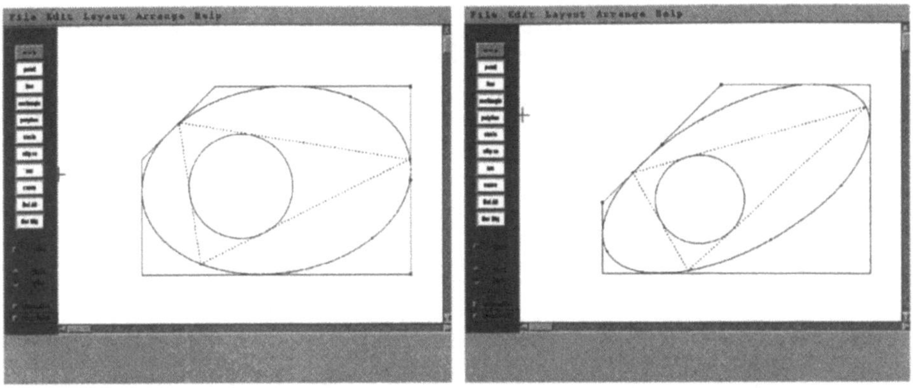

Figure 8(a,b). Move the lines without modifying the slopes, the geometric model changes as all the constraints are maintained

Constraint relaxation is also done by direct manipulation. You only have to drag a tangent point outside for the constraint to be relaxed. If the point is dragged with the left button of the mouse, the constraint is completely eliminated. When the geometric model is over-constrained during modification, the system gives you a message asking you to eliminate some constraints. The system allows you a free choice of the type of constraint to eliminate.

5. CONSTRAINT SOLVER

We have chosen in our approach the graph-constructive solving method. This method presents many advantages: in particular it is fast, more methodical and proven to be sound [Fud93]. Two steps characterise this method: *graph construction-analysis* and *constraint solving*.

- *Graph construction-analysis*: the user sketch, annotated with constraints, is translated into a graph whose vertices correspond to geometric elements: points, lines, circles, ellipses, splines, arcs, and whose edges represent the constraints between them. After the construction step, the graph is analysed and a constraint solving strategy is defined. The graph construction is performed only in case of drawing or modifying with constraint. In the example shown in Figure 3a, when we drag a point to the first line, the constraint *ellipse through five points* is maintained. In this case as soon as we select the point on the ellipse with the right mouse button, the graph as indicated in Figure 9a is generated. Graph analysis deduces the type of constraint to solve. In this case the constraint "*ellipse through five points*" is detected and solved at each modification of the first point. But when the first point is almost on the line, it is recognised, and the graph is reconstructed. A node representing the line is added (as indicated in Figure 9b). When this point is near the tangent point between the ellipse and the line, the point acquires the label *tangent,* which again modifies the graph, and a new constraint is detected corresponding to the new situation. This constraint becomes *"ellipse through four points and tangent to one line"*. The fifth point became the tangent point with the first line.

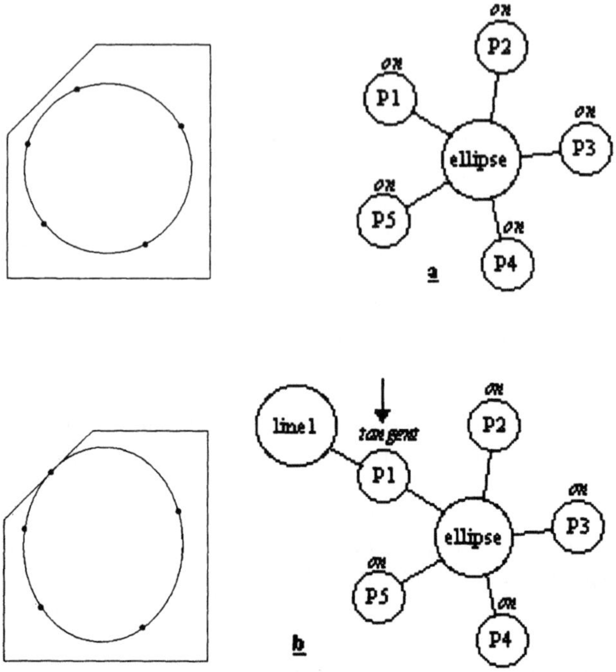

Figure 9(a,b). Geometric model and corresponding constraint graph.

- *Constraint solving*: when the construction steps involve ruler-and-compass constructions, only quadratic equations need to be solved. However we are more ambitious and some construction steps are permitted that are not ruler-and-compass; and in those situations the solutions are found numerically. In the example shown in Figure 8, the constraint which maintains the eight points dividing the elliptic perimeter into equal parts is solved using the elliptic integral of the second kind. Its resolution is done using Simpson's rule, with an error of less than 1/10000. The resolution of the tangent constraints between ellipse and other geometric entities use iterative methods, with the same error margin.

5.1 Well-constrained problems

A well-constrained problem can have many solutions. Thanks to certain heuristics we conduct the system to find a particular solution. The example shown in Figure 10 is well constrained (as the circle is defined by three points), yet can have up to 8 solutions. The solution which best matches the mouse position is used.

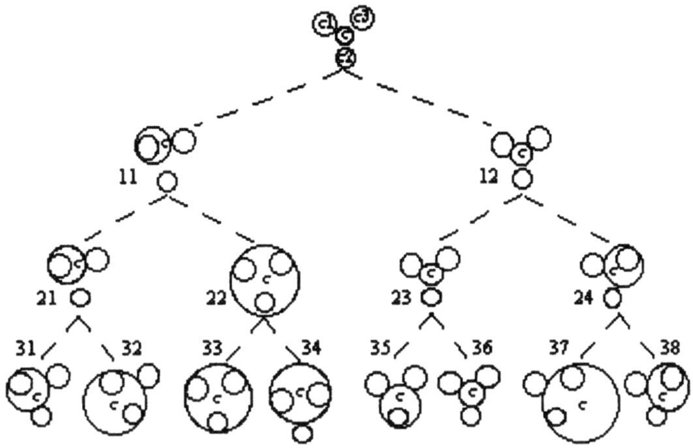

Figure 10. The eight possible solutions driven by the user decision thank to the mouse position

5.2 Under-constrained problems

Our approach supports the under-constrained problems. In this case the system is driven by the degree of freedom of the geometric elements. For example, by default, the circle has both of its parameters, the radius and the centre, frcc. Here, the user can display the degrees of freedom of the circle and can fix or release them. The degrees of freedom don't intervene in the well-constrained problems. In the example shown in Figure 11 we have a tangent between two circles *c1* and *c2*. In case of any modification we have several possibilities. If we would like to modify the radius of *c2* without moving the tangent point (case 1, Figure 11) we leave free both parameters of *c2*. If we would like to modify the centre of *c2* without modifying its radius, we fix its radius parameter (case 2, Figure 11). This modification can be done by dragging the circle *c2*, or numerically by using the numerical panel. In the first case we could put the exact radius wanted for *c2*. In the second case, we could put the exact angle between *c1* and *c2*.

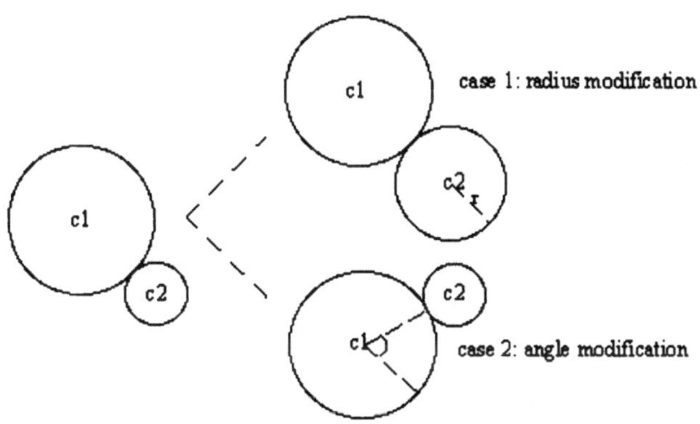

Figure 11. Different solutions driven by degrees of freedom

5.3 Conflict Management

Some conflicts can appear in the case of modifications. In Figure 12 we show an ellipse constrained by five points. If we drag one point, the new ellipse parameters are computed (Figure 12b). If the conditions $A\times C- B\times B>0$ and the eccentricity $e< 1$ are not satisfied (A, B, C and e correspond to the conic parameters), there is no solution. This situation corresponds geometrically to one point being inscribed in the polygon formed by the other four. In this case the system keeps displayed the last solution found (Figure 12c) until the conditions are satisfied again (Figure 12d).

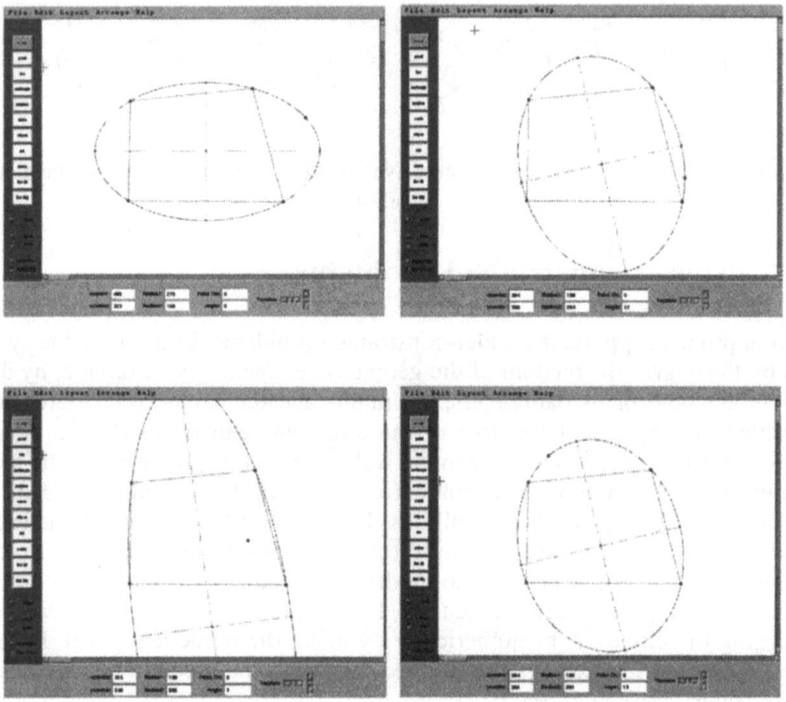

Figure 12(a,b,c,d). Conflict management

6. EXAMPLES

We have tested several examples which allow us to evaluate the performance of this approach. The example of Figure 13 shows how it is easy to work with elliptic forms. This geometric model is defined by an elliptic perimeter divided into eight equal parts. From each point a line perpendicular to the ellipse is drawn, these lines have the same length. It is possible to make a lot of variations of this model maintaining all the constraint: modify the ellipse orientation, its radii. Examples of buildings that have an elliptic perimeter include the *Lipstich Building* in New York designed by Philip Johnson and the *conference room of the Chamber of commerce* in Berlin designed by Nicholas Grimshaw & Partners.

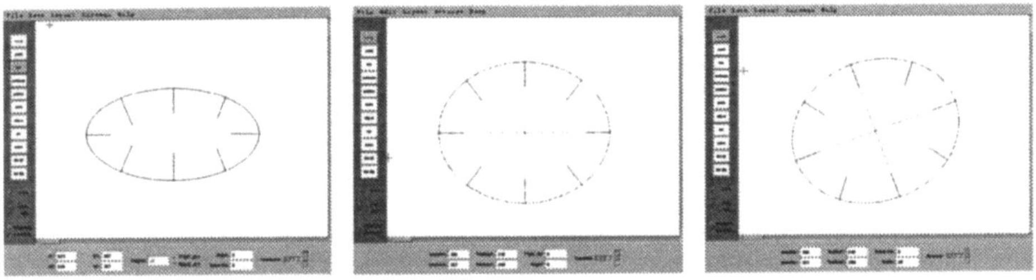

Figure 13. Example with an elliptic form

In Figure 14, the geometric model is defined by three circles and lines. Many variations of this model are presented: radius (Figure 14b) and circle orientation (Figure 14c).

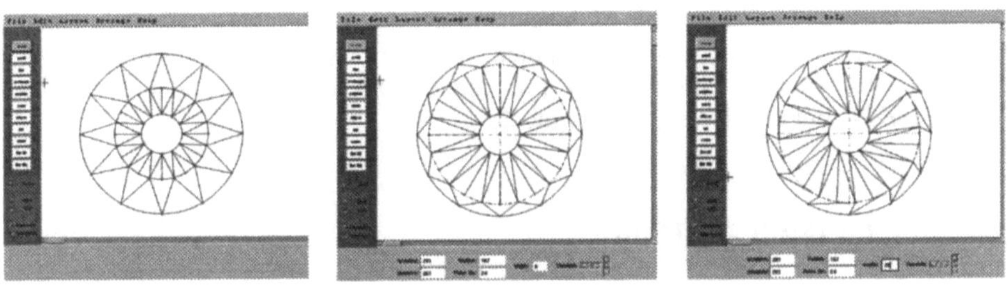

Figure 14(a,b,c). Example with circles and lines

In Figure 15, playing with circles is shown. Figure 15b shows the modification of the small circle just by a simple dragging. The Figure 15c shows an arbelos.

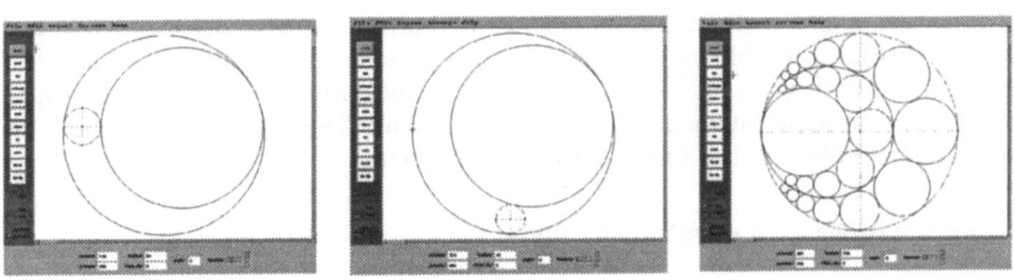

Figure 15(a,b,c). Example with circles

The last example (Figure 16) is a model with circles and an ellipse. Some variations are also shown.

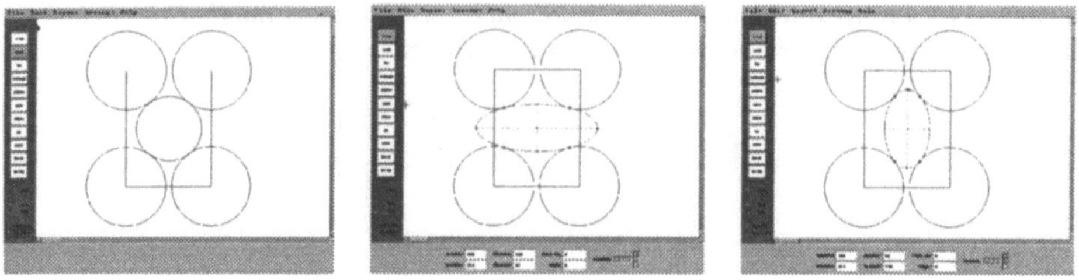

Figure 16. Example with circles and ellipses

7. IMPLEMENTATION

This application is developed in Java, JDK 3.0.1 of SGI's Java software (IRIX 5.3). This release is conformant with the behaviour of Sun's 1.1.3 JDK. Java is an Oriented Object Language that is based loosely upon C++. Java Applets can be transmitted over the Internet and executed, in theory by any Java-compatible Web Browser. We find in practice that this applet executes well under Linux and Windows/NT, but poorly on the Macintosh.

8. CONCLUSION

The approach presented in this paper show an interactive, simple and high-level modification of part geometry via numerical panel or by simple geometry dragging. Several aspects were discussed: the constraint and geometric object structure, the method of constraint solving, and interaction via the user interface.

Compared to previous work, we have enlarged the geometric object structure to ellipses and splines with constraint model enrichment. We have discussed well and under-constrained problems. The system manages constraint conflicts, without any instability. Another advantage of this approach is the particular user interface which makes this system interactive, simple and useful. This interface supports the resolution of both well and under-constrained problems.

Many extensions are presently under study:

- Constraint enrichment, particularly with respect to splines.
- A diagnostic tool for the over-constrained problem in order to guide the user to relax a specific constraint.
- Incorporating this system as a module in a CAAD package.

ACKNOWLEDGEMENTS

This work is sponsored by Informatix Inc. The author would like to thank Paul Richens, the Martin Centre director, for valuable help and comments about various aspects of this research project.

REFERENCES

Aish R, 1992, "Computer-Aided Design Software to Augment the Creation of Form", in *Computer in Architecture*, 97-104, Longman, London.

Alasdair S, et al., 1993, "Constraint Definition System: a Computer-Algebra Based Approach to Solving Geometric-Constraint Problem", *Computer Aided Design,* 25 (12) 741-750.

Aldefeld B, 1988, "Variation of Geometries Based on a Geometric-Reasoning Method", *Computer Aided Design*, 20 (3) 117-126.

Andre G P, et al., 1992, "Computer Aids For Design Development", in *Computer in Architecture*, 15-24, Longman, London.

Bruderlin B, 1986, "Constructing Three Dimensional Geometric Objects Defined by Constraints", *In workshop on Interactive 3D Graphics ACM*, 111-129.

Borning A H, 1981, "The Programming Language Aspects of ThingLab, a Constraint Oriented Simulation Laboratory", *ACM TOPLAS*, 3 (4) 353-387.

Borning A, Maher M, et al., 1989, "Constraints Hierarchies and Logic Programming", *In Proc. Of the 6th International Logic Programming Conference*, 149-164.

Boudon Ph, et al., 1972, *La geometrie chez l'architecte : Etude exploratoire* (UP-architecture-Nancy).

Bouma W, et al., 1995, "A Geometric Constraint Solver", *Computer Aided Design*, 27 487-501.

Buchberger B, 1988, "Algebraic Methods for Geometric Reasoning", *Annual Reviews in Computer Science*, 3 85-120.

Dimensional Constraint Manager, 1993, D-Cubed Ltd, 68 Castel Street, Cambridge, CB3 0AJ, UK.

Dohmen M, 1995, "A Survey of Constraint Satisfaction Techniques for Geometric Modeling", *Computer & Graphics,* 19 (6) 831-845.

Fudos I, 1993, "Editable Representations for 2D Geometric Design", *Master of Science Thesis*, Perdue University.

Fudos I, Hoffmann C M, 1993, "Correctness Proof of a Geometric Constraint Solver", *Technical Report* CSD-TR-93-076, Department of Computer Sciences, Purdue University.

Gao X S, Chou S C, 1998, "Solving Geometric Constraint Problems. I. A global propagation approach", *Computer Aided Design*, 30 (1) 47-54.

Gao X S, Chou S C, 1998, "Solving Geometric Constraint System. II. A Symbolic Approach and Decision of Rc-constructibility", *Computer Aided Design*, 30 (2) 115-122.

Heydon A, Nelson G, 1994, "The Juno-2 Constraint-based Drawing Editor", SRC Research Report 131a.

Kondo K, 1990, "PIGMOD: Parametric and Interactive Geometric Modeller for Mechanical Design", *Computer Aided Design*, 22 (10) 633-644.

Kramer G, 1992, *Solving Geometric Constraint Systems* (MIT Press).

Light R, Gossard D, 1982, "Modification of Geometric Models Through Variational Geometry", *Computer Aided Design*, 14 (4) 209-214.

Owen J C, 1991, "Algebraic Solution for Geometry from Dimensional Constraints", *In ACM Symp. Found. of Solid Modeling,* Austin TX, 397-407.

Requicha A, 1977, "Dimensionining and tolerancing", Technical report, Production Automation Project", University of Rochester, PADL TM-19.

Solano L, Brunet P, 1993, "A System for Constructive Constraint-based Modeling", *Modeling in Computer Graphics*, B. Falcidieno and T. Kunii editors, Springer Verlag.

Sunde G, 1988, "Specification of Shape by Dimensions and Other Geometric Constraints", *In Geometric Modeling for CAD Applications*, North Holland, IFIP, 199-213.

Sutherland I, 1963, "Sketchpad, a Man-machine Graphical Communication System", *In Proc. of the Spring Joint Comp. Conference*, 329-345, IFIPS.

Todd Ph, 1989, "A k-tree Generalization that Characterizes Consistency of Dimensioned Engineering Drawings", *SIAM J. DISC. MATH.*, 2(2) 255-261.

Van Sommers, 1984, *Drawing and Cognition* (Cambridge University Press, Cambridge).

Collaborative Design with NetDraw

Dongqiu Qian and Mark D. Gross
Sundance Lab for Computing in Design & Planning, University of Colorado, Boulder, CO, USA

Key words: synchronous collaboration, shared drawing.

Abstract: The paper describes NetDraw, a Java-based object oriented drawing program that employs a server-client architecture to provide a shared drawing environment for collaborative design. NetDraw goes beyond conventional shared whiteboard applications in its support for concurrency control, groups and constraints, and ephemeral gesture objects. Small and simple enough that users can learn it quickly, NetDraw is designed to run on small platforms such as handheld computers. We describe NetDraw's features and an early evaluation of its use.

1. INTRODUCTION

Design is team work. Architects and engineers work together to produce a design, and at times the design team sits face-to-face, usually around a table, drawing on a piece of paper or a whiteboard with a variety of colored pens. During those meetings designers draw concurrently or serially to show their ideas to others and they point to elements in the drawing as they discuss the design. One important tool for synchronous distance collaboration in design is a shared drawing surface where designers can discuss and mark up proposed designs. Although the Internet makes it possible to carry on these graphical conversations at a distance, computer support for this activity remains limited and primitive.

The idea is hardly new: Along with the development of personal computer and internet technologies in the 1980's and 1990's, a community of researchers has emerged that is developing tools and approaches for Computer Supported Cooperative Work, or CSCW. Using technology to support collaboration among distributed participants was the focus of the early Media Spaces experiment at Xerox PARC [Harrison, 1993; Bly 1988], which used video cameras in collaborators' offices to create a shared virtual workplace. Since then numerous efforts have sought to develop computer based drawing environments that designers working from different locations can share. Commercial Web browsers (e.g. Netscape Communicator) now include an option for a virtual whiteboard along with text and video conferencing. These products are primarily intended for general purpose meeting support, not for architectural design.

We describe NetDraw, a synchronous collaborative drawing program built in Java—initially as a learning project—that responds to this need. We begin with a review of collaborative drawing software, and locate NetDraw with respect to similar systems. We then outline NetDraw's basic functions, focusing on those that go beyond off-the-shelf solutions. We discuss the evaluation of NetDraw in which several designers used the program to carry out a simple design task. We conclude with lessons learned from building and evaluating NetDraw and our plans to modify NetDraw to take in to account these lessons.

1.1 Scenario

Here is an example of how a shared drawing program might be used in a design teaching context. Steve, (a student) working at home has encountered a problem with his site plan: He doesn't know how to deal with the circulation system. Steve decides to discuss the problem with his teacher (Theresa) using NetDraw. (See figures 1 and 2). Steve and Theresa connect their computers to the Internet. Theresa starts a NetDraw server and both NetDraw clients connect to the server. Steve loads his design draft, which he has saved or scanned in an image file, into NetDraw. The image appears on Theresa's screen so that both Steve and Theresa are looking at the same design.

In the chat window, Theresa types: "What's the problem?" Steve replies, "I need help on the entry and circulation system." Theresa asks, "Show me what you have in mind." Steve uses the rectangle tool to draw two rectangles representing two buildings. The two rectangles appear on Theresa's screen as well. Theresa says, "Your entry is in a difficult place. Look at the flow of pedestrian traffic." Theresa draws two circles to indicate the entries. Then she draws an arrow on the screen to indicate the pedestrian traffic flow. The gesture appears on Steve's screen, but over a short period of time, it disappears. Steve selects the two rectangles, groups them, and moves them to a better position. "Like this?" he asks.

2. COMPUTER SUPPORTED COOPERATIVE DRAWING

We examined commercial products for interactive conferencing where users draw on a real or screen whiteboard, as well as research developments in CSCW drawing tools. All these systems support synchronous design drawing and discussion. Some are bitmap based and do not allow users to manipulate the objects of the design. Others, though object-based, provide inadequate mechanisms to deal with concurrent object manipulation. Some products, like Netscape's NetConference and Microsoft's NetMeeting include simple whiteboard programs, but they provide only simple diagramming tools. On the other hand, more sophisticated solutions, such as Bentley Systems' Microstation/J, offer an object oriented model but require users to buy into an entire CAD package that is inappropriate for early or casual design discussions.

NetDraw tries to strike the middle ground between simple whiteboard programs and complex CAD package functionality. It provides basic 2D drawing program features including grouping and simple constraints, solid concurrency control in a server-client architecture. In addition, it provides an export function to write DXF files for AutoCAD. It also can communicate with our freehand sketching program (the Electronic Cocktail Napkin) over a network. NetDraw is a thin client, suitable for running on handheld computers that support Java, and we intend to use it also in mobile CAD applications [Gross et al., 1997].

We consider related research work with respect to several design issues: structured graphics versus bitmap drawing, Web based architecture versus point-to-point 'standalone' systems, the handling of concurrent operations, and how the systems support (or don't) multi-user awareness.

2.1 Structured vs. bitmap graphics

Two kinds of graphics structures are often used in computer based drawing tools: structured or object-based graphics and bitmap graphics. Many early collaborative drawing applications were based on bitmap graphics. For example, WE-MET [Wolf, 1992] is a pen-based sketching tool designed for small, informal work groups. It records all the meeting events, so during or after the meeting these events can be reviewed. GroupSketch [Greenberg, 1992] is a multi-user sketchpad: designers make and erase marks on a bitmap surface. It uses a pen-shaped cursor to indicate freestyle drawing mode. BoardNoter [Stefik, 1987] imitates the informal functionality of a chalkboard. It provides four tools: a piece of chalk, an eraser, a keyboard and a pointer. To draw a rectangle, the designer must first select the chalk to pick it up, then draw the four strokes of the rectangle.

Conversation Board [Brinck, 1992] is a shared drawing tool based on structured graphics. It provides multi-user sketching with various colored markers, editable multi-line text, circles, rectangles, lines, and connectors (a connector is a line or arrow between two other objects and it remains connected to them even when the objects are moved.) Tivoli [Moran, 1995] is an object-based system. As it is designed primarily for meeting support, it uses a set of built in objects for managing whiteboard text, for example, lists and tables of items. GroupDraw [Greenberg, 1992] is also object-based, users can create, move, resize and delete drawing objects. It also provides a scrollable work surface.

2.2 Web-based vs. 'standalone' application

Today the Web offers an 'enabling technology' for CSCW applications [Bentley, 1997]. BSCW (Basic Support For Cooperative Work) [Bentley, 1997] is an extension of a standard Web server using the CGI (Common Gateway Interface) programming interface. It provides shared workspaces for document management. Every shared workspace is a shared repository that can contain information such as documents, images, links to other pages or FTP sites, member contact information etc. Every authorized user has the same right to access this shared workspace. Users employ applications they are familiar with to create documents (e.g. Microsoft Word for text editing); then they can upload the documents to the shared workspace with the web browser. Other users then download the document from the shared workspace and make modifications.

Alliance [Salcedo, 1997] is a cooperative authoring application on the Web that allows geographically dispersed collaborators to produce and maintain documents together, asynchronously. Every user site runs an HTTP server and an instance of the Alliance software. Unlike BSCW, Alliance provides an editor for users to work on the shared document. In a collaborative session, every user works with a local copy of the shared document, but the modification of the local copy doesn't appear on others' screens automatically. The modifier must perform a 'save' operation to explicitly notify other users about the changes; and other users have the right to decide whether or not to have their screen updated. Alliance doesn't use HTML and a Web browser; it uses HTTP (Hyper Text Transfer Protocol) to communicate directly with CGI scripts running on the servers. These systems supported by CGI and HTTP technology are asynchronous, and they use HTTP, which is poorly suited to real-time conversations.[Dix, 1997]

2.3 Concurrent Object Manipulation

GroupDraw [Greenberg, 1992] uses 'coupling Status' to indicate an object is a 'private', 'public' or 'sharable' one. But this 'coupling Status' only indicates whether an object can only be viewed and manipulated by the creator (private), can only be viewed by others (public), or can be manipulated by all others (sharable). When two or more users want to manipulate an object at the same time, the system assumes optimistically that there will be no conflict. Conversation Board [Brinck, 1992] uses action verification to solve the concurrence problem. Before an action is initiated, the system will check whether it can be completed. For example, the resizing tool will verify that an object still exists before it tries to change the size of the object.

SimpleDraw [Brinck, 1993] uses a token to indicate who has the right to modify objects on the drawing surface. When one user holds the token, others can only watch. In BSCW, a user can set access rights to control the visibility of the information and the operations (e.g., rename, delete, edit, and replace) that others can perform. BSCW has no locking mechanism; it adds "a note" to an object (e.g. a file) to indicate the object's status. For example, it places a note on a shared workspace, "Fred is editing this document." Other users can see that, although they may have the right to edit the document, now is not the best time.

Alliance defines four roles for every document fragment (a fragment is a part of a document): Reader, Writer, Manager, and Null. The 'reader' role only allows a user to read the fragment; the 'writer' role allows a user to modify the fragment; the 'manager' role allows the assignment or change of roles and modification of a fragment; the 'null' role forbids the viewing of a confidential fragment. At the outset, the 'manager' role is assigned to the author who creates the document. Then the 'manager' can assign all these four roles to other users. At any time at most one author can edit each fragment. For example, if users A and B share a document with two fragments, both can view the whole document, but they cannot have the 'writer' role for the same fragment at the same time.

2.4 Multiuser Awareness

One feature that distinguishes a multi-user from a single-user application is that in groupware, users must be aware of who is around and what activity is happening. Several approaches have been explored. Some systems implement multiple cursors to identify different participants. In GroupSketch and GroupDraw, every participant's cursor is labeled with their name and the cursor changes shape according to different tasks. Every user in the session can see who is doing what on the shared drawing surface. Some systems, such as BoardNoter, implement a telepointer that is seen by all the users in the same collaborative session, while the individual cursors are not. To draw attention to a specific area of the shared drawing surface a user can pick up the telepointer and move it around the area.

BSCW uses different icons to support event and activity awareness. It uses five different icons to represent five event types: New, Read, Change, Move, and Touch. For example, when a document has been modified, the icon associated with this document changes to indicate that a 'Change event' has happened. The 'note' linked with a document also makes other users aware of a document's status.

Alliance also uses different icons to indicate different roles. Inside every user's local copy of the document, an icon included with every fragment indicates the user's role with respect to this fragment. When a user acts as 'reader' for a fragment, each time when a new version of that fragment becomes available, the icon before the fragment changes to notify the reader.

The icons and telepointers used in the systems described are static ; designers may find it difficult to use them to show their ideas to others. We agree with Stefik that when a collaborative group reaches a certain number of users, sharing cursors on a What You See is What I See model does not work: "The WYSIWIS display of cursors from multiple users is unacceptably distracting" [Stefik, 1987]. Therefore in NetDraw we explored the idea of using an ephemeral (disappearing) gesture for users to indicate to one another the subject of concern.

3. THE NETDRAW SYSTEM

NetDraw was implemented in Java 1.1 on a Windows PC; it runs as both an application and an applet on any Java-enabled browser. It uses a client-server architecture: Any number of users can log in to the NetDraw server and use their Web browsers to observe and participate in a graphical conversation. The server echoes drawing and editing actions to all clients. NetDraw provides a simple suite of drawing commands: As with standard Draw programs a designer can create and edit drawing objects, choose color, line weight, etc. NetDraw goes beyond standard Draw features in several ways:

• A designer can group and ungroup objects and apply geometric constraints to groups of objects.

• Beyond simple geoemetric shapes, NetDraw also provides a freehand sketch tool, and freehand sketches are curves that can be edited at control points.

• A 'gesture' command allows the designer to make a non-permanent graphical mark on the drawing to indicate parts of the drawing during a discussion (deictic reference).

• Designers can see the syncronisation status of objects because NetDraw uses colour to indicate which objects are currently being manipulated.

• NetDraw also records snapshots of the drawing, which a designer can play back to review the history of a design process. The snap shot records the design state (the actual objects, not merely a pixel map screen image), including the constraints, groups, and chat stream at that point. The design history feature can be set to record snapshots at regular intervals, or only to record them on command.

• Users can provide different methods for objects, allowing objects to have different appearances and behaviors.

• A MUD-like text based chat facility augments the graphical conversation.

• A locator for uploading graphic (GIF and JPEG) files from any location for use as an underlay on the drawing.

• Identifier and information association between graphic objects and their attributes.

• In addition, NetDraw can communicate with AutoCAD via file transfer, and with Cocktail Napkin network clients.

NetDraw provides two collaborative functions: Draw and Chat. NetDraw's drawing features include: object based graphical editing, gestures that fade away over time, graphical underlay, grouping and graphical constraints, and recording design history snapshots. NetDraw's text-based chat offers a substitute for face-to-face conversation or video links. Figure 1 shows the chat window. At the top is a one-line type-in area; on the bottom a multi-line text field displays the conversation transcript. Whenever one user types a sentence, it appears on the screens of all other users. NetDraw prefaces each sentence with the author's name.

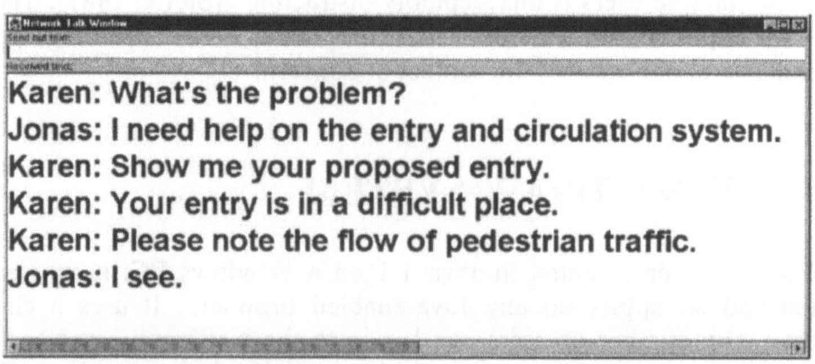

Figure 1. The text chat window displays the author's name with each comment.

3.1 Drawing Features

NetDraw is an object-oriented drawing program. Designers can draw lines, circles, rectangles and freehand sketches in different colors, and they can select, move, resize, and delete those objects. Designers can access those drawing and modifying tools quickly from a tool palette. Figure 2 displays NetDraw's drawing window. On the right is the shared surface where graphical objects are drawn. Every change made to the drawing area is immediately echoed on all designers' displays. On the left is a snapshot area where designers can capture an image of the current state of the session. By clicking on the saved snapshots, designers can quickly review a previous state of the design history.

NetDraw can also use a graphical image as the background of the drawing surface. The image may reside on the local machine or on any Web server. This image could be a topographic base map or a sketched design drawing imported from another program. The designer can link objects with descriptive text. The text can simply indicate the object's name, e.g. "table", "kitchen", or it can provide information about the object. Each user also controls whether to display the text associated with each object.

Besides the normal graphical objects offered by other drawing packages, NetDraw also provides a 'gesture' object. A designer can draw a gesture on the drawing surface; like other drawing objects, it will appear on every other designer's screen. However, unlike ordinary drawing objects, gesture objects are temporary; they fade away over a short period of time. Designers can draw attention to a specific area of the drawing surface with a gesture. A designer can use a gesture object not only to draw attention, but also to discuss their design concepts with others.

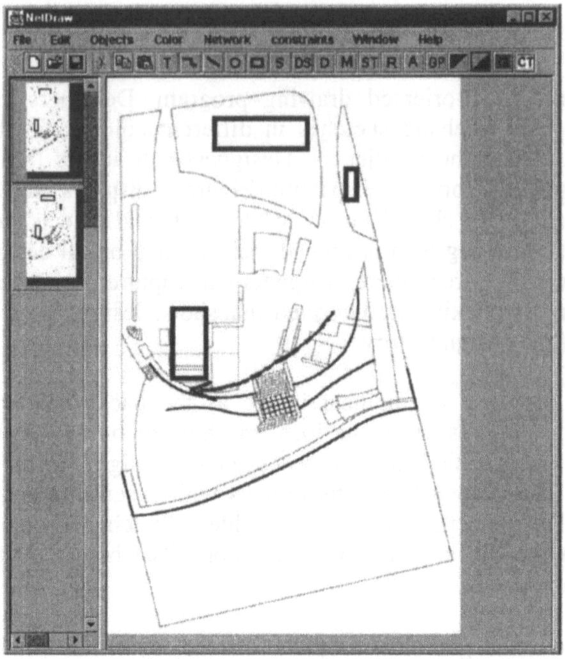

Figure 2. The NetDraw graphics screen with a scanned image underlay (actally a pointer to a Web site), geometric objects, and an ephemeral gesture (red arrow). Two snapshots from the design history are on the left; the command menu bar is on top.

3.2 Groups and Graphical Constraints

NetDraw provides a "Group" command with which designers can combine two or more graphical objects into a group. Instead of modifying every graphical object in a group one by one, designers can apply modifying actions to the group. NetDraw also provides geometric constraints that can be applied to groups. When a constraint is applied to a group, the NetDraw assumes that the first object in the group is the dominant object. For example, if a designer applies "Top = Top" (top alignment) to a group of objects, NetDraw will change the Top coordinates of all other objects to equal the value of the first (dominant) object's Top. Later, when a designer modifies an object in the group, NetDraw will make this object the dominant object and re-apply the constraints. For example, consider that objects A, C, B, and D are grouped in that order. When a designer applies "Top = Bottom" to make a vertical stack of the objects in the group, the system uses the following assignments to satisfy the constraints:

C's Top ← A's Bottom
B's Top ← C's Bottom
D's Top ← B's Bottom

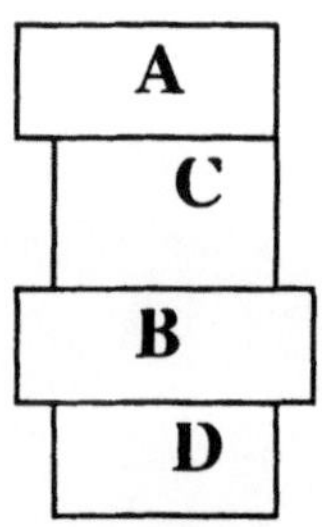

Later, when you modify object B, moving or stretching it, the system uses these assignments to satisfy the constraints:

C's Bottom ← B's Top
A's Bottom ← C's Top
D's Top ← B's Bottom

Designers can add one or more constraints to a group. However, NetDraw doesn't check for conflicts among the constraints. For example, if a designer first applies "Top = Top", and later adds "Bottom = Bottom" to the same group, constraint satisfaction proceeds as follows:

1) satisfy "Top = Top", to make all the objects in the group top-aligned.

2) satisfy "Bottom = Bottom" to make all the objects in the group bottom-aligned.

If all objects are not the same height, the "Top = Top" constraint will not be maintained. In the geometric constraints satisfaction process, NetDraw only moves objects; it will not resize them.

3.3 Concurrency Control

The NetDraw Server mediates and coordinates designers' actions. Before collaboration begins, one designer (the host) starts a NetDraw Server. Then all designers, including the host, can connect to the server. Once these connections have been made, the designers can begin their collaborative work. During the session, any participant can join and leave at any time as long as the server is running. When a new participant joins the session, the NetDraw server sends a message notifying all other participants.

It's likely that two or more designers may try to manipulate the same object simultaneously. NetDraw uses a locking synchronization mechanism to ensure that no two designers can modify one object at the same time. Every object NetDraw object has an unique identifying number. A central server distributes those unique numbers and conducts the object lock management. The NetDraw Server adopts a 'first come, first served' mechanism. When an instance of a NetDraw client starts, it connects to the NetDraw server to join the current collaborative session. At that time, the server assigns a number range to the NetDraw client. Then every time the client creates a new object, NetDraw picks a number from the number range and assigns it to that object. The NetDraw server maintains a list of currently locked objects. Before a client can change an object, the object must be locked: the client must ask the server for permission to modify the object. If another client has already locked the object then the server will queue the request and send a notification message back to the client. Once the object is unlocked, the server grants permission to the first request in the queue. The designer who made the request is notified and can start modifying the object. If the designer deletes an object, the server removes all pending requests to access this object.

NetDraw uses colors to indicate an object's lock status: yellow, pink and green. After the designer picks up a tool and selects an object to operate on, the object turns yellow indicating that it has requested permission from the server to lock the object. The designer cannot start the action until permission is granted. After the server replies, the object will turn either green or pink. If the object turns green, the designer was granted permission to modify, the object is locked, and the designer can begin to operate on the object. If the object turns pink, it indicates that another designer is editing the object right now, and the designer must wait. As the request is queued, once the object is unlocked, its color will turn to green indicating that modification can begin.

4. NETDRAW EVALUATION

In order to obtain relevant feedback for the continuing design of NetDraw, we conducted two pilot evaluations of the use of NetDraw. In the first 'formal' evaluation, described in detail below, two architecture students from the College participated in the study. In the second, 'informal' evaluation, we interviewed three researcher/faculty members from the College used NetDraw.

4.1 Setting

We used two PCs running Windows NT with 17" color monitors in a laboratory setting. We used keyboard and mouse as the input devices and the monitors as output devices. Both PCs are connected to the local area network, and then to the internet. Two participants sat back to back at a distance of 8 feet, so we asked them not to talk during the session. This setting allowed us to observe both participants at the same time. The two participants in the pilot experiment were recruited from the students of the College of Architecture and Planning at the University of Colorado who had experience with design and computer use.

4.2 Procedure

Participants in both the the 'formal' and 'informal' evaluations were given a 20 minute training session on the system functionality. Then after a 10 minute practice session, we gave the participants in the formal evaluation a design task and asked them to begin the collaborative session with NetDraw. The informal evaluators used the same task, but they were not assigned it. Here is what we told our formal evaluators:

"The task is to design the layout for a digital design studio. The studio must accommodate 12 students, each with a drawing table and a computer workstation. In addition, the studio should provide a small lecture area for the students to gather for presentations, and a shared scanning and printing station. The room for the design studio is 34'x37', with a corner removed by an angled wall. The plan of the room has been scanned into a image file, and the file can be downloaded from a Web server."

4.3 Preliminary Findings

The goal of our experiment was to learn how NetDraw performs in supporting a real design task. During the formal session we observed participants and made notes. We interviewed both the formal and informal evaluators after their design session. Our preliminary findings are drawn from all five evaluators' reported and recordedd experiences.

Although the system appeared to work well technically, no one completed the design task. All the drawings participants made are only conceptual designs. This is consistent with the experience reported in a larger experiment with collaborative CAD [Vera, Kvan, et al 1998]. There, designers responded to limitations in bandwidth by reducing low-level design exchanges and focusing on higher level discussions of design issues.

We found that the system worked well when participants were brainstorming general ideas about the design, but when it came to generate the details, one problem emerged: the participants had no idea of the size of the desk they had drawn. Here is the recorded discussion between the two participants, Paul and Mike:

(after working for a while)
Paul: I like the configuration. *agree with the configuration*
Mike: ok so how long is the desk? *get to the detail*
Paul: 4 feet? *confused*
Mike: we really need a ruler function. *identified the problem.*

NetDraw uses one window for graphical drawing and another window for text chatting, but the screen size is not big enough to keep two windows clearly visible at the same time. During the session, the two windows partially overlapped. This caused a problem: When Participant A made a change in the drawing, or sent a text message to Participant B, if B was working in the other window as A then B did not immediately notice the new text or change to the drawing.

We also found that "associating text with graphical objects" was useful for the text discussion. During the experiments, participants often added "associated text" to a graphical object to indicate its identity. For example, after one user associated "printer station" with a rectangle object, he referred to it by name:

Mike: how is the printer station? *referred to "printer station".*
Paul: it fits nicely in the corner. *understood which object referenced..*

During the experiment, participants often used the gesture object to show their ideas about a specific design task. For example, during a discussion about locating the 'lecture area', Mike used a gesture to show his idea of how to set up the video projector (Figure 3).

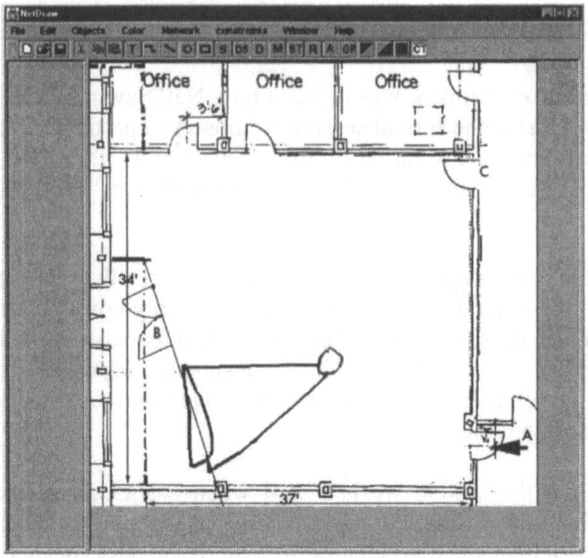

Figure 3: Using a gesture to indicate desired location of projector.

We conducted a post-experiment interview with the participants to obtain their general reactions and comments. The lack of ruler and scale was their largest concern: the participants had no idea whether the sizes of desks they drew were reasonable. Other issues were as follows:

Text scroll and wrap: Currently the text field used to display the incoming chatting scrolls, and each new sentence is added to the bottom of the window. Participants found it inconvenient to scroll to the bottom to read: the new sentence should be on the top of the window. When the sentence was too long, participants had to scroll horizontally to read it: Text wrapping should be implemented. Some participants found that changing between the text chatting window and the graphical window was tedious: integrating them on the same window would be helpful.

Text gesture: After finding the graphical gesturing helpful, our two formal participants proposed the idea of text gesturing. Similar to graphical gesturing, text gesturing would enable a designer to type directly on the drawing surface. The text would only stay on the surface temporarily. With text gesturing, participants can directly type something like this: "how about this printer station?" near the graphical objects that represent the printer station.

Gesture duration: Almost every participant who used the gesture object reported that the staying time of the gesture object is too short. They drew the same gesture object two or more times to make sure their partners understood their ideas. Some participants asked us to implement keyboard short-cuts for the commands. They had used single user CAD applications and they found keyboard short-cuts helpful: when you use one hand with the mouse, the other hand triggers commands.

5. CONCLUSION AND FUTURE WORK

Our goal has been to support collaborative design, so we designed NetDraw to imitate the functionality of a physical design meeting. But the experiments suggested something that we missed: Different design tasks or even different phases of the same design task impose different requirements for the software system. NetDraw can support group brainstorming to generate general ideas for a design task, but it fails to support generating design details.

There seems to be a gap between designers' ability to make judgements in computer supported drawing environments versus drawing in the physical world. In physical design meetings, although the whiteboard or tracing paper have no rulers, designers can somehow still "feel" the size of a specific object. But working on the computer screen with the scanned image, it seems difficult for designers to determine dimensions. Simply mimicking the functionality of a real world whiteboard doesn't work well.

The feature that was most clearly successful was the gesture object, which participants used often, and seemed to enjoy using. We intend to tune the time that the gesture object remains on the screen, and we are considering adding a textual gesture object as well, as called for by the two formal participants.

We also found that in as a short design session as we used in our evaluations, more complex features such as constraints were not explored. We are planning further, more focused evaluations, that examine the use of specific features for particular tasks.

ACKNOWLEDGEMENTS

This research was supported in part by the National Science Foundation under Grant No. IIS-98-19856. The views and findings contained in this material are those of the authors and do not necessarily reflect the views of the National Science Foundation. We thank Ray McCall for suggesting that we add the gesture feature to NetDraw, a feature that we had previously demonstrated in the Electronic Cocktail Napkin.

6. REFERENCES

Dix, A., 1997, "Challenges for Cooperative Work on the Web: An Analytical Approach", *Computer Supported Cooperative Work: The Journal of Collaborative Computing*, 135-156

Bentley R., Applelt W., Busbach V., Hinrichs E., et.al., 1997, "Basic Support for Cooperative Work on the World Wide Web", *International Journal of Human-Computer Studies*, 827-846

Bentley R., Horstmann T., Trevor J., 1997, "The World Wide Web as Enabling Technology for CSCW: The Case of BSCW", Computer Supported Cooperative Work: The Journal of Collaborative Computing, 1997, 111-134

Bly S.L., 1988, "A Use of Drawing Surfaces in Different Collaborative Settings", Proceedings of the Conference on Computer Supported Cooperative Work 1988, 250-256.

Brinck T., Gomez L.M., 1992, "A collaborative medium for the support of conversational props", Proceedings of the Conference on Computer-Supported Cooperative Work 1992, 171-178.

Brinck T., Hill R.D., 1993," Building Shared Graphical Editors Using the Abstraction-Link-View Architecture", Proceedings of the Third European Conference on Computer-Supported Cooperative Work 1993, 311-324

Greenberg S., Roseman M., Webster D., 1992, "Issues and Experiences Designing and Implementing Two Group Drawing Tools", Readings in Groupware, 609-620

Gross M.D., Do E., McCall R.J., 1997, "Collaboration and Coordination in Architectural Design: approaches to computer mediated team work", TeamCAD '97, 17-23

Harrison, S., 1993, "Computing and the Social Nature of Design." ACADIA Quarterly 12, no. 1, 10-18.

Ishii H., Kobayashi M., 1992, "ClearBoard: A Seamless Medium for Shared Drawing and Conversation with Eye Contact", Readings in Groupware, 829-836

Moran T.P., McCall K. et al., 1995, "Some design principles for sharing in TIVOLI, A whiteboard Meeting-Support Tool", in Greenberg, Hayne, and Rada, Groupware for Real-Time Drawing -- a designer's guide, McGraw-Hill, 24-36

Stefik M., Bobrow D., Foster G. et al., 1987, "WYSIWIS:Early Experiences with Multiuser Interfaces", Readings in Groupware, 585-595

Salcedo M.R., Decouchant D., 1997, "Structured Cooperative Authoring for the World Wide Web", Computer Supported Cooperative Work: The Journal of Collaborative Computing, 1997, 157-174

Vera, A., Kvan, T., et al., 1998, "Expertise, Collaboration, and Bandwidth", Proc. ACM Human Factors in Computing (CHI '98), 503-510.

Wojtowicz J., 1995, Virtual Design Studio, Hong Kong University Press

Wolf C.G., Rhyne J.R., 1992, "Communication and Information Retrieval with a Pen-based Meeting Support Tool", Proceedings of the Conference on Computer-Supported Cooperative Work 1992, 322-329.

The Computer's Role in Sketch Design: A Transparent Sketching Medium

Michael Trinder
The Martin Centre for Architectural and Urban Studies, University of Cambridge Department of Architecture

Key words: sketching, sketch design, user-interface, transparency, immersion, computer-aided design

Abstract: Starting from an analysis of the current unsuitability of computers for sketching, three key requirements are identified, in particular the notion that re-drawing or over-drawing are more important than editing and tweaking. These requirements are encapsulated in the broad concept of Transparency, understood both literally and metaphorically. Two experiments in implementing aspects of Transparency are described. One subverts the Macintosh window manager to provide windows with variable transparency, so that tracing between applications becomes a practical possibility. The other implements a graphical interface that requires no on-screen palettes or sliders to control it, allowing uninterrupted concentration on the design in hand. User tests show that the tool can be learnt quickly, is engaging to use, and most importantly, has character.

1. INTRODUCTION

Computers are being used by architects almost exclusively for the later stages of design: one UK practice has been quoted as using CAAD for upwards of 90% of their issued drawings while early presentation work and sketches are made with pencil and paper. Under the temporal, financial and aesthetic pressures of the design office any tool will be used only for tasks that designers perceive are within its capability. Whether their current perception is based in prejudice, preference or past experience, designers avoid using computers for sketching. The software currently available for sketch design is plainly not working.

Of course, the lack of useful software is not a hindrance to design. Buildings have been and are built without any need for information technology; design is famously as at home on an envelope, a cigarette packet or a cocktail napkin as it is on the drawing board or MacPaint. However, architects using CAAD to produce hard line drawings, but only pencil and paper to sketch, experience real problems when moving between these contexts. Electronic drawings must be printed out before they can be reworked. Alterations must then be copied tediously back into the computer. If the computer could support sketch design, the architect could work in either context, as they see fit.

This paper presents a small part of a PhD investigation into how to design interfaces that support sketch design. Our approach to the problem is to understand that sketch design and the computer as having some fundamental differences that need to be recognised before they can be overcome. We see the computer's role in sketch design not as some design companion but as presenting a fluid and engrossing sketching medium, stage managing the transition of drawings from the back-of-the-envelope to hard line drawings, renderings and models and back again.

2. RELATED WORK

Existing work that is relevant to the transparent medium and sketching tool presented here can be divided into two broad categories: projects concerning issues of transparency and those investigating sketching. The amount of work dedicated to the latter far outweighs the former, and much sketching/painting research has little relevance to the proposals for a transparent interface. However, a few projects which tackle the sketching discipline have used aspects of transparency, and it is on these that we concentrate.

2.1 Sketch design software

The Cocktail Napkin [Gross 1994, 1996] concentrates on gesture recognition to interpret sketches into structured diagrams. The marks made cannot be separated or recombined in new ways, nor do they lend themselves to re-interpretation or ambiguity. PerSketch [Saund, Moran 1994] attempts to rectify this shortcoming by giving the user 'access to visual objects as they are perceived by the human visual system.' This is achieved by retracing existing objects to work with emergent forms.

Neither the Napkin nor PerSketch fit our approach to sketching. Piranesi [Schofield 1994], [Schofield, Richens 1995] comes closer in an attempt to extract the rendering process from the parametric, trial and error realm and allow selective, interactive rendering by using geometry derived from a 3D image's Z buffer.

Finally, a realised building project, the Groningen Museum by Coop Himmelblau makes an interesting example of what architects *expect* from sketch design software. [Rattenbury 1995] describes the process: 'an original cryptogram, overlaid three times, was converted into a model and then scanned straight into a computer using sophisticated software and transformed into plans "since we didn't care too much about the functions of the space." The computer in Coop Himmelblau's office was then linked to a computer in Groningen's redundant dockyards where the gallery's huge steel panel components were welded together by the shipbuilders and then transported to the site for assembly.'

Coop Himmelblau may be interested in an automatic building solely for rhetorical reasons, but they see computers as being able to cope with the shifts of interpretation necessary to design buildings through drawings. Most of all this demonstrates that it is possible, with some effort, to produce architecture within the digital medium.

2.2 Uses of Transparency

[Colby & Scholl 1991] provide some important groundwork, showing how transparency and blur can be used to prioritise information within a complex display and relating scales of each to visual impact. [Staples 1993] examines the 'desktop' GUI as an artistic artifact, drawing analogies between the use of perspective and transparency in painting and their application to interface design, noting that the latter confers supernatural and special status on objects in addition to allowing multiple images to overlay. Although this is a purely theoretical paper, it does contain a mocked up image of a transparent window which comes closest to the medium we describe here.

PAD [Perlin, Fox 1993] and The Macroscope [Lieberman 1994] use transparency to allow 'semantic zooming'. This involves slowly replacing abstract information with more detailed versions as the user zooms in closer.

[Kamba *et al* 1996] use transparency to allow multiple interface widgets to occupy the same screen space so as to free up the small displays of most hand-held PDA devices. Unfortunately for their experiment, they imagine their widgets as being above the information on display, yet render them as if they were below it. What they show is that transparency does not work when the user's perception of depth differs from that maintained by the interface. [Harrison *et al* 1995a, 1995b] and [Harrison, Vicente 1996] examine the user of transparent interface widgets in extensive detail in order to determine the optimum levels of transparency for these elements. Their intention is to prevent dialog boxes, menus and icons from taking up valuable screen space in graphic intensive applications such as 3D design rather than to exploit overlaid images.

[Bier *et al* 1993,1994] present a compelling use for such transparent interface widgets. Their Toolglass has been the foundation for a large number of papers which form the most extended investigation into Transparency undertaken so far, passing images up and user interaction down through a stack of toolglasses to allow the user to construct an interface out of the overlapping elements. Importantly, the documents that the user is working on are always shown as opaque. The transparent medium that we present here makes *every* window transparent, removing the distinction between interface tools and documents.

[Carter 1993] and [Wellner 1993] employ a combined overhead projection system and camera to set up a visual feedback loop similar to [Kreuger 1991]'s installation work. Their DigitalDesk system overlays a digital desktop onto a real one, allowing hand drawn images to be instantly digitised and copied or used as the basis of digital transformations. DigitalDesk is the most literal, and successful, expression of the transparent computer.

Finally, [Kramer 1994, 1996] implements a sketching environment from a stack of freeform transparent sheets of drawings and, as such, provides the closest parallel to our work on a transparent medium.

3. SKETCHING AND SKETCH DESIGN

3.1 Sketching Theories

Studying sketching is fraught with potential confusions. The biggest problem is linguistic - the words *sketch* and *drawing* refer both to the act as well as its end product, the act of sketching is seen to be different from the image that it produces. To help distinguish these, *sketch* will be used to refer to drawings that are being made and *image* to the final result. The image is made up of *marks* and sketching can be seen as the act of *mark making*. Quite why these distinctions are important to sketch-design software becomes clear after digging a little deeper into the nature of sketching.

Today design is inexorably entwined with some sort of drawing. Whether this is because drawing is a natural design medium, a historical accident or even a device to exclude the layman is not really an issue. What is more important is that there has been a tendency for theories of sketch design to take this relationship as read and concentrate solely on sketching as an act - looking at the marks made - and the images produced. Whilst the *inter*relationship between sketching and design is not in question, failing to assess this relationship leads to serious confusions in understanding just what is so important about sketching.

It is usually understood that the image *is* the design: perspectives, for example, are presented to both clients and planners as objective tests of a design's validity. [Pérez-Gómez 1983] shows how this relationship between architectural drawing and design developed, in the mid seventeenth century, from a functional view of reality. The proposal that the world can be represented in an objective, self consistent manner developed at this time underlies our contemporary appreciation of visual truth: images are to be judged by how well they describe the world. [Coyne 1995] outlines the two most prevalent explanations of how images are able to encode reality, the Correspondence and the Constructive theories, before going on to propose a theory based in Phenomenology.

Phenomenology sees everyday experience and action as activities of complete immersion, of being thoroughly engaged in doing. The intellectual stances of subject and object represent a breakdown of this engagement. Applying this theory to architectural representation, [Veseley 1993] concludes that it is meaningless to search for information in sketches as the images only reveal design rather than encompass it. Architects can be surprised by their sketches and find new ideas in them, but this is not because the idea is actually *in* the image. Rather, the act of drawing helps to reveal the idea, revelation that is not possible *without* the image but entirely separate from it. They are reacting less to what they see than what they have thought about whilst doing.

3.2 Computer-based sketching

Drawing is a pre-reflective, immersive state that is culturally based and culturally biased. As [Veseley 1993] explains, any drawing act is made as part of the established cultural context that allows us to create, understand, share and experience space without being able to represent it in its entirety. However, if the sketching experience is one of undifferentiated involvement, then any intervention into this, occasions when *things* suddenly intrude, times of breakdown, will halt the drawing prematurely. The decision to treat a drawing objectively must only be that of the sketcher and not one imposed by a data structure unable to support drawing or an interface that gets in the way. Sketching depends on lack of constraint and on fluidity.

If the phenomenological viewpoint is correct, the computer must be interfering with the pre-reflective immersive state that the sketching act requires. Although the computer tries to present an electronic version of your desktop, this desktop is subtly different in a number of important ways:

- Every action is limited in some way. There are very few fundamental limits, things that the computer cannot do at all, but every application imposes restrictions on what the user can accomplish.
- A computer application is an encapsulation of its programmer's understanding of a task. It is not possible to rotate text in Microsoft Word, you cannot blur a drawing in a CAD system, you cannot draw a continuous diagonal line in Photoshop. Whilst object-oriented software and integrated 'office' packages attempt to get around the problem it is important to recognise that the limitations still exist.
- Whilst the modern GUI has helped users learn how to use applications more quickly by presenting their interfaces in the same way, this visual similarity unfortunately implies that the underlying applications function in the same way.
- Furthermore the interfaces get in the way, eating into the limited screen area available to the user. Even when most of the interface is placed into floating palettes a drawing can be almost completely obscured.
- Electronic documents are infinitely mutable – there is no record of a drawing's evolution beyond what is explicitly recorded by the computer or user. Whilst this is a definite advantage for drafting, drawings are no longer definitive records.
- Most CAD applications use a selection and manipulation tool (usually represented by an arrow) as the default tool. After drawing an object, the current tool automatically switches to selection. This promotes editing and manipulation of a drawing over creating new marks – the user is encouraged to switch from drawing to analysis of the drawing. Such applications naturally interrupt any immersive state and place emphasis on the product not the act.
- The tendency to slip into continuous editing can also seriously compromise the quality of work carried out electronically: an investigation by [Penz & Wright 1994] to find differences between paper and digital design shows that design students evaluated fewer alternatives at a basic level when computers were involved. This tendency, termed the 'Icon Trap' in [Lawson 1990], by which a single image or set of images come to *be* the design, blinds the designer to any investigation other than that dealt with in these drawings. [Jansson, Smith 1991] and [Purcell, Gero 1996] make specific studies of this 'design fixation' and show how images given to test subjects can shape the way they subsequently design, even reproducing faults inherent in the examples. The cure lies in first realising the iconic status of a drawing (usually an early concept statement) and setting it aside for new work to start. This is less easy to do in a digital medium that is seamlessly

mutable and pitched toward editing existing documents rather than re-creating them.

- Computer displays have an extremely low resolution when compared to that of paper and pencil.
- The mouse is a terrible drawing instrument, closer to a brick than a pencil. It was designed purely to give two dimensional positional information - for *pointing*. Its design restricts movement of the hand and arm to allow only this pointing movement. Its standard shape forces the hand and arm into a pointing gesture with forefingers extended to press buttons and the wrist locked in position, a straight line running from elbow to finger tip. It is at best extremely difficult to move a tool fluidly when in this position, let alone use of the fingers, wrist and elbow to vary the orientation and pressure with which that tool is used. [Zhai, Milgram, Buxton 1996] show that devices that take input only from the arm or hand instead of the fingers are inherently less flexible and accurate.
- Finally, modern graphic interfaces use opaque windows. Architects don't use opaque media very often – they use translucent and transparent media such as tracing paper and polyester film to copy, elaborate, annotate and evolve drawings. The only opaque drawings usually to be found in an architectural practice are mechanical reproductions.

3.3 The Computer's Role: Transparency

We propose that the concept of *transparency* has significant potential for the development of creative electronic media. The term transparency implies much more than a purely optical effect, extending to ideas of ubiquity, unobtrusiveness and interchangeability. It has been an ideal of the Modernist movement in architecture, only truly realisable with recent innovations in glass technology. Taking the broader meaning, *transparency* can be used to investigate four separate domains:

Computer Transparency
The degree to which the physical form of a computer's input and output devices promotes intuitive interaction.

Interface Transparency
The degree to which the user interface promotes the experience of immersion.

Data Transparency
The degree to which user actions are freed from the limitations imposed by a particular data structure.

Media Transparency
The degree to which conceptually separate groups of data can be combined visually.

4. A TRANSPARENT MEDIUM

With user experience of windows being based entirely on a stacking illusion, it would seem to be a trivial extension of the desktop metaphor to let any covered images retain their presence. Turning the standard Macintosh document window into a transparent medium introduces transparency within an easily understood context, adding new functionality to an existing interface element rather than re-inventing the entire interface.

4.1 Simulating Transparency

There are a wide variety of transparent and translucent media in use by architects although these are usually referred to by only three names: drafting film ('Film'), tracing paper ('Trace') and, more rarely, detail/layout paper. Superficially these papers all appear very different but, in use, significant characteristics become clear that are much more consistent. Transparent media are used not only to bring images together as one, but also to place images in some sort of hierarchical relationship: the partial transmission of the image diminishes its visual significance letting the upper drawing read as the new, revised or more important version. Degrading images is important to the sketching process, letting earlier images fade in order to create new ones, as is the ability to distinguish between drawings so that, when simply copying from one sheet to another, it is possible to determine what is still left to transfer. These media degrade and separate images by reducing their contrast, altering their colour balance and by subtly blurring them.

Creating a window that has similar properties turns out to be more complex than it would at first seem. All computer simulations of transparency are generated by some sort of blending function that produces a single output, the colour to be displayed, from a pair of inputs, the images to be overlaid. At its simplest level, pixels are taken directly from one or other of the images according to a binary mask. This is exactly the technique adopted by [Perlin, Fox 1993] for their PAD interface. In this way images with non rectangular boundaries or containing 'holes' may be displayed by only transferring the relevant pixels. Extending this mask from a single bit per pixel to, generally, eight bits per pixel allows a mixture of colours within a pixel to be expressed. This extra information, the alpha channel, is usually conceptualised as the opacity of the upper or foreground image.

[Colby and Scholl 1991] stated that there is little perceptual difference between a physical transparency and a rendering in which one image can be seen to continue through another, in which the two images are blended by linear interpolation. There are however situations in which the results of alpha compositing are wholly different from that experienced using physical transparent media. The standard linear blend function is symmetrical; the designation of foreground and background is only meaningful in that the value α is defined in terms of the opacity of the 'foreground' image. To deliberately associate this value with a particular image, the notation $f\alpha$ will be used to represent the alpha value associated with the foreground image **F**. This foreground image will always be the uppermost of the two.

The standard linear blend formula shows how the output pixels are actually calculated:

$$\mathbf{I} = \mathbf{F}.f\alpha + \mathbf{B}(1 - f\alpha)$$

where **F** = Foreground pixel, **B** = Background pixel, fα = Foreground alpha value and **I** = Resultant pixel for display

The problem with this standard blend is that the result of compositing two images with fα =1/4 is identical to that of compositing them with fα =3/4 with the notional stacking order is reversed (fig 1). Put another way, the stacking order we perceive may be different from that maintained by the computer. As [Metelli 1974] shows, we will read the stronger image as the foreground.

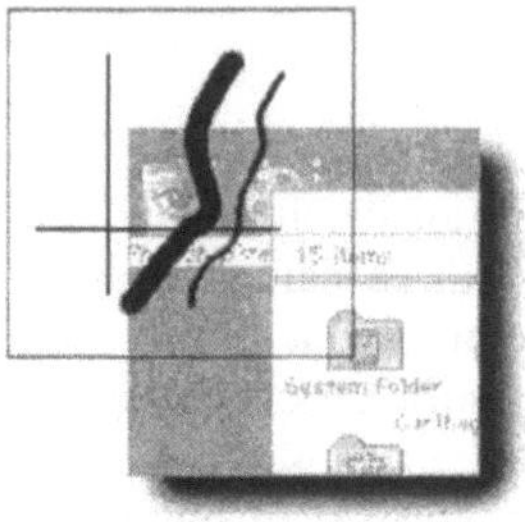

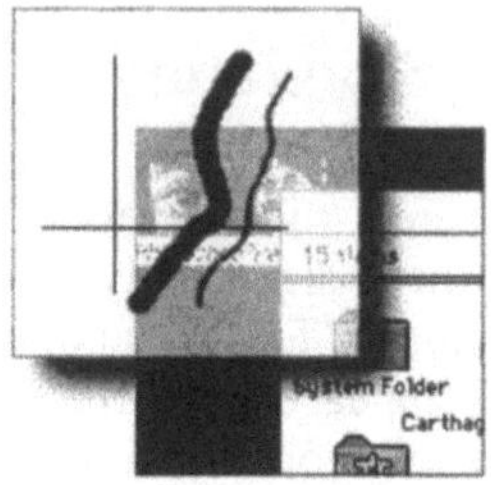

Figure 1. fα =1/4 fα =3/4

Instead, we must use two successive blends. The first reproduces the effect of the medium, toning down and obscuring the images below based on one alpha value, mα. Altering mα changes how much of the background image is visible behind the foreground, whose transparency remains constant. The second blend introduces the marks made upon the medium by blending the composite background image with the image the application wishes to display.

A series test blends were carried out using a typical piece of the Macintosh desktop and the foreground marks shown in fig 1. The first experiments simply mixed the background image with solid white (**W**) or a colour (**M**) according to mα using pattern dithers or alpha blending. Of the four experiments, the coloured alpha blended images were the most effective (fig 2), but all four suffered from the same flaw. At 100% transparency, when mα is zero, the background image is passed through untouched making it very difficult to distinguish. The perceived stacking order reverts to being based on the relative strength of the images. It is impossible to determine the actual stacking order of the images without trying to move one of them. Whilst the user may well wish to combine images perfectly in this manner at some stages in their working process, this tendency toward absolute transparency is an unfortunate one for a proposed sketch design environment - old drawings would continue to have the same impact as new ones.

The solution adopted was to reverse the behaviour of the medium's colour. Rather than have the medium gain colour as it becomes more opaque, it instead starts out tinted and slowly fades to an opaque white sheet. Fading the colour of the medium as transparency drops may not reflect any physical experience of transparent media but the result is that the opaque composite is indistinguishable from a standard opaque window whilst increasing transparency lets in a colour to help separate foreground and background (fig 3). As the sheet becomes more transparent its white surface fades away to reveal a tinted image beneath.

The final blend adopted went one stage further in degrading the underlying image by first passing it though a simple blur filter. Although only a minimal amount of blurring was used, it immediately separates the two images, the contents of the window and the images that can be seen though it, into foreground and background. Thin marks in the foreground can be clearly distinguished as they float slightly in front of their backdrop.

In the final interface the medium's transparency is altered by a slider in the corner of each window. Moving the slider pops up a small preview frame to give real-time feedback of the current transparency setting (fig 4).

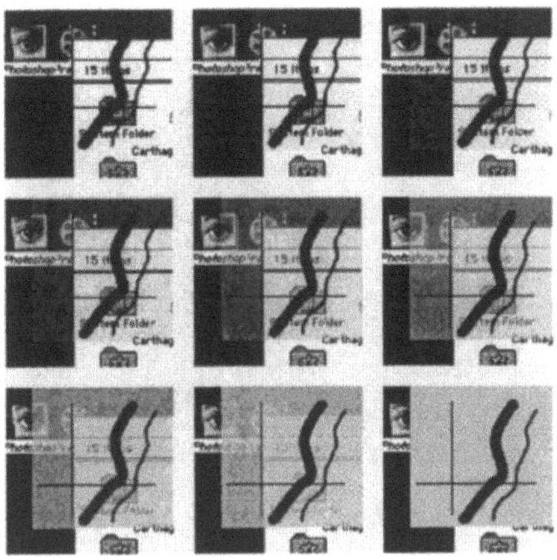

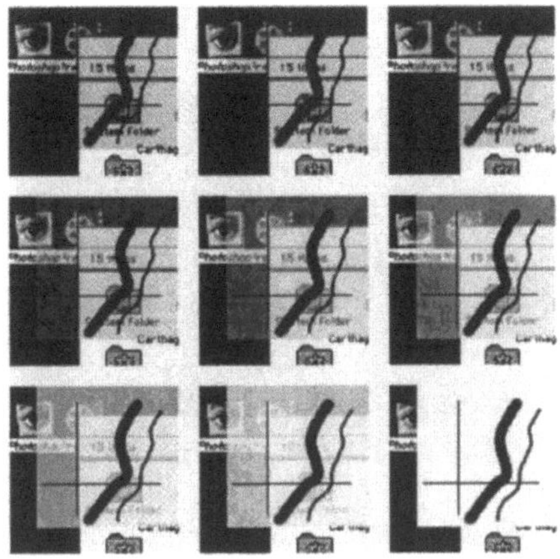

Figure 2 left. two successive alpha blends: $\mathbf{I} = \mathbf{F}.f\alpha + (1- f\alpha).(\mathbf{M}.m\alpha + (1-m\alpha).\mathbf{B})$
Figure 3 right. Reversed colour blend with blur: $\mathbf{I} = \mathbf{F}.f\alpha + (1- f\alpha).(\mathbf{W}.m\alpha + (1-m\alpha).f(\mathbf{B}).\mathbf{M})$
($\mathbf{W}$ = (1,1,1) RGB, $\mathbf{M}$ = medium colour, $f()$ is a simple blur convolution kernel)

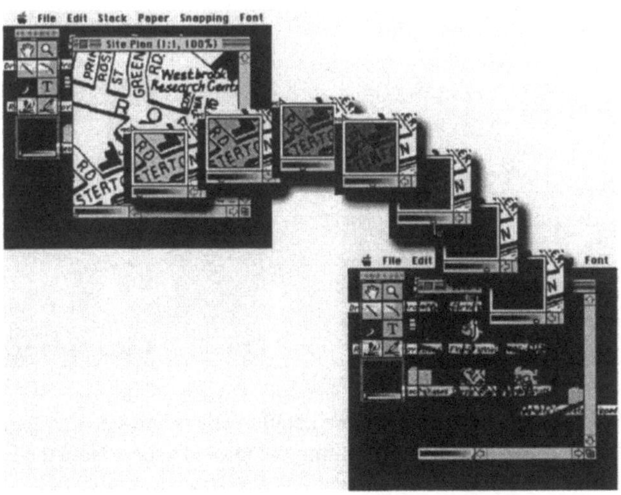

Figure 4. Altering a window's transparency in real time using a preview frame.

4.2 Using Transparent Windows

Even at their most basic level, simply overlaying one image onto another, transparent windows allow the user to work in novel ways that reflect the way physical drawings are used. An initial sketch can be overlaid, drawn again and again until the idea becomes clearer. Significantly, as more windows are overlaid, the original image is increasingly obscured as it fades, blurs and is covered by new marks.

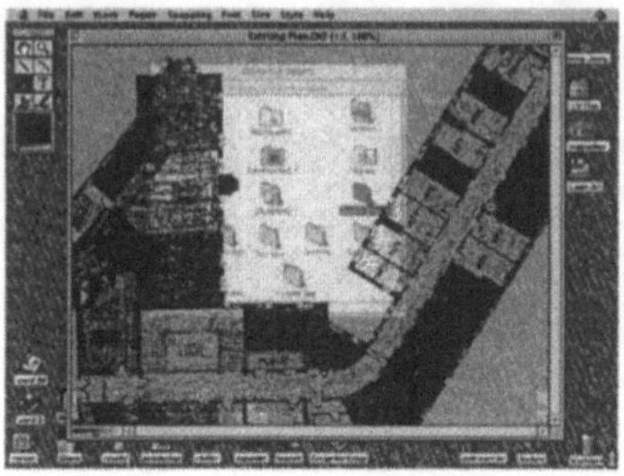

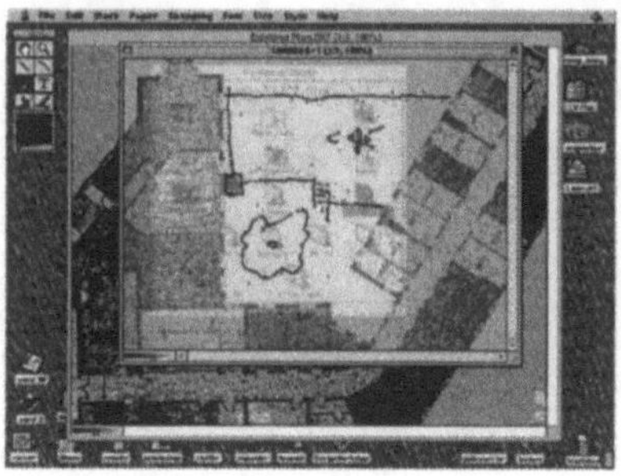

Figure 5 left. A CAAD plan with the Finder beyond
Figure 6 right. The plan overlaid by a new sketch of a proposed cafeteria

Combining the windows of different applications takes the idea much further (figs 5-9). A CAAD drawing can be sketched over to rework a detail, or a hard line drawing can be developed using a sketch as a guide. Text can be appropriated from a word processor (fig 9), images directly from scanning software or even a video monitor.

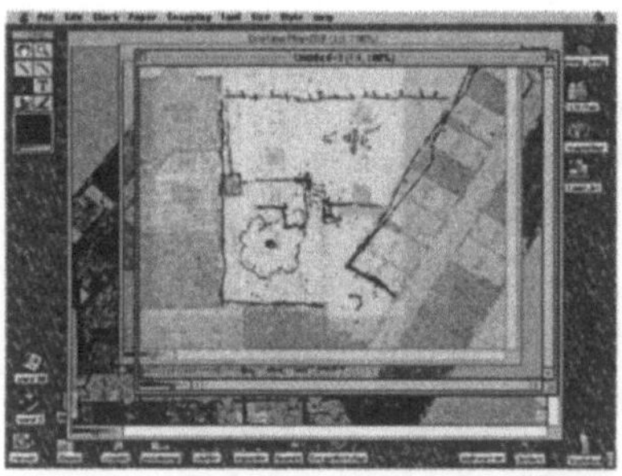

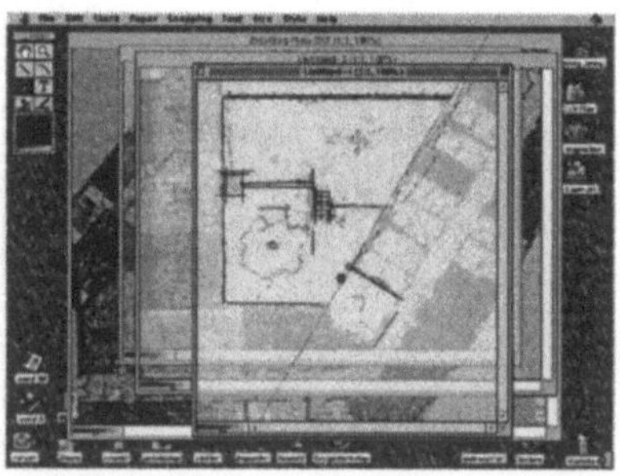

Figure 7 left. Another sketch with more detail overlays and starts to obscure the original
Figure 8 right. A two dimensional drafting window is added. Note the full screen snapping lines.

Simply overlaying windows in this fashion requires no alterations to existing applications as long as they do not draw directly to the screen. The transparent window exploits the fact that all applications must present their data as a pixel map in order to exist within a GUI. A custom Window Manager redirects application's display into an off-screen cache from which produces a composite. This leaves the user to interpret the images as they see fit – something the human brain is extraordinarily accomplished at doing.

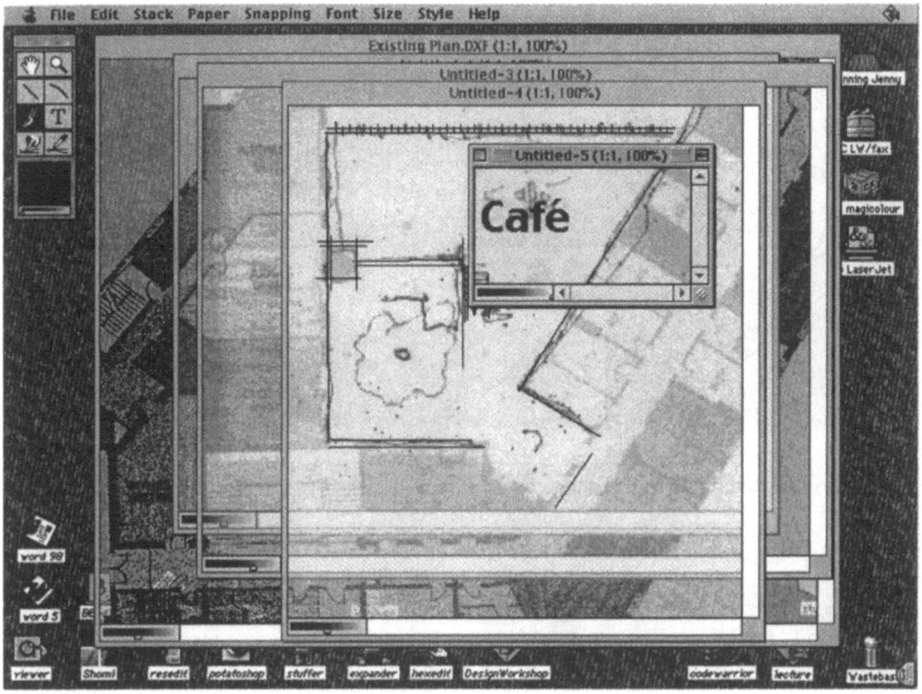

Figure 9. A text window is positioned to add an annotation.

With some simple additions to an application, the transparent medium can become much more than way of compositing images. As long as an application responds to requests from the window manager for the nearest geometrically significant point to a given co-ordinate, geometric snapping can take place between windows maintaining any data structure. Passing two-dimension geometry between windows is achieved by using the display as a standard co-ordinate system. Since snapping is no longer limited to a single window, construction lines must be drawn across the whole screen. Finding the intersection between vectors in two different windows within the co-ordinate space of a third is perfectly possible (fig 10).

Two special tools, designated *Push* and *Pull*, allow pixel map images to be pushed between windows that will accept pixel data. Most obviously, this allows the composite image produced by a stack of windows to be recorded by pulling the various images up into a new window that covers them all. A perspective rendering can be populated by opening stock images of people and rubbing over them with the push tool, as if they were dry transfers. Intermediate windows can be used as masks to limit the transfer to specific areas (fig 11). This is just not possible with opaque windows. The translucent windows allow the user to simultaneously see the source and destination images in alignment and carefully control the process.

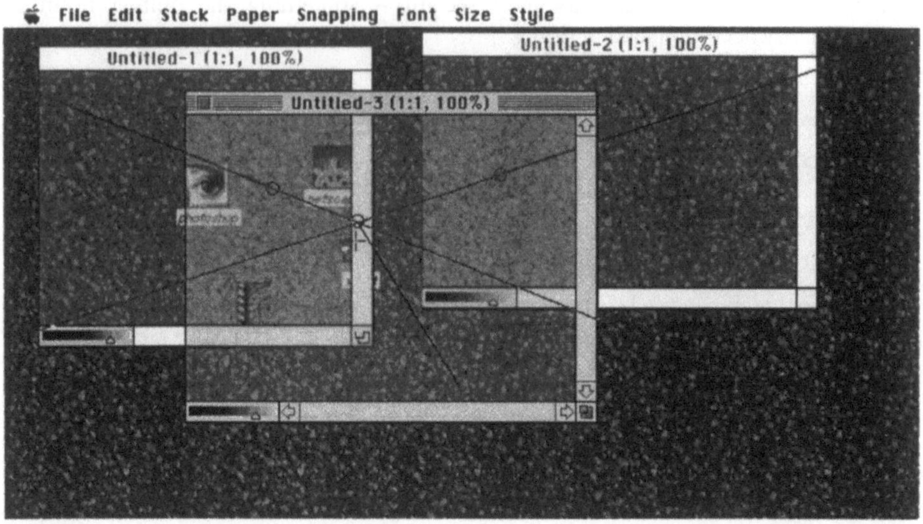

Figure 10. Snapping to the intersection of two discrete lines in a third co-ordinate space.

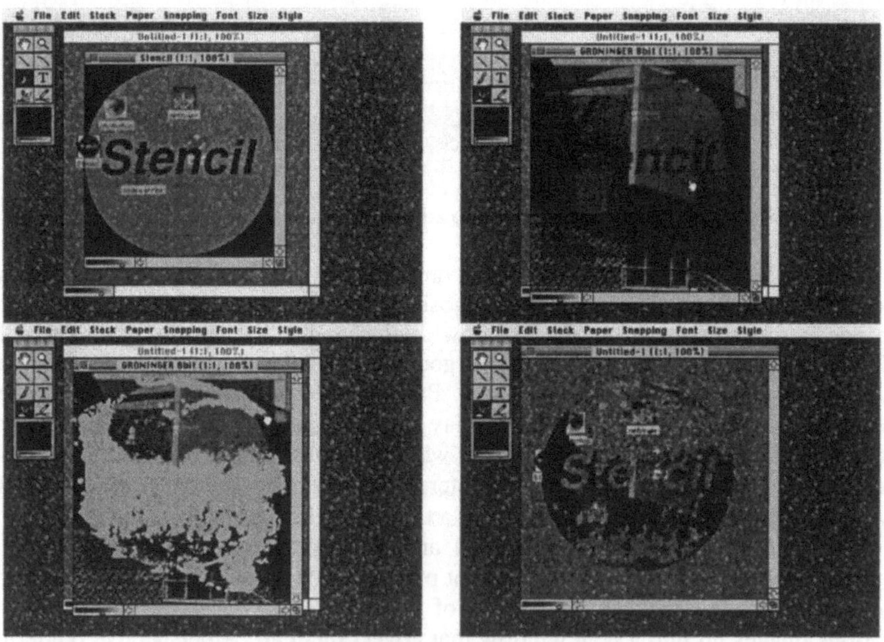

Figure 11. Transferring an image through a stencil: *Top left* The stencil in place over the receiving window. *Top right* The source image placed ready for transfer. *Bottom left* Transfer in progress. *Bottom right* The results in the lowest window

Finally, using a live video feed as a source gives the user a constantly changing palette of images and colours with which to paint, requiring skill and timing to use successfully. A novel creative medium can be built by the user by combining simple applications whose hybrid functionality lies outside the scope of any of those applications. The example image (fig 14) was created from live television broadcasts, moving the video window to an appropriate position and then rubbing through the image before it disappeared. The unpredictable and uncontrollable nature of the source images mean that concentration is essential and that continuous drawing and redrawing is the only possible strategy.

5. A TRANSPARENT SKETCHING TOOL

Having already concluded that the mouse has, frankly, all the drawing ability of a brick, we chose a pressure-sensitive drawing tablet as the minimum level of hardware necessary for sketching. Not only does this afford a greater level of Computer Transparency by presenting the user with a familiar drawing instrument and a more natural hand position, the pressure-sensitive tablet offers an extremely rich stream of input. Table 1 shows how extensive this input can be.

Table 1. Stroke characteristics available with different input devices

Mouse	Tablet	Wacom UltraPad
X	X	X
Y	Y	Y
Button	Tip	Tip
	Pressure	Side button
		Pressure
		Plan orientation
		Tilt
		Tool used
Velocity	Velocity	Velocity
Acceleration	Acceleration	Acceleration
Bearing	Bearing	Bearing
Curvature	Curvature	Curvature
	Rate of pressure change	Rate of pressure change

Table 2. The variety of marks made by the sketching tool:
left using just pressure, *right* using pressure, velocity, bearing and acceleration.

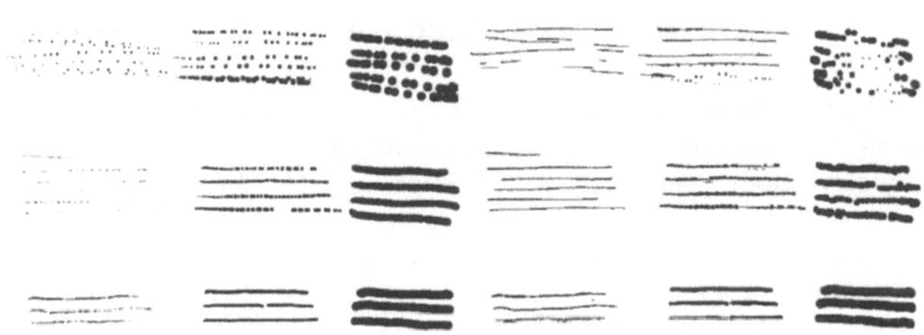

	low	medium	high	low	medium	high
High						
Medium						
Low Velocity						
			pressure			

This sketching tool is a deliberate experiment in Interface Transparency. The tool is designed to be entirely controlled by the manner in which the stylus is used, making clean and 'dirty' lines, brush marks, hatches, area fills and splattery marks without any need for the user to select options or adjust sliders in a tool palette. The traditional palette of brush shapes and characters can be just as adjustable but the characteristics must always be adjusted before making a mark. The user must predict the marks they wish to make and use the stylus sometimes to draw, sometimes to select options. The sketching tool can be constantly adjusted and remain a drawing tool at all times.

The marking algorithm uses both a circular brush mark and a line drawn directly between sample points. The brush is scaled proportionally to pressure and inversely to velocity, so as to make it possible to hatch. Hatching involves repeated high velocity strokes using a thin line and an arcing motion of the hand – the velocity and hand motion combine to make fine pressure control more difficult. Reducing the brush size at speed compensates for this.

The brush is offset from the sample points according to acceleration: the acceleration value tends to be relatively constant during a stroke, but rapidly reaches extremes at the start and end of the stroke and at sharp changes in direction. Perturbing brush marks with acceleration lends rapid, sloppy strokes a noisier appearance than carefully controlled strokes and also produces blots and blotches at the ends of lines if the user stops dead instead of 'following through'.

Connecting sampled points by a straight line not only copes with widely spaced samples but also has the useful effect of separating the mark into two parts: line and shading. At low pressures, when the added effect of velocity often scales brush sizes down to a single pixel, the tool generates curved lines that faithfully follow the stylus. As velocity increases, the sample points become more widely spaced, the level of approximation increases and the lines straighten out. Increased pressure in both cases thickens the line into a brush stroke and the tool becomes a shading rather than drawing device. In fact, at high pressures connecting sample points by a line detracts from the tool's shading. Fast, high pressure strokes result in occasional brush marks strung along a thin line. To rectify this unfortunate situation, the line is not drawn above a 50% pressure level.

6. USER FEEDBACK

6.1 The Transparent Medium

The transparent medium has been demonstrated to two broad groups of people during its development: programmers involved in interface design and professional architects. Both groups suggested improvements, but there is an appreciable difference between the comments made by the two groups. When the software was demonstrated to those involved in programming and interface design, the general reaction was an appreciation of the visual effect and the implications this had for the interface, but few could see any use for it. It looked nice, but the main question was simply 'Why?' When design professionals were shown the same software their response was much more immediate: the overwhelming question was 'When?'

This difference of opinion is not just a reflection of different attitudes to the computer. Of course a programmer would question the need for such a complex solution to an apparently trivial problem and of course an architect would empathise with something deliberately based on their working process. What is important is that designers could immediately see *how* to work with transparent windows. They recognised familiar approaches to design that their own computers would not support. Whilst this is reassuring and supports the theory that transparency will have positive benefits for sketch design, it also indicates that these techniques are not universally applicable. Some activities do not become any easier simply because the window is translucent. The computer programmer has little use for overlaid information or visual ambiguity. The large processing and storage overhead the translucent window entails is hard to justify in these circumstances.

6.2 The Sketching Tool

The proposed sketching tool has potential flaws. Its design was based on a theoretically strong but unproved assessment of the marks made during sketching. For it to be used with confidence and easily learnt, it should react in a controllable and predictable manner to the sort of movements made by the hand whilst sketching. If the user cannot control the marks made, does not discover how to make certain marks or creates marks that they had not intended then the sketching tool will not provide an immersive sketching experience. Its success depends on exactly the same criterion as the transparent medium – is it useful?

Twelve test subjects were taken from a cross section of the architectural profession, from first year students through to professionals with several years of post qualification experience. They were presented with a sketch executed with the prototype tool (fig 12) and asked to practice using the tool until they felt confident to make a copy of the sketch. Drawings were recorded as pixel map images as well as raw streams of time-stamped data from the tablet, allowing any session to be replayed in real time. Subjects' comments were recorded on tape.

Three measures were derived from the data – the time taken before the subject 'felt confident' with the tool (taken as being the time before an attempt was made to copy the sketch), the amount of control the subject had over the stylus (measured by looking for distinct groupings of pressure and velocity) and the fluency of the reproduction (how many types of mark they discovered). Of these, the time measurements show the clearest results (fig 13). The subjects fall into two surprisingly distinct groups when their contact times – the time spent with the stylus in contact with the tablet – are considered. Most went straight into copying the test image after only cursory experiments with the pen, whilst four spent nearly 2.5 minutes trying out different marks and lines. The faster group continued to make discoveries about the pen whilst drawing the final image: subject 4 for example only discovered the relation of speed to line weight right at the end of the test. The slower group produced a wide range of marks during their practice. The only obvious correlation between type of subject and the contact time is that all the students fall into the faster group.

Although this was an informal pilot study with too few subjects to allow statistical analysis, there are some interesting observations to be made from the way the subjects drew and the way they talked about these drawings. In fact recording their comments whilst drawing proves more useful than any of the analytic methods.

The overwhelming topic of conversation was the sensitivity of the pen to pressure, making marks too thick, and to acceleration, causing large splatters. The version of the tool used for the tests was deliberately extremely sensitive when used at low velocities - part of the experiment was to see if users would discover that line weight became more controllable with increased velocity. Unfortunately, because the test subjects were unfamiliar with the drawing tablet - a topic raised by four of them whilst drawing - and because they were copying the test drawing very carefully, they tended to draw at low speeds. When the user concentrates on getting one aspect of a mark correct - as these test subjects were - he or she is doing so to the detriment of the other variables. Furthermore, where a small amount of acceleration noise textured a line the subjects were pleased with it, but where there was too much splatter or where they couldn't get rid of the splatter they were less happy. Only a few discovered that drawing faster gave them more control over the lines. Whether this is because it is not intuitive or because of the care with which they were copying the sketch is not possible to say. Those that did work out the velocity control tended to say so out loud at the moment of discovery - 'Aha, that's what you want, it's thinner if it is quicker' or '...actually...the faster you move it, the lighter the lines you make with it'.

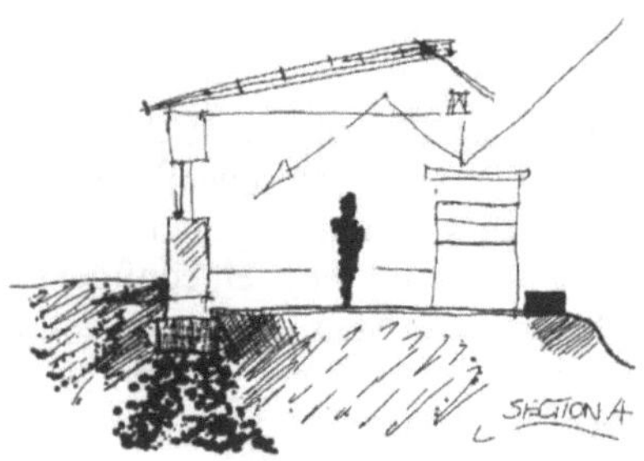

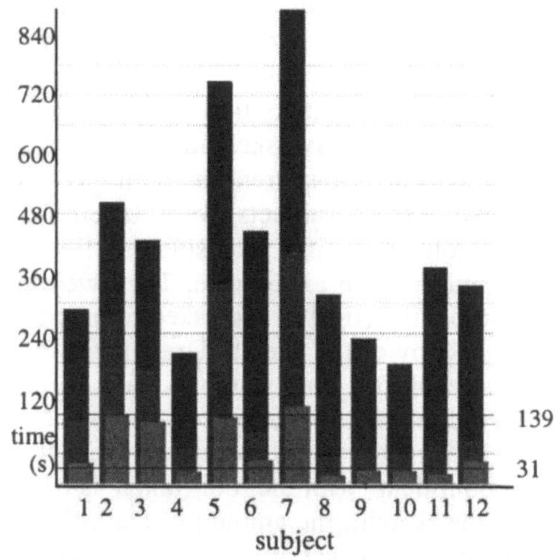

Figure 12 left. The sketch to be copied.
Figure 13 right. Total time (black), learning (grey) and actual contact times (offset).

Finally, some subjects had real problems orienting their strokes, coping with the dislocation between the stylus tip and the marks on screen and with the low resolution of their sketches. These are all problems of Computer Transparency that the sketching tool tested does not address.

The way subjects talked about their drawing proves to be almost as interesting as what they actually said. Whilst drawing, they would speak in incomplete sentences, only expressing fully formed ideas when they stopped to discuss something. These moments are excellent indicators of 'breakdown'. When subjects talked about the experience, there was a definite tendency to address their complaints to the stylus itself rather than the computer (or the monitor). There were also a number of comparisons between the pen and physical drawing tools - analogies to typewriters, Rotring pens, Sumi brushes, Croquil pens and Gerald Scarfe were made. They were seeing the *stylus*, and not the software, as having these characteristics – it felt to them like a real tool with definite character.

That the subjects were able to produce engaging sketches whilst noting that the pen was sometimes 'out of control' indicates the demand that designers place on their drawing tools. Figure 15 shows the quality of image that some were able to produce. The test tool was extremely sensitive to acceleration, causing splatter, and to pressure at low velocities, causing blotches on lines. Whilst the latter is a genuine problem that needs to be addressed in future versions of the tool, the former seems better adjusted as a 'preference' by an interface widget, a slider perhaps. The degree of splatter is not a variable that needs constant adjustment, reflecting as it does a personal preference rather than the emotional content of a particular stroke. Adjusting the overall level of splatter with a slider shouldn't present the same sort of interruption to drawing that adjusting line weight by a similar mechanism causes.

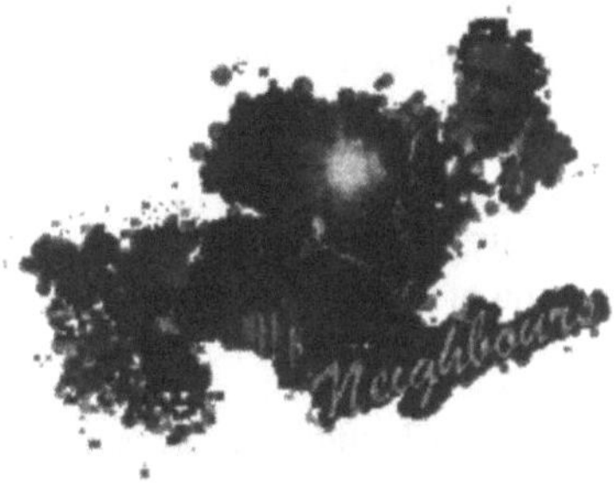

Figure 14 left. A composite image created from live television broadcasts.

Figure 15 right. Fern leaves drawn with the sketching tool after 15 minutes practice by a landscape architect.

7. CONCLUSIONS

The transparent medium opens up the computer interface to ambiguity. Each application sees no change to its environment, and yet the user sees a completely new way to work with the computer. The user can bring together applications with different data models and effectively fuse their capabilities into a new working environment. An application can be exploited, for purposes beyond those defined by its authors, as part of an interface constructed by the user. The sketching tool presented here is just one potential tool for use within this interface.

For a computer to support sketch design it must allow the user to freely combine images and other data, it should promote creation of new images over the editing of existing drawings and it should not interrupt the user whilst she is sketching or otherwise force her to make a decision about the content of the drawing. The transparent medium describes not only the sketching interface, but also the computer's role in the interaction. The computer should literally be a transparent medium.

This research was funded by the Engineering and Physical Sciences Research Council and supervised by Paul Richens. It forms part of a PhD entitled 'The Potentials of Transparency in Sketch Design Interfaces' submitted to the University of Cambridge in July 1998.

REFERENCES

Bier E, Stone M, Pier K, Buxton W, DeRose T, 1993, "Toolglass and Magic Lenses: The See-Through Interface", *Computer Graphics (Proc. SIGGRAPH)* August 1993 73-80

Bier E, Stone M, Fishkin K, Buxton W, Baudel T, 1994, "A Taxonomy of See-Through Tools", *Proc. CHI 94*, Boston, MA 358-364

Carter K, 1993, "Computer Aided Design: Back to the Drawing Board", *Proc. Creativity and Cognition 93*, Loughborough

Colby G, Scholl L, 1991, "Transparency and Blur as Selective Cues for Complex Visual Information", *SPIE Image Handling and Reproduction Systems*, vol 1460 114-118

Coyne R, 1995, *Designing Information Technology in the Postmodern Age* (MIT Press, Cambridge, MA)

Gross, M D, 1994 "The Fat Pencil, the Cocktail Napkin, and the Slide Library", *ACADIA '94*, Association of Computer Aided Design in Architecture 103-113

Gross, M D, 1996 "The Electronic Cocktail Napkin - a computational environment for working with design diagrams", *Design Studies*, 17 (1) 53-69

Harrison B, Ishii H, Vicente K, Buxton W, 1995a, "Transparent Layer User Interfaces, An Evaluation of a Display Design to Enhance Focused and Divided Attention", *Proc. CHI 95*, Denver, CO 317-324

Harrison B, Kurtenbach G, Vicente K, 1995b, "An Experimental Evaluation of Transparent User Interface Tools and Information Content" *Proc. UIST 95*, Pittsburgh, PA 81-90

Harrison B, Vicente K, 1996, "An Experimental Evaluation of Transparent Menu Usage", *Proc. CHI 96*, Vancouver 391-398

Jansson D, Smith S, 1991, "Design Fixation", *Design Studies*, 12 (1) 3-11

Kamba T, Elson S, Harpold T, Stamper T, Sukaviriya P, 1996, "Using Small Screen Space more Efficiently", *Proc. CHI 96*, Vancouver 383-390

Kramer A, 1994, "Translucent Patches: Dissolving Windows", *Proc. UIST '94*, Marina del Rey, CA 121-130

Kramer A, 1996, "Dynamic Interpretations in Translucent Patches", *Proc. AVI 96*, Gubbio

Kreuger M, 1991, *Artificial Reality II* (Addison-Wesley)

Lawson B, 1990, *How Designers Think. The Design Process Demystified* 2nd Ed. (Butterworth Architecture)

Lieberman H, 1994, "Powers of Ten Thousand: Navigating in Large Information Spaces", *Proc. UIST '94*, Marina del Rey, CA 15-16

Metelli F, 1974, "The Perception of Transparency", *Scientific American* , April 1974

Penz F, Bourne M, Wright P, 1992, "Tools for Design: a controlled experiment comparing computer work with traditional hand drawings", *Penz F (Ed.) Computers in Architecture: Tools for design* (Longman, London) 3-14

Pérez-Gómez A, 1983 *Architecture and the Crisis of Modern Science* (MIT Press)

Perlin K, Fox D, 1993, "PAD: An Alternative Approach to the Computer Interface", *Computer Graphics (Proc. SIGGRAPH)* Anaheim, August 1993 57-62

Purcell T, Gero J, 1996, "Design and other types of Fixation", *Design Studies*, 17 (4) 363-383

Rattenbury K, 1995, "In Glorious Technicolour", *Building Design*, Jan 27, 1995 12-14

Saund E, Moran, T P, 1994, "A Perceptually-Supported Sketch Editor", *Proc. UIST '94*, Marina del Rey, CA 175-184

Schofield S, 1994, *Non Photorealistic Rendering: A Critical Examination and Proposed System*, Doctoral Dissertation, Middlesex University

Schofield S, Richens P, 1995, *The Piranesi System*, (The Martin Centre, Cambridge, England)

Staples L, 1993, "Representation in Virtual Space", *Proc.INTERCHI 93*, Amsterdam 348-354

Vesely D, 1993, "Architecture, Simulation and Reality", *Scroope Five*, Cambridge Architecture Journal 5 26-28

Wellner P, 1993, "Interacting with Paper on the DigitalDesk", *Communications of the ACM* 36 (7) 87-96

Zhai S, Milgram P, Buxton W, 1996, "The Influence of Muscle Groups on Performance of Multiple Degree-of-Freedom Input", *Proc. CHI 96*, Vancouver 308-315

Making Designs Come Alive: Using Physically Based Modeling Techniques in Space Layout Planning

Scott A. Arvin, Donald H. House
Visualization Laboratory, Department of Architecture,
Texas A&M University, College Station, TX. USA

Keywords: physically based space layout, physically based design, responsive design, space layout planning, computer-aided design, human-computer interaction.

Abstract: This paper introduces the concept of responsive design. It elaborates this concept as an approach to free form, adaptable, automated design applying physically based modeling techniques to the design process. Our approach attempts to bridge the gap between totally automated design and the free form brainstorming designers normally employ. We do this by automating the initial placement and sizing of design elements, with an interactive engine that appears alive and highly responsive. We present a method for applying these techniques to architectural space layout planning, and preliminary implementation details for a prototype system for developing rectangular, two-dimensional, single-story floor plans.

1. INTRODUCTION

During schematic building design, an architect develops an overall aesthetic motivation for the building, while trying to arrange individual spaces so they meet the adjacency and area requirements specified in the functional program. The process is often thought of in terms of sculpture, where a design is molded, formed, and massaged. Architects view a building design as constantly evolving and rearranging while they modify existing designs and discover new ones. There has been much research in the past few decades into ways to automate the design process, but most suffer from a lack of intuitive user interaction and a discontinuity with the design language of architects. Architects typically resist using these tools because they tend to be inflexible, prescribing designs rather than allowing the architect to discover designs.

Now, imagine a drawing of a floor plan that is constantly moving, shifting and changing as the designer works on it. If a room is moved out of the building envelope, the building closes in and repairs the hole. If the room is replaced in a different location, the rooms around it adjust to make space for the new addition while remaining the size they need to be. The drawing responds to these changes by managing adjacency and area requirements, allowing the architect to concentrate on design. One could call this *responsive design.*

As we envision it, responsive design is a process that allows a designer to easily make decisions whose consequences immediately propogate throughout the design. Such a responsive design process would be automated, natural, intuitive, flexible, and interactive. It would be automated in that given a set of design intentions, a design solution can be produced. It would be natural in that design elements and intentions are specified in the working language of the designer rather than a low-level representation. It would be intuitive in that the individual elements and the overall design act in expected ways. It would be flexible in that designs are easily modifiable and the design process does not specify the "best" design, but enables designs to emerge. Finally, it would be interactive in that responses to user inputs happen in real time.

We are attempting to create a methodology for a responsive design process by applying physically based techniques to architectural space planning. Although we believe that responsive design can be applied to many graphic design domains, space planning is a relatively simple and well-studied area in which to begin to understand and apply the complexities of this methodology.

In our approach the architect defines programmatic objectives in the usual manner, which are then modeled as physical objects and forces used in a dynamic physical simulation. We model spaces as masses, with adjacencies between spaces modeled as springs connecting the masses. Objectives specified in the architectural program are translated into forces applied to the masses. A dynamic simulation proceeds allowing the mass-spring system to quickly reach equilibrium. The designer then modifies and adds objectives by directly manipulating the graphic model rather than by re-specifying design objectives in the language of the underlying system. The mass-spring representation allows the graphic model to immediately adapt to those changes. It is important to note that we do not intend to simulate the actual behavior of building elements, but to simulate the way architect's may view and interact with design elements during their conception. One way to think about this approach is that it simulates the design behavior of voids rather than the actual behavior of solids.

It can be argued that given enough computing resources, it may be possible for most design methods to be made responsive. The mathematical and computational requirements in physically based modeling are much more extensive than those required by 3D graphics, so why use this approach? First, physical laws are ingrained in all of us, whether or not we know the mathematics behind them. Our minds and bodies know the physics of structure and motion, otherwise we would not be able to survive. Even though we do not think of spaces and walls as being alive and moveable, spaces that move based on design intentions and that act like physical objects should be intuitive to the user. Second, working with a model whose motion is a result of mass-spring interaction should give somewhat the same feeling as working with a deformable substance such as clay. Third, it is conceptually simple from the user's point of view. Rather than having to define a long list of highly specific knowledge-based rules, the designer can think in terms of where a space *wants* to be, which is translated by the system into a force that moves the space. The underlying dynamic computation is complex, but the conceptual interface can be fairly simple and intuitive. Finally, it provides a means for finding a *locally* optimum design solution.

Our approach combines the advantages of automation and manual flexibility by producing a locally optimum space layout solution from an initial arrangement of spaces, *and then letting the designer design*. An approach that provides a globally optimum solution can be inflexible and may not be as useful to designers.

Preliminary results of using physically based techniques on simple floor plans indicate that this responsive approach is extremely promising. We have to date been focusing our efforts on a pilot study, developing this approach for single story, rectangular spaces. But even on simple models it feels natural and fun to move spaces around to try out many designs. The ease and speed of generating and manipulating new designs allow new and more interesting ideas to emerge that would be more difficult to discover with non-responsive approaches.

2. BACKGROUND

There exists a wide body of research into automated space planning methods. We present here just a few of those methods that contrast with ours. We then describe the concept of physically based modeling and review some of basic components used in physically based simulations.

2.1 Automated Space Planning

Architectural space layout problems tend to be ill-defined (Yoon, 1992, p. 8) and over-constrained. Problems that are not well-defined are ill-defined (Simon, 1973), in that the initial constraints on the problem are not fully formulated. Resolving ill-defined problems is a process of searching for and refining a set of design constraints. Problems that are over-constrained have too many possible solutions (Balachandran and Gero, 1987). Automated space planning systems need a method of providing a good solution from a large set of possible solutions, and a method of allowing the designer to modify the set of design constraints to continuously refine the problem definition.

Some approaches to automated space layout planning use an iterative process of constructive initial placement, in which spaces are positioned one at a time (Liggett and Mitchell, 1981; Flemming and Chien, 1995). In these approaches, an ordering function is needed to determine which space to position first.

Some approaches are generative in nature, in that they seek to produce all or a large number of the possible designs within a design space. Some examples are evolutionary design techniques (Jo and Gero, 1998; Gero and Kazakov, 1998), and shape grammars (Flemming, 1987).

Methods of producing optimal space plans have been the focus of many approaches such as Liggett and Mitchell (1981).

Constraints have been used in architectural design (Gross, 1986) in, for example, three-dimensional solid modeling (Tobin, 1991; Martini, 1995), and space layout planning (Yoon, 1992).

One of the most notable and extensive research projects in recent years is the "Software Environment to support the Early phases in building Design," or SEED (Flemming and Woodbury, 1995). SEED partitions the schematic design problem into a variety of modules, one of which is SEED-Layout (Flemming and Chien, 1995). SEED-Layout supports design space exploration through an iterative, constraint based approach that can be either manual or automated. It also supports a case-based approach where previous designs can be used to produce new designs.

2.2 Physically Based Modeling

Physically based modeling is a subfield of computer graphics and visualization. It attempts to represent dynamic motion and changes in geometry by modeling objects as mechanical elements that behave according to the laws of physics. Dynamics are most often derived by the use of forward numerical simulation over discrete time intervals. In a forward simulation, the system is moved from its state at the current time to its state at the next discrete time step, using forces to determine accelerations, and thus changes in velocity during the time step, and velocities to determine translations. The process of making this forward extrapolation is called numerical integration. An excellent introduction to the concepts of physically based modeling and a practical guide to the implementation of these concepts in the computer is given in Witkin and Baraff (1997).

Physically based modeling has been used to model the realistic behavior of rigid bodies (Barzel and Barr, 1988; Baraff, 1989) as well as deformable models (Terzopoulos et. al., 1987). Others have recently begun to use dynamics in geometric design. Qin and Vemuri (1998) and Mandal, et. al. (1997) use physically based techniques to interactively manipulate smooth surfaces of arbitrary topology. In their approach, a user defines the points of an initial control mesh, which are manipulated by applying synthesized forces until the desired shape is achieved.

Harada et. al. (1995) use a physically based approach to allow for discrete and continuous manipulation of generalized floor planning problems. They demonstrate their approach on architectural floor plans represented as rectangular dissections, as well as circuit board layout and page layout.

There are three key concepts in physically based modeling needed to understand the rest of this paper. These are the spring-mass-damper modeling element, and the notions of colliding and resting contact between objects.

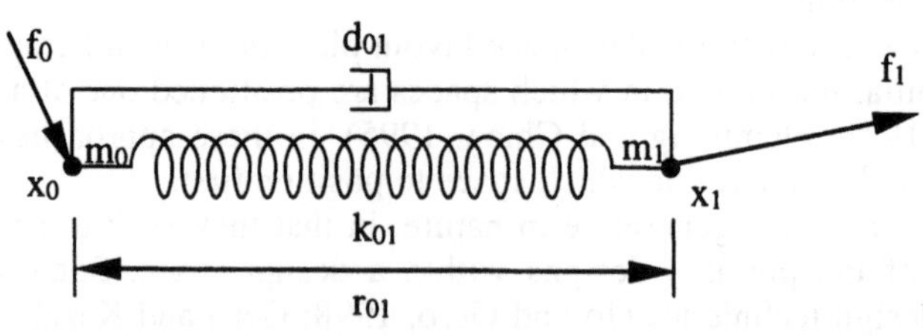

Figure 1. Two masses connected with a spring and a damper.

Figure 1 shows a simple spring-mass-damper system. It consists of two points with mass m_0 and m_1 connected by a spring with spring constant k_{01} and a dashpot with damping constant d_{01}. The spring exerts forces on the two masses with magnitude proportional (with proportionality constant k_{01}) to the difference between the rest length r_{01} of the spring and its current length. The direction of the force will be along the line connecting the point masses. As the masses move farther from each other, the spring applies a force to try to move them closer, and as they move closer the spring tries to separate them. The dashpot is attached in parallel with the spring, and acts like the hydraulic piston on a screen door closer. It damps the motion of the masses by producing forces proportional (with proportionality constant d_{01}) to their relative velocity towards or away from each other, thus reducing the kinetic energy introduced by the spring forces.

Colliding contact occurs between two objects at that instant in time when they touch each other, and have a non-zero relative velocity towards each other. *Collision detection* is the process of determining if and when two objects collide. *Collision response* is the process of determining the result of a collision. (Moore and Wilhelms, 1988) Two bodies are in *resting contact* if they are touching each other, the relative velocity between the two bodies is zero, and the relative acceleration of the bodies is zero. (Witkin and Baraff, 1997)

3. PROBLEM

Our problem is how to create a responsive design system within the domain of architectural space planning. Space layout planning can itself be broken down into two problems, determining topological or qualitative properties, and determining geometric or quantitative properties (Jo and Gero, 1998; Flemming, 1989). The topological problem is one of determining the relationship between individual spaces, without regard to the dimensions of any building elements. The geometric problem is one of determining the physical dimensions of the building elements. Both of these problems need to be solved in any approach to space planning.

Our solution to this problem arises out of taking a physically based viewpoint. We treat each space as a solid with mass. Design objectives are treated as forces that are applied to these spaces, and collision detection is used to keep the spaces from overlapping. One of the primary advantages of using a physically based approach is that it is possible for all design objectives to have an effect on the final location of plan elements.

4. IMPLEMENTATION

In applying physically based techniques to space planning, the first problem is how to represent the elements of a space plan such that forces can be made to act on them. The second problem is how to represent architectural design objectives as forces that can be applied to these elements. It is useful to think of design objectives as wants or needs. For example, space A "wants" to be next to space B, or space C "needs" to be 200 square feet in area. It then becomes easier to understand and define the physical forces needed to accomplish the design objectives. *Figure 2* shows a single space with forces acting on its elements. Arrows labeled with *a* represent forces applied to the space location, and that may change the way this space relates to another. Arrows labeled with *b* represent forces applied to the polygonal edges of the space boundary, and that may change the geometric position of the edges.

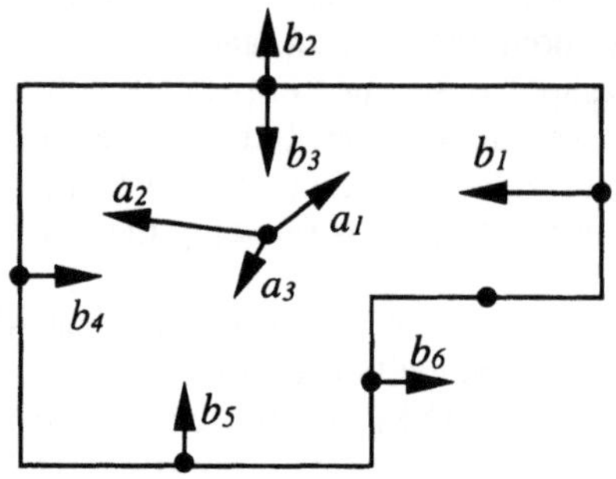

a: Forces applied to the space location.

b: Forces applied to the polygonal edges of the space boundary.

Figure 2. Simple space with forces acting on its elements.

In this section we present the details for implementing a physically based space planning system. We discuss how to represent spaces with polygonal boundaries and nodes that can be used in a dynamic simulation, how to represent design objectives as forces that can be applied to nodes, and the process for resolving a space plan from a given set of design objectives.

4.1 Node types

A node is a point in space on which a force can be applied. The data structure representing a node in our implementation contains values for mass, position, and velocity, as well as a force accumulator and other geometric information that may be required for each node type. Each unique node type has its own graphic representation. Nodes are typically connected to other nodes by springs. The type of the node determines how its movement, and the movement of the node to which it is connected, is constrained. The two types of nodes we use are the point node and the line node.

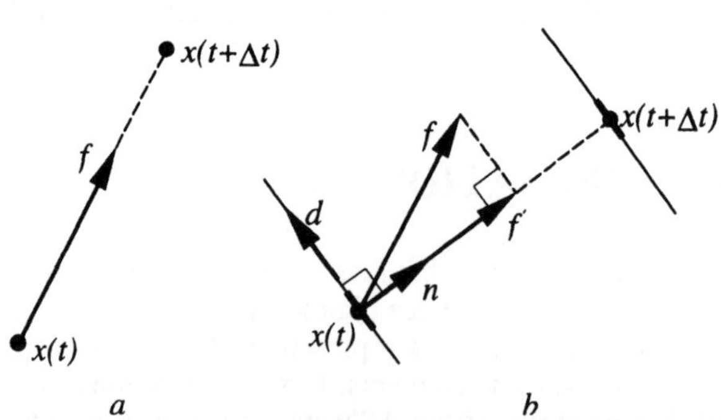

Figure 3. Node types.

Point Node. A point node is the simplest type. The data structure for a point node stores typical node information, and it is displayed as a dot. A force applied to it is not constrained in any way. *Figure 3a* shows a point node with position $x(t)$ at time t, with a constant force f applied to it, and no initial velocity. $x(t)$ is accelerated in the direction of f so that at time $t+\Delta t$ it is at position $x(t+\Delta t)$.

In our implementation we use point nodes to define the center of spaces.

Line Node. A line node defines an infinite line passing through a point. Any forces applied to a line node are constrained to act perpendicular to the line it defines, thus preserving its orientation. The data structure for each line node contains unit direction and unit normal vectors in addition to the typical node information. A position and a direction are all that are needed to define a line; the additional normal vector is stored so as not to repeat its calculation. A line node is displayed as a dot with a short bar going through it parallel to the direction vector. *Figure 3b* shows a line node with unit direction d, unit normal n, and position $x(t)$ at time t, with a constant force f applied to it, and no initial velocity. f is constrained to act along n by

$$f' = (f \cdot n)n,$$

where the $\cdot$ operator is the dot product of two vectors. $x(t)$ is accelerated in the direction of n so that at time $t+\Delta t$ it is at $x(t+\Delta t)$.

In our implementation we use line nodes to define the polygonal edges of space boundaries. They could also be used to define linear spaces such as corridors.

4.2 Polygonal shapes

Shapes are limited to polygons with non-crossing edges and are defined as an ordered list of line nodes, each connected to a common center node. *Figure 4* shows an arbitrary n-sided polygon, with center node c, edge line nodes e, and vertices v. Each vertex v_i can be found as the intersection of the direction vectors of e_i and $e_{i\oplus 1}$, where the operator $\oplus$ is addition modulo n.

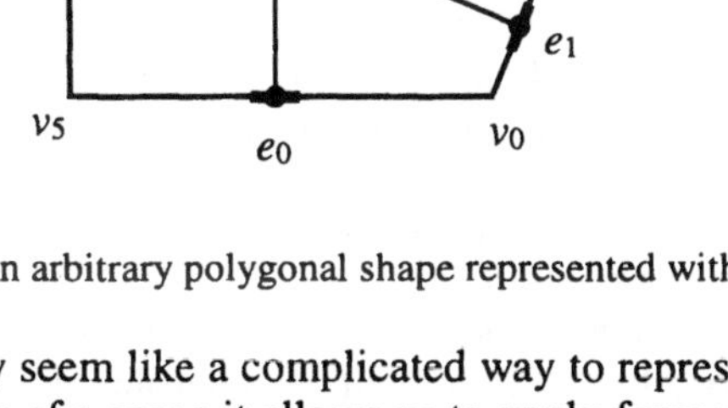

Figure 4. An arbitrary polygonal shape represented with edge line nodes.

Although this may seem like a complicated way to represent a shape, when used to define the boundary of a space it allows us to apply forces to individual segments without having to worry about maintaining their orientation. The typical task in space planning is to move a wall, not to move the end points of a wall. If a shape were represented with point nodes at its vertices, in order to maintian orientation any force applied to one vertex would have to be separated into components that are applied to its surrounding vertices. For non-orthogonal shapes, computing the necessary components could get unnecessarily complicated. *Figure 5a* shows a rectangle represented with vertex point nodes. A force applied to one of its vertices as shown results in the non-orthogonal polygon shown in *Figure 5b*. *Figure 5c* shows a rectangle represented with edge line nodes. Forces f_1 and f_2 applied to its edges are constrained to yield f_1' and f_2' ,which act normal to each edge, resulting in the maintained rectangular shape shown in *Figure 5d*.

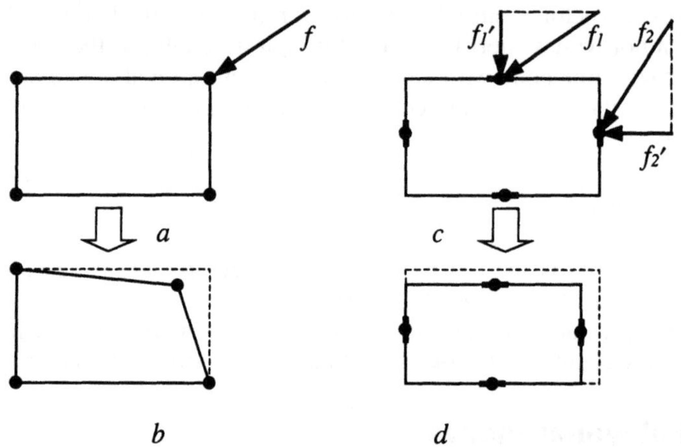

Figure 5. A rectangle represented with vertex point nodes and edge line nodes.

4.3 Spaces

A space defines any arbitrary area or volume. The data structure representing a space contains a common center node to define its location, and a polygon to define its boundary. These values are optional, to be able to define any generalized space. For example, a space used to represent *the outside* would not need a defined position or shape, and could be used when a building space needs to relate to the outside.

A space may also contain any number of child spaces. Child spaces may, in turn, contain their own set of child spaces. In this way, a hierarchy of spaces of arbitrary depth can be defined, similar to that described by Flemming and Chien (1995). A parent space and its child spaces defines a self-contained physical system, and the relationship between the parent and its children is defined by the parent boundary. If a parent boundary exists, as in *Figure 6a*, the system of child spaces needs to be contained within that boundary. If a parent boundary does not exist, as in *Figure 6b*, the solution of the system of child spaces will define the parental boundary. This spatial structure allows the definition of a wide variety of spaces, and allows for spatial hierarchies of arbitrary depth.

The location of the center node relative to the edges is not important to the polygonal representation, but is important in determining space adjacencies. For simple shapes the geometric center or center of mass is fine, but for more complicated shapes a more appropriate center may need to be defined by the designer.

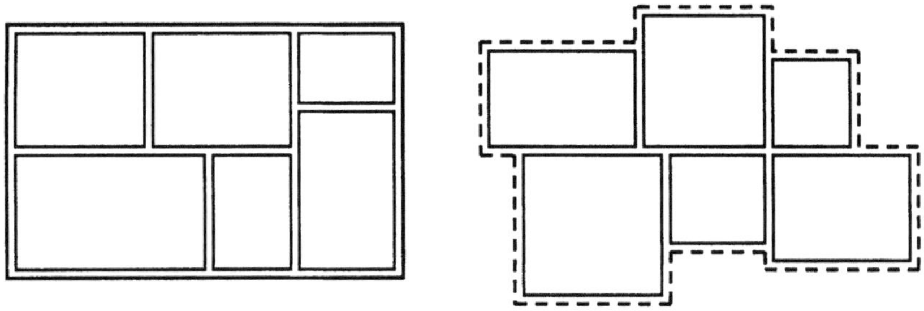

a *b*

Figure 6. Parent space with and without a defined boundary.

4.4 Objectives

We define a design *objective* as something that a designer wants to happen, specifically, any intention that influences geometry. It could be argued that any intention that determines the position of any object could be called an objective, but for the purposes of our approach we limit objectives to those intentions that affect the position of one space relative to another or that affect the location of walls.

We use the term objective instead of *constraint* to avoid confusion with the term used in physically based constrained dynamics. In constrained dynamics, a constraint is a condition that, once met, continues to be met throughout a dynamic simulation (Barzel and Barr, 1988; Witkin and Kass, 1989). One example is an area constraint, where any applied forces that may act to change the area of a polygon will be counteracted to ensure that the area remains the same. Constrained dynamics is not a process of *constraint satisfaction*, but one of *constraint maintenance*. If an area is not as it should be, constrained dynamics will *not* act to change it to the correct area.

Design objectives need to be translated into forces that act on the nodes. They are separated into two categories, topological objectives and geometric objectives, in keeping with the two problems in space planning noted earlier. Topological objectives apply forces to the center of a space, and geometric objectives apply forces to the polygonal edges of space boundaries. Any design objective that affects the position of any element in an architectural plan can probably be translated into force objectives. Here, we will only give details for the few that we found to be necessary to demonstrate the concept of responsive physically based space planning.

Adjacency Objective. An Adjacency Objective is a type of topological objective. It connects the center nodes of two spaces with a spring, and applies forces to those nodes depending upon the distance between them. Using circular spaces as an example, the rest length of the spring may be the sum of the radii of each space. If the spaces are not located next to each other, the spring will produce forces on each space node that attempt to move them together. The spring constant defines the strength of the adjacency, and is initially set on a scale of 0.0 to 1.0 relative to other adjacency spring constants. *Table 1* shows how descriptive terms for adjacency relationships might be mapped into numerical values for spring constants.

Table 1. Adjacency Spring Constants

Adjacency requirement	Spring constant
Immediate	1.0
Important	0.7
Convenient	0.3
Unimportant	0.0

Area Objective. An Area Objective is a type of geometric objective. Recall that a space may contain a polygon that defines the space boundary, and that a polygon is defined as an ordered list of line nodes. An area objective acts to maintain the area of the boundary polygon. The center node of the space becomes the center node of the polygon, and a spring connects this node with each edge line node. The rest lengths of the springs are used to maintain the correct area of the boundary polygon and are set to the perpendicular distance from the center node to the line node.

4.5 Dynamic Simulation

Once a set of spaces and objectives has been defined, a dynamic simulation is allowed to run to produce a layout solution. In our current implementation, the initial location of each space's center node is chosen at random. The dynamic simulation itself proceeds in two phases due to the separate tasks of the general space layout problem. First, the relationship between spaces needs to be resolved, and then the position of walls separating the spaces needs to be resolved. In our approach they cannot be solved at the same time. For this reason, the first phase is to run the simulation with only topological objectives being applied. Once a topological simulation has reached equilibrium, the second phase is started, in which geometric objectives are applied. Once a geometric simulation has reached equilibrium, the designer can begin to analyze and interact with the design by modifying existing objectives and adding new ones.

Although this may sound like the two steps are turning objectives on and off, in our current approach all of the objectives are being applied at all times. The only difference between the two modes is in the boundary shape used in collision detection and response, as described below.

4.5.1 Topological resolution

The first phase in solving a space layout is to determine the location of each space relative to all other spaces. Any force acting on space centers will affect the position of the entire space. Using the objectives we have descibed previously, adjacency objectives are the only forces that have an effect. For collision detection, boundary shapes are treated as circles and are able to slide around each other if necessary. If polygonal boundary shapes were used, boundary corners may catch each other and keep one space from being able to move to the other side of another. The dynamic simulation runs until the system is in equilibrium, which is defined as the point in time when all velocities are zero.

Collision detection and response during this phase can be simulated with the use of stiff springs between overlapping spaces. Although this method is inaccurate, in that some spaces may still overlap, it is adequate for determining topological relationships.

4.5.2 Geometric resolution

Once the dynamic simulation for space resolution has reached equilibrium, space boundaries are switched from a circular to a polygonal representation. Collision detection and response then act to keep spaces from overlapping, resulting in an arrangement that is very close to a recognizable building floor plan.

Collision detection and response are much more complicated for geometric resolution than for topological resolution. Overlapping spaces are unacceptable, so accurate collision response is required. Especially important and problematic is dealing with resting contacts. Although it would seem that resting contacts would be easy to deal with, their handling is one of the more difficult problems in physically based modeling (Witkin and Baraff. p. D49).

To simulate resting contacts, forces must be applied to each object that is in contact with another object so that the contact will be maintained during the next time step. Although colliding contacts can be solved in turn for each individual contact, all resting contacts must be solved together. For a body in resting contact with two other bodies, the resting forces need to account for both contacts.

In our approach we use constrained dynamics to maintain resting contacts (Witkin and Baraff, 1997). Once two spaces are determined to be in resting contact, a distance constraint is applied between the parallel edge line nodes of the contiguous boundary edges. As noted in section 4.4, this constraint acts to maintain the distance between the two nodes. Constraint forces are then solved as a complete system.

4.6 User Interaction

Currently, the definition of spaces for a design project is accomplished with a text file. In the future we will implement an interface to interactively input and edit design spaces and objectives.

The method of interacting with objects in a physical simulation is different from that used to edit static graphic objects. In a typical CAD application, a user moves an object by directly changing the position of the object. In a forward physical simulation, the inputs to a system are forces, and the outputs are positions and velocities. Directly changing the position of an object is inconsistent with this notion of simulation.

One method of moving objects in a physical simulation is to attach a temporary zero length spring between the mouse's cursor position and the object being moved. This spring then applies a force on the object in the direction of the mouse's position. This approach is not usually satisfying to the user, because the cursor position and the object position are not the same during the editing process.

A second method, and the one that we favor, is to temporarily set the mass of the edited object to infinity, and then directly change the position of the object. All dynamic calculations using mass involve multiplication by the mass inverse, so the value of each object's mass inverse is stored instead of its mass. To set a mass to infinity, we simply set the mass inverse to zero, making the object unresponsive to any forces acting on it. Thus, while the user is moving a point it cannot be moved by the dynamics. Once the user lets go of the object the mass inverse is set to its previous value and the simulation proceeds as usual.

4.7 Implementation Details

Software development was done on a Silicon Graphics O2 workstation running Irix 6.3, with a 200 MHz R5000 processor, and 192 MB RAM.

Programming was done in object-oriented C++, using OpenGL for the graphics, and the FLTK user-interface toolkit (FLTK, 1999).

Ordinary Differential Equations were solved using Runge-Kutta fourth order numerical integration.

Table 2 lists some of the constants we used in our force computations.

Table 2. Dynamic constants.

Adjacency spring constant	0.0–20.0
Shape spring constant	500.0
Edit spring constant	500.0
Spring dashpot	2.0
Coefficient of Restitution	0.0
Viscosity	10.0

5. RESULTS

During the development of the dynamics for our physically based approach to space planning, we used the architectural program shown in *Figure 7*, from an example by Karlen (1993). During the early stages of development, we needed a program that was fairly small but which contained many adjacency requirements. It needed to be small because the emphasis in early development was in getting the dynamics between individual spaces to work. It needed to have many adjacency requirements so that many locally optimal solutions were possible. A small number of adjacency requirements would yield a small number of solutions.

University Career Counseling Center	Sq.Ft.
Reception	250
Interview Station	220
Director	140
Staff	180
Seminar Room	300
Rest Rooms	200
Work Area	120
Coffee Station	50
Guest Apartment	350

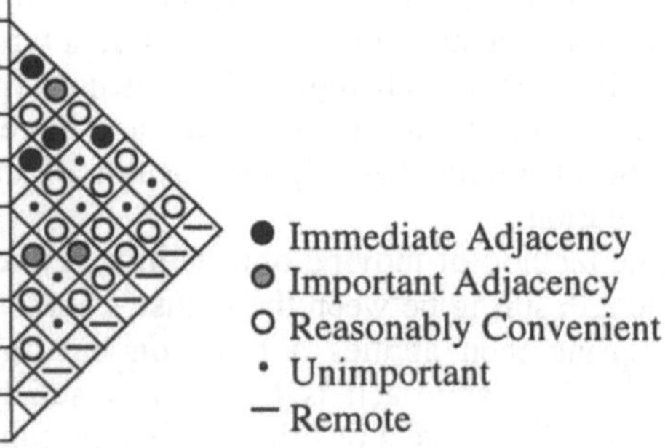

Figure 7. Sample Adjacency Matrix [Redrawn from (Karlen 1993), p. 22]

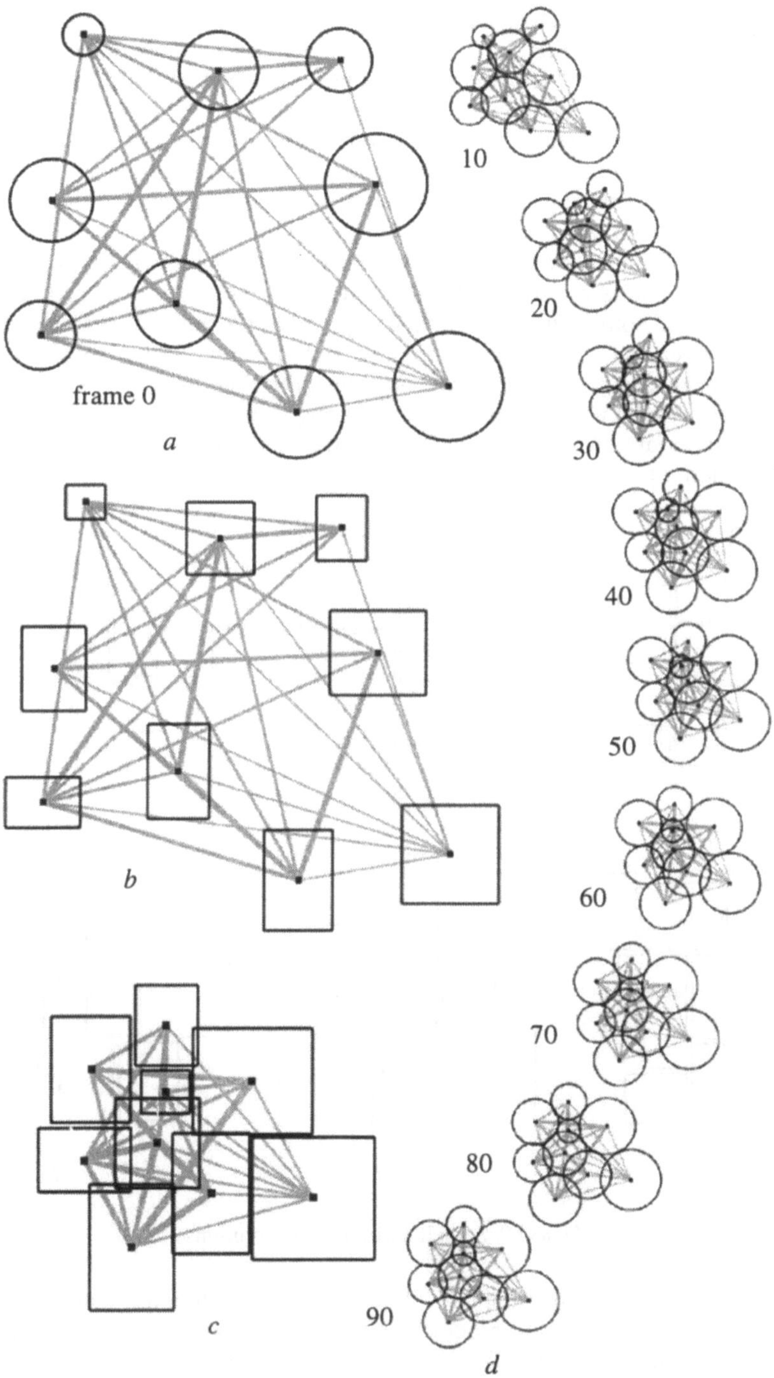

Figure 8. Sample Topological Resolution

Figure 8 shows a sample topological resolution using the program in *Figure 7*. *Figure 8a* and *8b* show each space boundary drawn with its required area and with random initial positions, *Figure 8b* displaying boundaries drawn as rectangles with random proportions, and *Figure 8a* displaying boundaries drawn as circles. Recall that during topological resolution, circles are used in collision detection. It is difficult to show the dynamic movement with static images, but *Figure 8d* shows every tenth frame from the dynamic simulation, with frame 90 showing the spaces in equilibrium. The entire sequence took three seconds to compute and display, so the illusion to the user is of smooth natural motion. Notice that most of the movement occurs between frames 0 and 10 when the spaces are coming together, and that any movement after that is a result of the spaces rearranging themselves and coming to equilibrium. With some initial positions it is possible for the system to almost be at equilibrium when one space manages to move onto the other side of another, and the whole system rearranges itself. *Figure 8c* shows the final topological solution with the boundaries drawn as rectangles again. Although some of the boundaries overlap, this in not important during topological resolution and the overlaps will be resolved during geometric resolution.

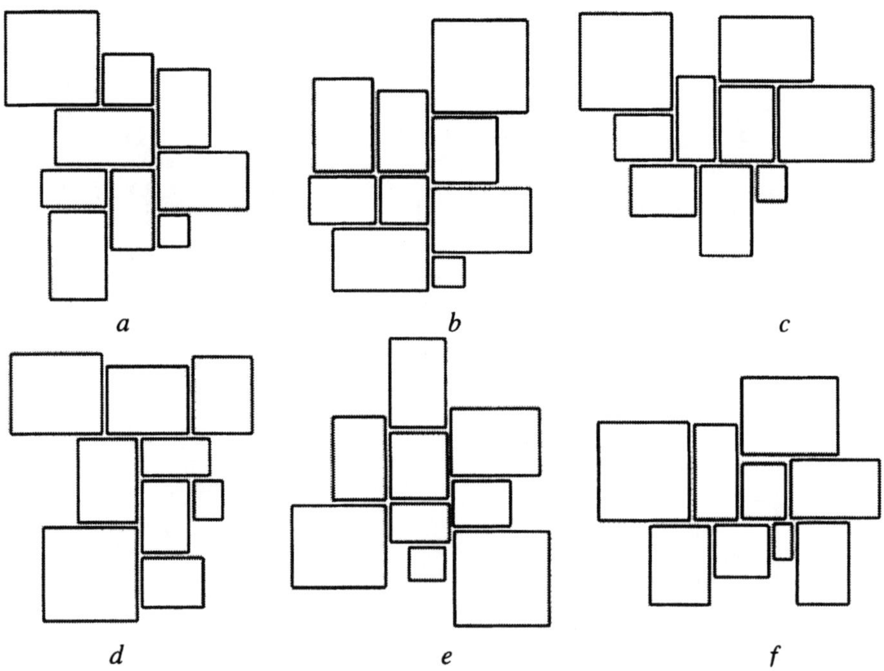

Figure 9. Sample Geometric Resolutions

Figure 9 shows six samples of geometric resolutions using the same architectural program. Initial proportions for each space were maintained from sample to sample, but initial positions were randomized. Final topological relationships were not edited manually. For these results, topological resolution was *not* performed before geometric resolution, so these do not represent locally optimal topological solutions. Note in *Figure 9f* that some wasted space is possible. This problem can be overcome by finding the shortest distance between parallel lines across the opening that needs to be closed, and attaching a spring between the line nodes for those lines.

Note the variety of designs produced from a simple set of objectives. The only objectives active in producing these samples are adjacency and rectangular area objectives. The addition of other objectives such as non-rectangular shape, parental shape, and alignment objectives, among others, should allow the architect to have a great amount of control over the design of space plans.

6. DISCUSSION

We feel that our prototype system provides a convincing demonstration of the attractiveness of the responsive design approach and the use of physically based methods in implementing this approach. It is difficult to convey, in a paper, the experience of working with our system. But, it really does evoke the feeling that one is working with a "living" design, one that responds to the user in ways consistent with programmatic objectives while still providing a high degree of intuitive designer control. We are surprised and pleased at the variety of designs that present themselves from random initial positions. Even when the number of spaces is small (3 or 4), it is enjoyable to rearrange them to see what happens. The fact that the physically based objectives do most of the work allows one to easily explore many design possibilities.

We proposed earlier that a responsive design process would be automated, natural, interactive, intuitive, and flexible. We feel that using a physically based approach to creating a responsive design system meets these criteria very well. Our results show that our approach is clearly automated and flexible. Design solutions are created automatically from a small set of design objectives, and the system is very flexible in that design objects can be easily manipulated, allowing new designs to emmerge. Our approach is for the most part intuitive in that design elements act in expected ways, although some of the springyness of objects may be a little unexpected. Our current implementation is highly interactive and results are produced in real time, but whether or not this will hold for more complicated designs is an open question. The last criterion for responsive design is that designers work with elements and intentions in their natural design language rather than through a low-level representation. How our approach fits this criterion is also an open question. An interactive interface that enables designers to specify objectives directly on the graphic elements of a space plan should provide for natural interaction. However, requiring the designer to have an extensive understanding of the underlying physical model will hinder the feeling of natural interaction.

Within our prototype system, the quality of interaction when rearranging spaces is, for the most part, as we envisioned it to be. When editing a space plan, the designer is able to move spaces around easily to quickly try out new designs. Because of the use of springs and forces, the spaces are very springy feeling. We feel that this can be attractive, in that some springyness adds to the feeling that the plan is alive, but that when it is too noticable it interfers with the interaction. This problem is, however, easily solved by the use of appropriate spring and damping constants.

The responsive feeling of our system depends upon the rate at which calculations can be done and the display updated. The *frame rate* is the rate at which one compute and display cycle is completed. During the final stages of geometric resolution, this frame rate can get a little slow, sometimes approaching non-interactive rates. This is due to two factors. First, accurately detecting and responding to collisions requires that the simulation clock be stopped at the time of collision and the response calculated. Finding the exact time of colliding contact requires a binary search, each stage of which requires a solution to the system's dynamic equations. Second, the handling of resting contacts requires the calculation of constraint forces at each resting contact, which requires the solution of a large set of simultaneous equations describing the constraint dynamics. However, although each of these problems is computationally intensive, their time complexity is at worst a low order polynomial. This makes us hopeful that increasing computer speeds, along with the application of appropriate approximation methods will allow us to achive responsive performance, even for complex building designs, in the reasonably near future.

It is in the nature of the space planning problem is that it is NP-complete, but we do not claim to be able to produce solutions that are globally optimal. Instead, we look for local optima and depend on the designer to recognize when a solution is weak and to make appropriate changes by hand to guide the system into a more optimal configuration. We also feel that we can get some reduction in time complexity by the use of a hierarchical approach. When using hierarchical spaces, such as those discussed in section 4.3, solving for the location of a space is dependent only on the parent and sibling spaces, and is not dependent on every other space in the hierarchy, thus reducing the global problem into a set of smaller, simpler problems.

7. FUTURE WORK

Since the responsive design approach to doing space planning is a previously unexplored concept, it raises many new questions and presents many opportunities for future work. Some of these are obvious and require answers and elaboration in order to make this approach truly useful to space planners. Other questions are of a more fundamental nature.

In the immediate short term, we plan to explore:

- What other kinds of design intentions can be translated into physically based design objectives? Some examples may be daylighting, view, non-rectangular shapes, etc.
- How can this approach be modified to handle circulation? Might the use of line nodes to define corridor space "centers" be a promising solution?
- How can this approach be extended to three dimensions in order to handle multi-story plans?
- What are appropriate values to set for the spring constants? Is it possible to find a range of values that apply to all design situations, or will the designer have to know how to set their own values?
- What are the interface issues in inputing and manipulating design elements?
- What is an effective way of displaying the various forces on a space so that they are meaningful to designers and allow them to understand the conflicting nature of a set of design objectives?

To validate and more firmly ground our work we plan to explore the following fundamental questions:

- Does our physically based approach truly fit the intended characteristics of responsive design?
- How does this approach fit into the architectural programming process?
- How does this approach fit into the overall architectural design process, and how can it be integrated with other aspects of schematic design and design development?
- How does this approach affect the quality of designs that a designer produces? Is it better than existing methods in allowing designers to discover new designs? Will it aid designers from getting fixated on a single design?
- Although this approach is not intended to produce globally optimum solutions, it should benefit the designer to be able to start from a more optimal solution than those we are able to currently provide. How can optimization methods be applied to this approach?
- How much of the underlying physical model should be revealed to designers? Do they need to know about the underlying forces and the details of the physically based system, or is it better for them to understand how it works through the use of another metaphor?
- How can this approach be applied to other design domains?
- What other benefits are there in using a physically based approach? The different forces applied by each objective may be used as a measure of the effectiveness of a particular design, to be compared to other design schemes.

8. CONCLUSION

We have introduced the concept of *responsive design*, as a new paradigm within which tools for architectural design may be developed. We have also demonstrated, with a simple prototype system, that the concepts of physically based modeling provide concrete principles around which to construct such tools. Our preliminary results with the prototype indicate that useful physically based responsive design tools can be built, and that these tools will be practical, enjoyable to use and integrate well with the exploratory nature of the design development process.

ACKNOWLEDGEMENTS

This work was supported in part by an O. N. Mitchell, Jr. Endowed Fellowship in Construction Management, the Visualization Laboratory and the Department of Architecture in the College of Architecture Texas A&M University.

REFERENCES

Balachandran M., and Gero J. S., 1987, "Dimensioning of architectural floor plans under conflicting objectives", *Environment and Planning B*, 14:29-37.

Baraff D., 1989, "Analytical methods for dynamic simulation of non-penetrating rigid bodies", *Computer Graphics*, 23(3):223-232.

Barzel R., and Barr A. H., 1988, "A Modeling System Based on Dynamic Constraints", *Computer Graphics*, 22(4):179-188.

Flemming U., 1987, "The role of shape grammars in the analysis and creation of designs", *Computability of Design*, edited by Kalay Y. E. (John Wiley & Sons, New York):245-272.

Flemming U., 1989, "More on the representation and generation of loosely packed arrangements of rectangles", *Environment and Planning B*, 16:327-359.

Flemming U., and Chien S. F., 1995, "Schematic Layout Design in SEED Environment", Journal of Architectural Engineering, 1(4):162-169.

Flemming U., and Woodbury R., 1995, "Software Environment to Support Early Phases in Building Design (SEED): Overview", Journal of Architectural Engineering, 1(4):147-152.

FLTK, 1999, *The Fast Light Tool Kit Home Page*, <http://www.fltk.org>, accessed 1 March 1999.

Gero J. S., and Kazakov V. A., 1998, "Evolving design genes in space layout planning problems", *Artificial Intelligence in Engineering*, 12(3):163-176.

Gross M., Ervin S., Anderson J., and Fleisher, A, 1986, "Designing with constraints", *Computability of Design*, edited by Kalay Y. E. (John Wiley & Sons, New York):53-68.

Harada M., Witkin A., and Baraff D., 1995, "Interactive Physically-Based Manipulation of Discrete/Continuous Models", In *Computer Graphics* Proceedings, Proceedings of SIGGRAPH 95, 199-208.

Jo J. H., and Gero J. S., 1998, "Space layout planning using an evolutionary approach", *Artificial Intelligence in Engineering*, 12(3):149-162.

Karlen M., 1993, *Space Planning Basics* (Van Nostrand Reinhold, New York).

Liggett R. S., and Mitchell W. J., 1981, "Optimal space planning in practice", Computer-Aided Design

Mandal C., Qin H, and Vemuri B. C., 1997, "Dynamic Smooth Subdivision Surfaces for Data Visualization", *Proceedings of IEEE Visualization '97*, 371-377.

Martini K., 1995, "Hierarchical geometric constraints for building design", *Computer-Aided Design*, 27(3):181-191.

Moore M., and Wilhelms J., 1988, "Collision detection and response for computer animation", *Computer Graphics*, 22:289-298.

Qin H., and Vemuri B. C., 1998, "Dynamic Catmull-Clark Subdivision Surfaces", IEEE Transactions on Visualization and Computer Graphics, 4(3):215-229.

Simon H. A., 1973, "The structure of ill-structured problems", Artificial Intelligence, 4:181-201.

Terzopoulos D., Platt J., Barr A., and Fleischer K., 1987, "Elastically deformable models", *Computer Graphics*, 21(4).

Tobin K. L., 1991, "Constrain-Based Three-Dimensional Modeling as a Design Tool", *Reality and Virtual Reality*, 1991 ACADIA Proceedings, edited by G. Goldman and M. Zdepski, 193-209.

Witkin A., and Baraff D., 1997, *Physically Based Modeling: Principles and Practice* (ACM SIGGRAPH 97 Course Notes, Course 19).

Witkin A., and Kass M., 1989, "Spacetime constraints", *Computer Graphics*, 22:159-168.

Yoon K. B., 1992, *A Constraint Model of Space Planning* (Computational Mechanics Publications, Southampton, UK).

An Interpolation/Extrapolation Process For Creative Designing

John S. Gero and Vladimir Kazakov
Key Centre of Design Computing and Cognition
Department of Architectural and Design Science,
University of Sydney, NSW, 2006, Australia
email: {john,kaz}@arch.usyd.edu.au

Key words: Creative design, computational exploration, design combination

Abstract: This paper introduces a new computational operation that provides support for creative designing by adaptively exploring design state spaces. This modification is based on the re-interpretation of the crossover operation of genetic algorithms as an interpolation and its generalization to extrapolation. Examples of the results of the application of the process are presented.

1. INTRODUCTION

Creative designing can be defined as that class of design activity when all the variables which define the structure and behaviour are not known a priori nor necessarily are all the processes needed to produce them. The implication of this conceptualisation of creative designing is that the focus is on processes for the introduction of new variables into the design and their integration into the existing variable structure. It is suggested that this is one basis for the production of potentially creative designs.

For a given set of variables and processes operating within a bounded context any model will construct a bounded state space. Creative designing can be represented in such a state space by a change in the state space. Non-creative or routine designing does not change the state space, it simply searches within it.

Exploration is the label we give to the class of processes that changes the design state space. However, in designing we work with three state spaces: function state space, behaviour state space and structure state space. It is customary to consider only the structure state space as the space for change. The behaviour state space is dependant on the structure state space. It is possible to have two classes of state space exploration. In the first class the state space maintains the same dimensions, ie. the same defining variables, but changes over time either additively or substitutively as in Figure 1.

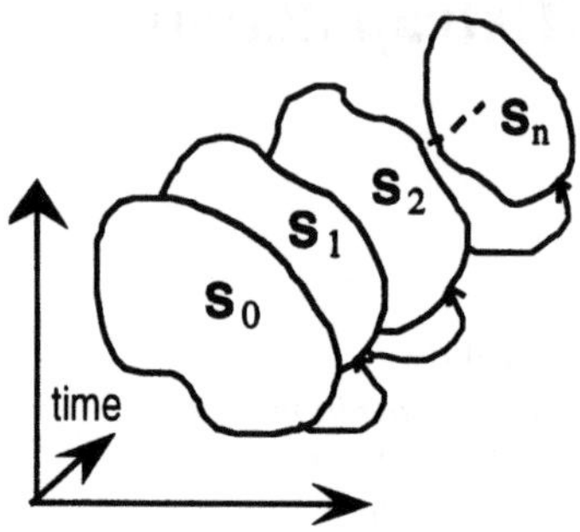

Figure 1 The change in the design state space over time. Here the state spaces (S_0, , , , S_n) keep the same dimensions since their defining variables are unchanged.

In the second class the dimensions of the state space also change with time with the change of defining variables as in Figure 2. In most current creative design systems the behaviour state space is characterised by the first class whilst the structure state space is characterised by the second class.

In this paper a process capable of playing a role in designing potentially creative designs will be introduced. A number of such processes have been developed which fall into one of the following categories: combination, mutation, analogy, first principles and emergence (Gero, 1994). We wish to commence with a process, which is able to be used in routine designing but which, under suitable conditions can be modified to produce non-routine designs. In another words, we would like to construct such a computational process that can behave either as a search or as an exploratory process, depending on the designs that are used to initiate it.

2. COMBINATION AS DESIGN TOOL

Combination is one of the well-known computational processes, which can be used to produce designs (including non-routine designs). One standard computational mechanism for combination is the crossover operation of genetic algorithms, where the genetic representation (genotype) of one design is combined with that of another. This results in a design which is different to either of its parents but which exhibits some characteristics of both. We take this genetic crossover as our starting point. However, crossover does not result in potentially creative designs since all the designs, which can be produced are already encapsulated in the genome. The basic assumption behind the genetic crossover construction is that the genetic representation is fixed and static, implicitly defines all the designs that can be described/generated using it. In terms of the concepts in Figure 1, this means that all design state spaces are the same during design process, which is based on genetic crossover combination process ($S_0 = S_1 = \ = S_n$), Figure 2.

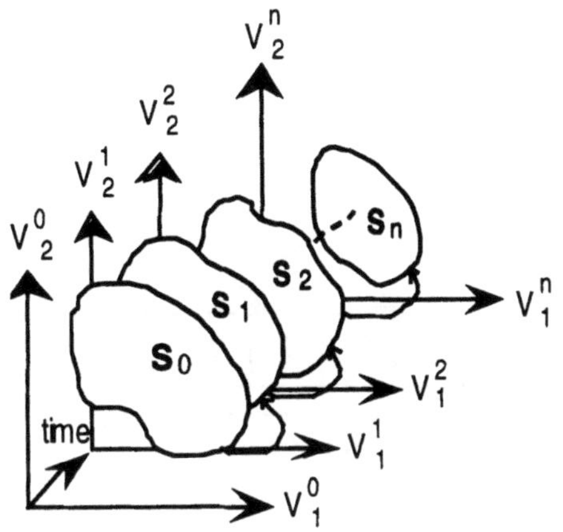

Figure 2. The change in the design state space over time. Here the state space is characterised by changing dimensions since its defining variable change over time.

In our derivations we will follow the following steps. We commence with a standard genetic crossover. Then we re-cast it as a particular type of interpolation operation. The next step is a generalization of this particular version of interpolation to a general form of interpolation. Finally, we further generalize this operator by replacing interpolation with a more general type of operation – extrapolation. As a result we construct a computational combination operator from the genetic crossover operator, that is capable of generating designs, which cannot be generated using a crossover based combination. Since each of these generalization steps includes genetic based crossover as a particular case, it is possible that both initial genetic crossover-based and generalized interpolation/extrapolation combination operators give the same results as the genetic crossover. But in the general case, they will produce different combinations from the same initial designs. In terms of state space evolution, this means that if designs to be combined belong to the state space S_0 and design-combinations produced by genetic crossover-based combination belong to the state space S_1 then the design-combinations that are produced by interpolation/extrapolation combination belong to its super-space $\mathbf{S}_1$: $S_1 \subset \mathbf{S}_1$. $\mathbf{S}_1$ can include additional dimensions compared to S_1.

3. GENETIC CROSSOVER–BASED COMBINATION

In genetic algorithms (GAs) all designs are represented in two forms: in an encoded form of genetic strings (*genotypes*) and in the decoded form of *phenotypes* – the design structures. The genetic crossover operation is the major computational engine of GAs. As a rule, when a fixed-length genetic representation is used, the crossover operation mixes two parental genotypes in component-wise fashion, when each gene of the resulting design comes from one of the parents, Figure 3.

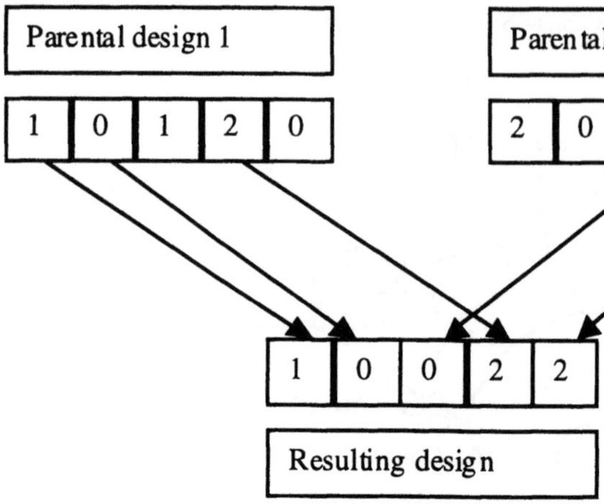

Figure 3. The standard genetic crossover function

This operation of mixing genetic components can be re-interpreted as the operation of choosing a point on a line segment between two points representing both parental designs in the space of genotypes, Figure 4.

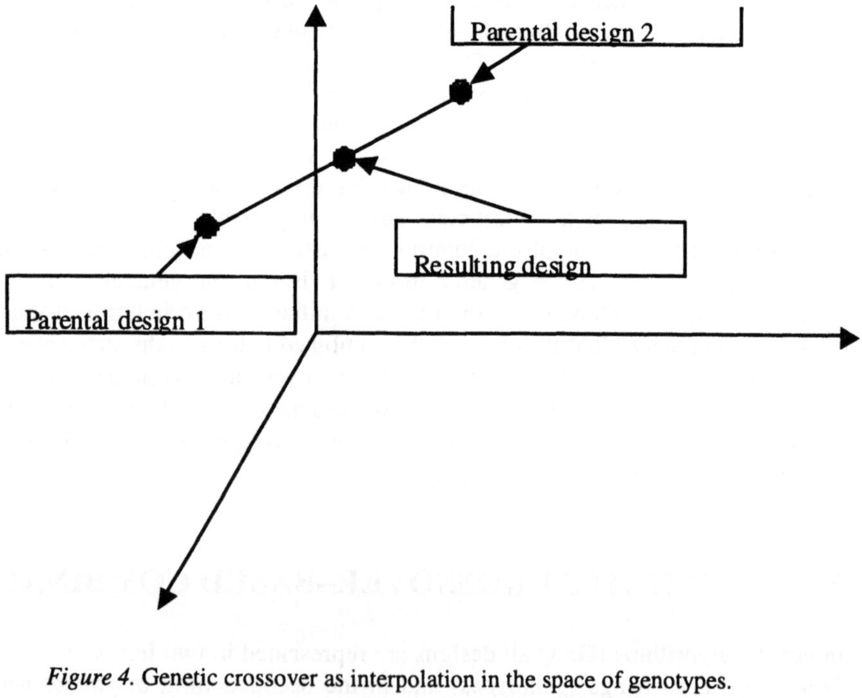

Figure 4. Genetic crossover as interpolation in the space of genotypes.

The distance, which determines what is a line segment here, is defined as a sum of the Hamming distances plus some penalty for the number of genes from different parents that are located side by side in the resulting genotype. The form of this penalty determines what version of crossover operator is used (one-point, two-point, uniform, etc.). In this paper we will assume that the one-point crossover is used. The mapping of this line segment in genetic space onto the space of actual designs produces some other type of trajectory that is, in the general case, much more complex, Figure 5.

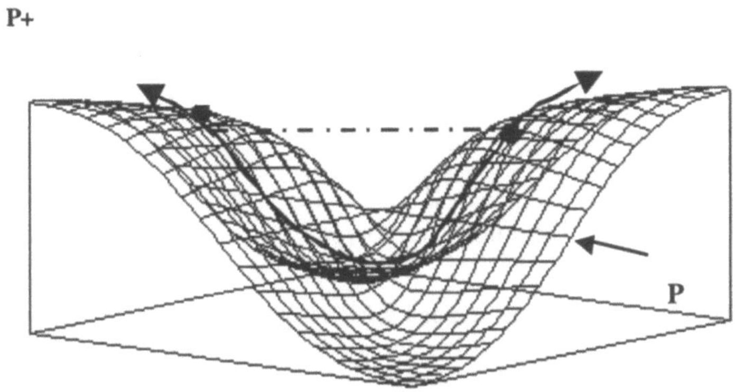

Figure 5. The illustration of the crossover-induced interpolation in P (the original structure space) and direct interpolation in enlarged space P+. The enlarged space P+ represents the complete three-dimensional space and the set P represents the surface in it. The solid line represents an interpolation in P (the mapping of the line segment in Figure 4 in genotypic space), whilst the dotted line represents an interpolation in P+

Because a genetic crossover has been constructed to produce genetic combination, its operations in genetic space can be easily understood, interpreted and described. But in design structure space (space of phenotypes) these are much more difficult tasks. Even the problem of expressing genetic crossover completely in terms of phenotypic space becomes a formidable one. The only exception is an extreme case when the genetic description becomes homomorphic with the phenotypic one. This case is illustrated in Figure 6, where a simple shape grammar is used to build a shape by adding elementary cells to one of 4 positions adjoining the currently labeled cell.

Ultimately, because we want to extend our structure space, we want to generalize our genetic representation by relaxing some of the constraints that are built into the genome. We will do that partial relaxation by replacing the original specialized "rigid" genetic representation, capable of representing only the designs from the fixed class, with a non-specialized representation that is homomorphic to the structure space representation. Once this is done, we will follow the above-described plan: derive the expression for genetic crossover in this representation, re-represent this expression first in an equivalent form and then construct various generalizations of it.

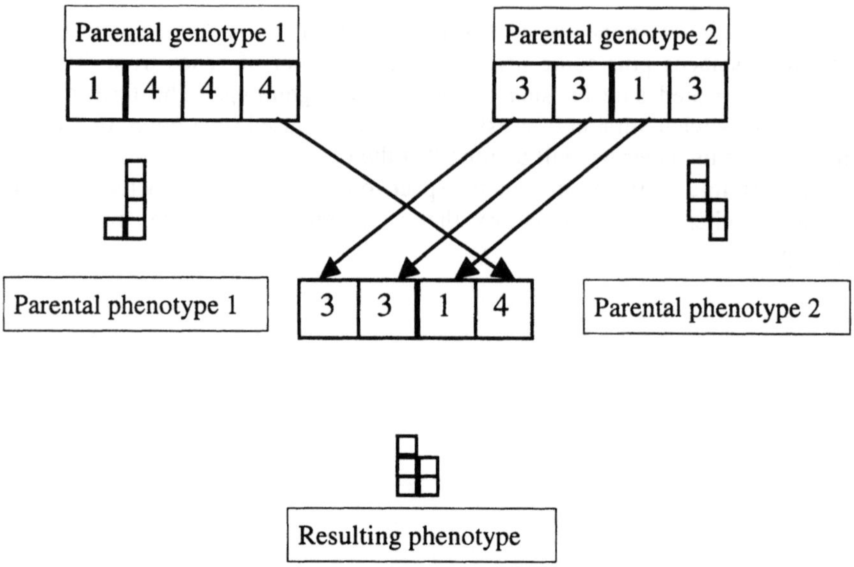

Figure 6. The case when both genetic and structure spaces are homomorphic and genetic crossover can be easily expressed in structure space terms.

4. GENETIC CROSSOVER AS A PARTICULAR CASE OF INTERPOLATION

As an example, in this paper we consider phenotypes (design structures) which are 2-d or 3-d shapes or objects. These designs are represented using an F-representation (Pashko et al., 1995) as real valued functions $F(\mathbf{x})$ such that $F(\mathbf{x})>0$ is inside the object, $F(\mathbf{x})=0$ is on its boundary and $F(\mathbf{x})<0$ is outside of the object. Here $\mathbf{x}$ is a 2-d or 3-d vector with a defined feasible bounded region D: $\mathbf{x} \subset D$. Assume that our design structures are represented using these F-representations and let us try to express genetic crossover in terms of this representation. Since the crossover yields a combination of two input designs, both designs here are represented as functions and a well known combination operation in space of functions is interpolation, it is natural to try to attempt to construct such expressions for genetic crossover as a function interpolation. Let us consider two design structures $F_1(\mathbf{x})$ and $F_2(\mathbf{x})$ and the transformation:

$$F_c(t,\mathbf{x}')= t\, v(\mathbf{x}')\, F_1(\mathbf{x}') + (1-t)\, w(\mathbf{x}')\, F_2(\mathbf{x}'),\ \mathbf{x}'(t,\mathbf{x}): D\rightarrow D, \qquad (1)$$

where t is a scalar which ranges from 0 to 1, $v(\mathbf{x})$ and $w(\mathbf{x})$ are non-negative scalar functions (called modulating functions) of $\mathbf{x}$, and $\mathbf{x}'(t,\mathbf{x})$ is a coordinate transformation which determines a homomorphism from (D,t) to D for any $t \subset [0,1]$ and depends continuously on t. Usually the mapping $\mathbf{x}'(t,\mathbf{x})$ is chosen to provide a correspondence between the positions $i_j(t)$ of the similar features (points, line-segments, etc.) in the two shapes: $\mathbf{x}'(t,i_j(0))=i_j(t)$. The mapping $\mathbf{x}'(t,\mathbf{x})$ can be constructed using the algorithm proposed by Fujimura and Makarov (1997). The only condition which is necessary for this formula to define an interpolation between $F_1(\mathbf{x})$ and $F_2(\mathbf{x})$ (that is, $F_c(0,\mathbf{x})=F_1(\mathbf{x})$ and $F_c(1,\mathbf{x}'(1,\mathbf{x}))=F_2(\mathbf{x})$) is the positivity of functions $v(\mathbf{x})$ and $w(\mathbf{x})$ for all $\mathbf{x} \subset D$. Hence, the goal is to find such positive functions $v^c(\mathbf{x})$ and $w^c(\mathbf{x})$ (we shall call them crossover modulating functions). Such functions must satisfy the requirement that the interpolation path $F_c(t,\mathbf{x})$ fits the mapping of the genetic interpolation $\mathbf{g_c}(t)$ induced by a crossover in genotypic space onto phenotypic space. With such functions, the GA search can be formulated completely in terms of phenotypic space without any explicit references to genotypic space (except implicit information built into functions $v^c(\mathbf{x})$ and $w^c(\mathbf{x})$). Whether or not this can be done (that is, whether or not $v^c(\mathbf{x})$ and $w^c(\mathbf{x})$ exist) needs to be investigated for each particular system. Note that for this problem phenotype interpolation turns out to be the problem of interpolation of real functions.

Now, that we have re-interpreted GA crossover as an interpolation operation in both genotypic and phenotypic spaces, we are in a position to produce a generalization of GA crossover which yields exploration, that is, enlargement of the phenotypic space, which is to be subsequently searched. In the next section we will do that and establish when this generalization will be productive, that is, when it could yield this enlargement.

4.1 Construction of generalized crossover using a general interpolation

Since we have already derived an equivalent formulation of genetic crossover as interpolation, equation (1), using the crossover modulating functions $v^c(\mathbf{x})$ and $w^c(\mathbf{x})$, we can now use the same formula with some other modulating functions that are more general than the crossover modulating functions. Essentially, this means that instead of choosing between various intermediate designs that have a fixed, given genetic structure (and therefore belong to a fixed given class of structures), we choose between intermediate designs that share only the ability to be represented (described) using the same framework. This is clearly a much weaker requirement.

4.2 Further generalization of crossover by replacing interpolation with extrapolation

The next obvious step in the generalization of genetic crossover is an attempt to extend the process of building combinations between designs, by starting from one design then building combinations that gradually transform it into the second design and allowing this process to continue beyond this second design. Mathematically, this means the replacement of interpolation with extrapolation. Intuitively extrapolation (if restricted to the states which are close to the parental points) seems to keep only marginally less information from these points but is capable of going outside the current space. Hence, we can further increase the potential exploratory power of our generalized genetic crossover in this way by completely replacing the crossover operation with the interpolating formula, equation (1), where the original range of t values [0,1,...,n], is extended, for example, over [-k+1,-k,...,-

1,0,1,...,n,n+1,...] to encompass extrapolation. Geometrically this case corresponds to the one when the extensions of the crossover-based interpolating path between two parental points on the surface **P** leave this surface, the heavy line in Figure 5.

5. CHOOSING THE FORM OF PHENOTYPIC INTERPOLATION

When we constructed the computational operations which could form the basis of our algorithm, we assumed that the non-crossover based interpolating modulating functions v(**x**) and w(**x**) in our example are somehow chosen. Their choice is critically important for the algorithm to be productive. Three strategies can be employed for finding such "good" interpolating operators. In the first approach ("blind extension"), some general parametric forms of these operators are chosen a priori, ie v(e,**x**) and w(e,**x**), where e is the vector of parameters $e \subset E$ within the feasible range E. Then when a combination is constructed additional search is carried out with respect to different values of e.

The second approach to the choice of interpolating formula (that is, modulating functions in our example) is based on the following strategy. First, the set of properties required of the states in an enlarged space in order to be valuable are established. Which properties should be preserved from the original space in the enlarged space need to be decided upon. Second, an interpolating operator with this set of properties is constructed. Various shape transformation algorithms which preserve different properties of the end point designs have been invented which assist in this process (eg Beier and Neely, 1992, Wolberg, 1990). This process is still largely experimental.

In the third approach the same set of properties required from the states in the enlarged design space are used as a "viability" test that is applied to each new design generated by generalized crossover. Only those new points which pass it are considered as viable combinations and are subjected to further processing. Of course, this is a search where there is no guarantee that such points can be found or even exist.

6. EXAMPLE

As an example, we consider the problem of designing a cross-section of a beam. The structure space consists of cross-sections of the pre-defined shape with a fixed area, Figure 7. This shape is determined by 4 parameters (the width and height of the top and bottom rectangular flanges and the width and height of the middle rectangular web) with the same range [1,100]. The area of the cross-section is fixed and is set equal to 16. The fulfillment of this area constraint is guaranteed by scaling the shape. The problem has a two-component behaviour (fitness) function **F** which consists of the moment of inertia ,**I,** and the section modulus, **Z.**

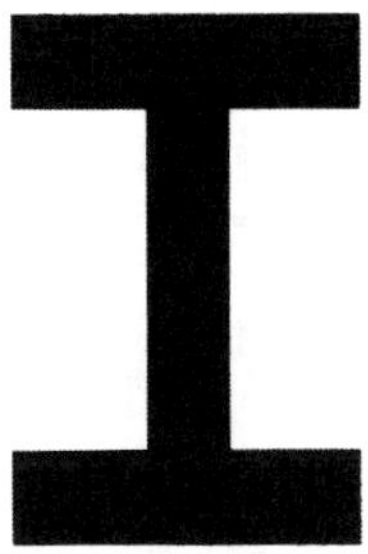

Figure 7 The cross-section template which defines the original structure space.

The standard genetic algorithm found the following Pareto-optimal designs in the original structure space, Figure 8. Because they all have the same genetic structure, they all have the same characteristic form.

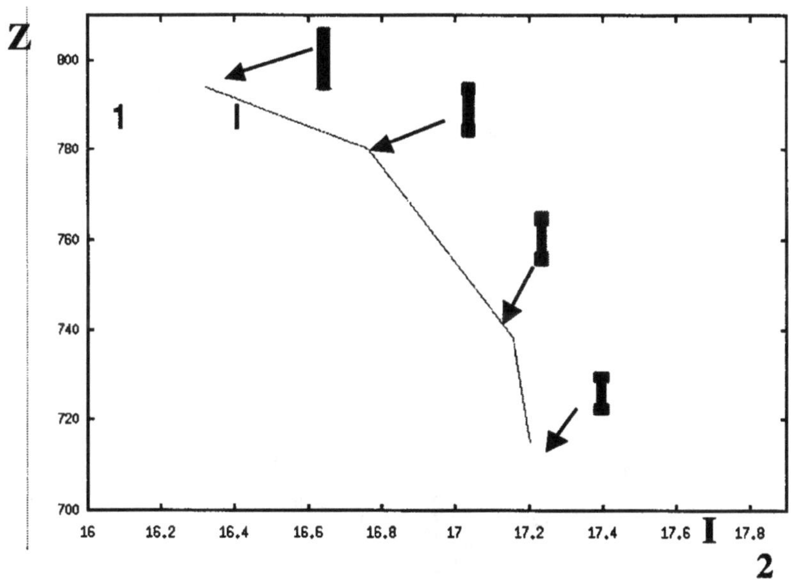

Figure 8 The Pareto set for the initial structure space with the corresponding shapes

Then these designs from the Pareto front are re-represented in F-representations, as real functions F(x), and their combinations produced using two types of interpolations – "linear" interpolation, when both modulating functions are always equal 1 in equation (1), and non-linear interpolation, when these modulating functions are chosen from a fixed class of (trigonometric) functions and "blind" search over modulating functions from this class is performed each time the combination is generated. The results are shown in Figures 9 and 10. It is quite clear that even the use of the linear interpolation leads to the structure space being extended, and that non-linear interpolation leads somehow to a more powerful exploration process which becomes extremely potent when extrapolation is switched on, Figure 11.

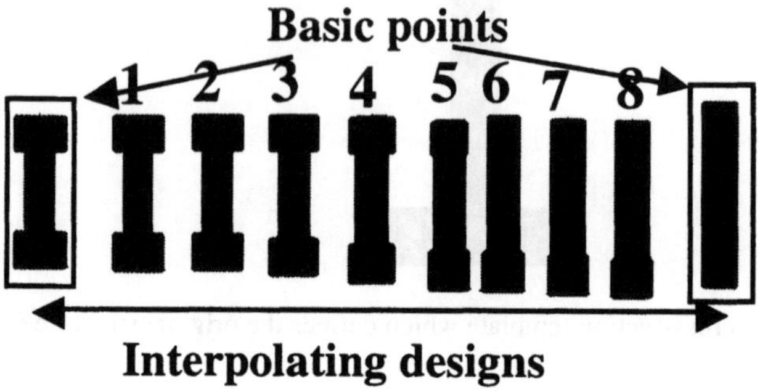

Figure 9. Combinations produced by the "linear" structure-structure interpolation

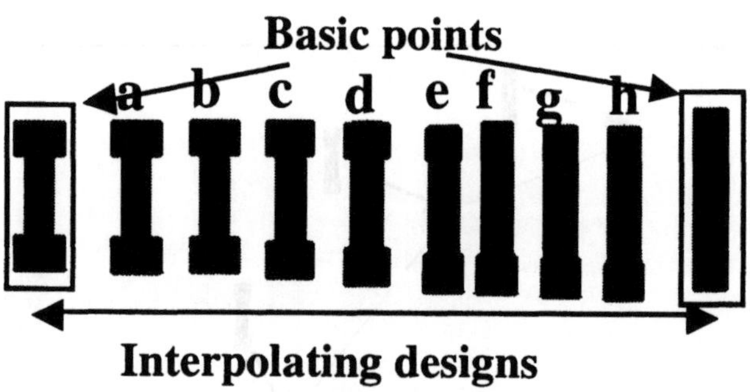

Figure 10. Combinations produced by the "non-linear" structure-structure interpolation

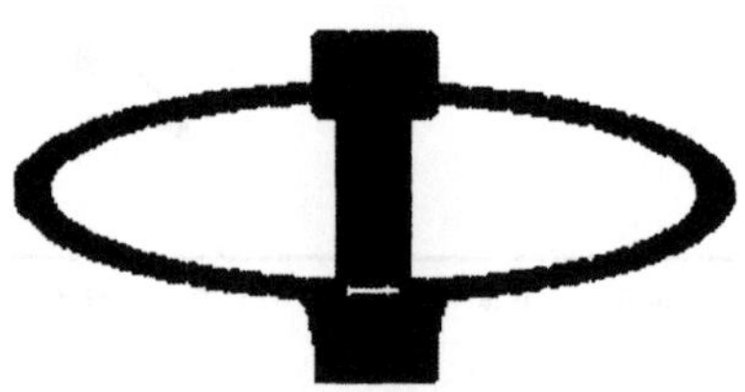

Figure 11. Unexpected combination produced by structure-structure extrapolation

Designs generated from both the linear and non-linear interpolations exhibit characteristics which are not inherited from their parents, as would be the case with genetic crossover. These novel characteristics are a demonstration of the capacity of this process to expand the state space of possible designs.

We can plot these novel designs on the Pareto-optimal front and we note that some of them dominate previous Pareto-optimal designs, Figure 12. From this example, we can see that this generalized combination produces both diversity and the potential to obtain designs which perform better than those produced by genetic combination alone.

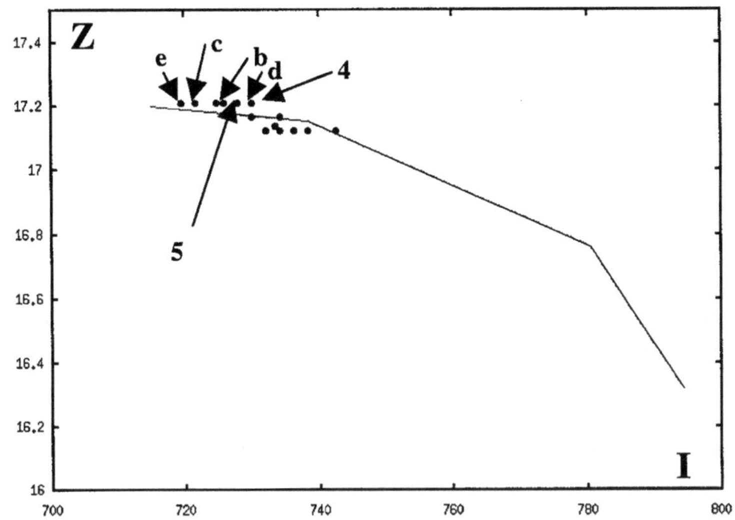

Figure 12. Improved design produced by generalized combination

7. DISCUSSION

One of the well-established notions related to creative designing processes is that an important means of characterising them is to determine whether they have the capacity to expand the state space of possible designs - exploration (Gero, 1994). Of the three state spaces used to describe designs (function space, behaviour space and structure space), only the structure space has the capacity to directly produce novel designs as it is expanded, although the other spaces can also be expanded and hence produce either novel interpretations or indirectly force the expansion of the structure space (Gero and Kazakov, 1998). There are two classes of expansion or modification operators applicable here. The first class contains those operators which rely largely on external knowledge which is applied to the values of the existing space and as a consequence expand it. The second class contains those operators which make use of emergent features in the design space and use those to expand the design space. The extrapolation operator described here belongs to the first class.

One of the difficulties in deriving operators that are capable of expanding a design space is the need for the operators to have the capacity to add variables to the range of variables used to describe the space originally. There have been a number of approaches adopted previously. These include splitting a single variable into two variables (Aelion et al., 1992), importing variables from other design spaces using combination or analogy, and emerging new features and reverse engineering new variables to describe them. The approach adopted here distinguishes itself from these and other approaches in that it opens a range of possible new variables. It does so by recasting the structure space from being restricted to the surface defined by the variables and the original processes of genetic algorithms to being a larger potential space. Which parts of that larger space are used is dependent on the interpolation/extrapolation functions involved and the values of the variables used when the functions are applied. The range of applicable functions may well be unlimited. Each function potentially produces a different trajectory outside the original surface and each trajectory represents potentially different structure variables which are required to describe the resulting design.

As can be seen from the example the resulting designs are unpredictable in the sense that they are unexpected given only knowledge of the original designs and of the interpolation/extrapolation functions. In this sense the process matches well the meaning of exploration both in the technical sense used in this paper and in the natural language sense. The designs produced by the system demonstrate both the novelty and unexpectedness of what can be generated.

This paper can be seen as a generalization of a number of related approaches (see for example, Graf and Banzhaf, 1995) where cross-image interpolation has been used to replace a GA's crossover operator. Our results are more general because we use a more general type of phenotype-phenotype (cross-object and not cross-image) interpolation which can be reduced to cross-image interpolation in a particular case only.

Acknowledgments

This work is supported by a grant from the Australian Research Council. Computing resources are provided by the Key Centre of Design Computing.

References

Aelion V, Cagan J and Powers G, 1992, "Input Variable Expansion - An Algorithmic Design Generation Technique", *Research in Engineering Design,* 4 101-113

Beier T and Neely S, 1992, Feature-based image metamorphosis, *Computer Graphics* (Proc. of SIGGRAPH), 35-52

Fujimura K and Makarov M, 1997, "Homotopic shape deformation", *International Conference on Shape Modeling and Applications,* Aizu-Wakamatsu, 215-225

Gero J S, 1990, "Design prototypes: a knowledge representation schema for design", *AI Magazine,* 11 (4) 26-36

Gero J S, 1994, "Computational models of creative design processes", *in* T. Dartnall (ed.), *Artificial Intelligence and Creativity,* (Kluwer, Dordrecht), 269-281

Graf J and Banzhaf W, 1995, "Interactive evolution in civil engineering", *in* J. S. Gero, M.L.Maher and F. Sudweeks (eds), *Preprints Computational Models of Creative Design,* (Key Centre of Design Computing, University of Sydney, Sydney, Australia), 303-316

Pasko A A, Adzhiev V D, Sourin A I and Savchenko V V, 1995, "Function representation in geometric modeling: concepts, implementation and applications", *The Visual Computer,* 11 (8) 429-446

Wolberg G, 1990, *Digital Image Warping,* (IEEE Computer Society Press)

Feature-Based High Level Design Tools

A Classification

Henri H. Achten, Jos P. van Leeuwen
Eindhoven University of Technology, The Netherlands

Key words: Features, Feature-Based modelling, architectural design, design process, design support

Abstract: The VR-DIS project aims to provide design support in the early design stage using a Virtual Reality environment. The initial brief of the design system is based on an analysis of a design case. The paper describes the process of analysis and extraction of design knowledge and design concepts in terms of Features. It is demonstrated how the analysis has lead to a classification of design concepts. This classification forms one of the main specifications for the VR-based design aid system that is being developed in the VR-DIS programme. The paper concludes by discussing the particular approach used in the case analysis and discusses future work in the VR-DIS research programme.

1. DESIGN INFORMATION FOR ARCHITECTURAL DESIGN

Design information is used throughout every phase of the design process. It develops along with the design from high level abstract concepts such as symmetry, co-ordination, spaces, function, etc. to low level concrete concepts such as dimensioned window frames, light sockets, and finishes. Traditional product modelling has been successfully applied to the phase in the design process when the early, formative phase of design has been concluded. For an adequate support of early design phases, they have proven to be too rigid and set. In particular, product modelling approaches lack flexible definition of object types nor can these be extended when required.

Design information at the start of a design process is quite different from design information at the phase of preliminary design. Design, as a problem-solving process, involves activities of searching information, analysing, manipulating, and structuring information, generating new information, and evaluating and communicating information. These are not sequential activities, but take place in cycles (Markus 1969), (Maver 1970), Roozenburg and Eekels (1995). (Lawson 1990) argues that designers tend to switch in an ad hoc manner between different activities, resulting in concurrency of activities with no predictable sequence. The dynamic nature of design should be supported by design aid systems.

In the Design Systems group at Eindhoven University of Technology, a research programme has been initiated called VR-DIS, meaning Design Information System and Distributed Interactive Simulation in Virtual Reality. The goals and projects of this programme have been reported in (de Vries et al. 1997). The possible advantages of the VR-DIS programme compared to conventional CAD systems are discussed in (Achten et al. 1998c). They propose that VR technology shows the best performance in the early design stage, using tools to create and evaluate (abstract) design models based on a three dimensional dynamic representation, and that it has the most potential in those areas where traditional CAAD has a poor performance.

In order to establish such a design support system, it is necessary to develop a functional brief. The performance specification should be formulated in terms that can easily be transformed computationally. Feature-Based modelling, an approach currently under investigation and development in the group, provides such a formalism. Research in the VR-DIS project on the possible uses of Feature-Based modelling in a VR environment concern conjoint measurement of user behaviour (Dijkstra et al. 1996; 1998; 1999; Dijkstra and Timmermans 1997a; 1997b; 1998), Case-Based Reasoning (Groot et al 1999; Mallory-Hill 1998; Achten et al. 1998d), and user interface (Coomans 1999).

The development of a design support tool in VR needs to be based on an understanding of the required tools and design information needs. For this purpose, a design case drawn from practice is studied. The case describes in terms of Features the steps in the design process represented by drawings. For this purpose, a description technique derived from previous research (Achten 1997a; 1997b; Achten et al. 1998b; Achten and van Leeuwen 1998) is extended and used for analysis of the drawings that are made during the design process. Each drawing is analysed and described in terms of Features. The transitions between drawings are also described in terms of Features. This description provides a formal basis for developing design tools that can be used in early design.

1.1 Feature-based modelling in the research context

Feature-Based modelling (FBM) originates from areas of Mechanical Engineering. The background and history of these techniques have been discussed and summarised in early papers by (Cunningham 1988), (Shah 1991; 1994), and (Bronsvoort 1993; 1996). FBM has been reviewed for its relevance to architectural design in (Van Leeuwen et al. 1996; 1997). The main conclusions from the latter reviews are that concepts of FBM are very relevant for modelling architectural information in a broader sense. In the VR-DIS programme, the following definition of the term Feature is employed (Van Leeuwen 1998a):

"A Feature is a collection of high-level information, possibly emerging during design, defining a set of characteristics or concepts with a semantic meaning to a particular view in the life-cycle of a building."

This definition reflects four important aspects of Feature modelling in the architectural context:

1. A Feature has high-level information with semantic meaning.
2. Both physical and non-physical characteristics and concepts can be defined.
3. Definition and use of emerging Features during design is supported.
4. A Feature relates to a particular view in the life-cycle of a building.

(Van Leeuwen 1998a) provides a Feature modelling framework for the development of information modelling systems for support of architectural design. The framework defines how Features are to be modelled. Feature models are flexible in that they support alteration of specific Feature Types during the design process. They are extensible through support of defining new Feature Types and Feature Instances. Also, it is possible to define relations between Feature Instances that have not yet been foreseen at the Feature Type level. This dynamic character of Feature modelling seems to be in accordance with the dynamic nature of design.

For Feature modelling, a tool has been developed in the group to define Feature Types and Feature Instances and to manipulate them. In time, Feature manipulation of the design is envisioned to be an integral part of the VR environment. Work by (Coomans and Achten 1998; Coomans and Timmermans 1998; Coomans 1999) is aimed towards this development. For now, the Windows-based Feature tool is used for Feature definition. Features can be represented in a graphical way (Van Leeuwen 1998a) or in a textual way, the Feature Type Definition Language (Van Leeuwen 1998b). In this paper, we will be using the textual representation.

2. THE CASE STUDY

A case study has been done on an actual building design, which is analysed in terms of Features (Achten and van Leeuwen 1998a). It concerns an actual design executed by an architect's office. In the office, AutoCAD is used from the very start of the design work. Drawings during the design process are made as new copies rather than changing or revising old drawings. In this way, key phases of the design process are available for analysis.

The case study is based on 30 drawings made during the design process. Each single drawing is taken as a step in the design process for which a Feature model can be established. The transition from one step to the next therefore, represents the design decisions taken from that phase to the next.

- We define as a phase: one single drawing. Notation: phase n, $n \in N$.
- We define as a step: the transition from phase n to phase n+1.
- We define the brief as phase 0.

The following figures represent the first six phases in the design case.

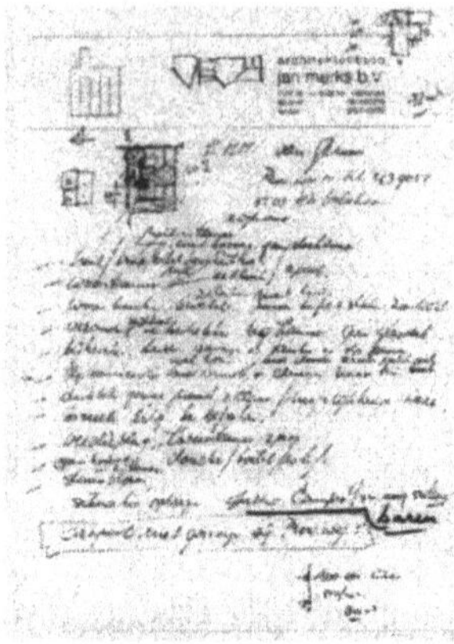

Figure 1. Phase 0 of the design case

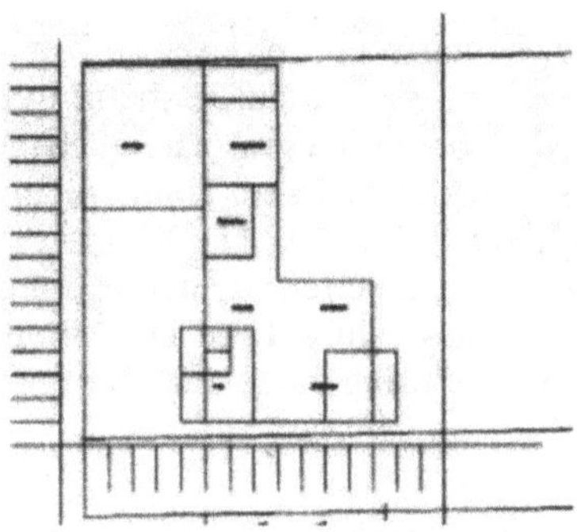

Figure 2. Phase 1 of the design case

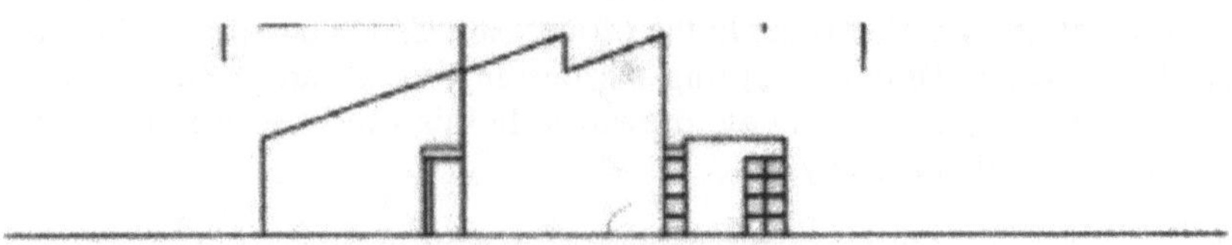

Figure 3. Phase 2 of the design case

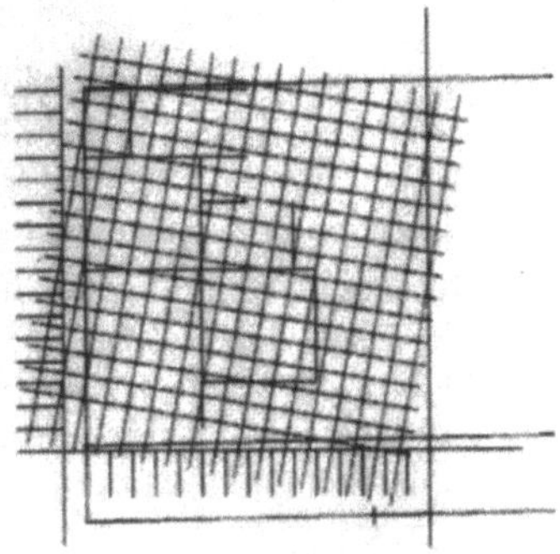

Figure 4. Phase 3 of the design case

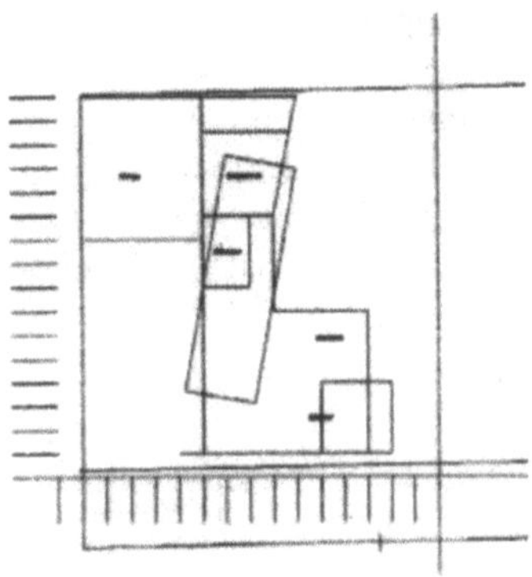

Figure 5. Phase 4 of design case

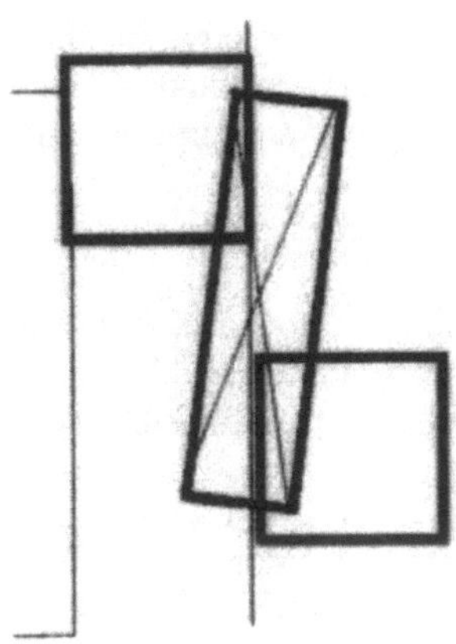

Figure 6. Phase 5 of design case

The analysis of the design case is carried out as follows:

1. In phase n, identify the elements that have to be defined as Features. These elements are concepts used in the design, and typically are nouns in the text or graphic units in the drawings.
2. If the elements are new, define a complex Feature Type or simple Feature Type for each new element. An element is defined as a complex Feature Type when it cannot be defined as a simple Feature Type (string, integer, real, Boolean, or enumeration). If the element in question already has been defined as a Feature Type in any previous phase, record which changes have taken place and determine whether these should be included at the Type level or the Instance level.
3. In the case of new elements, establish Feature Instances based on the new Feature Types defined in step 2. In the case of existing Feature Types, record changes in Feature Instances. Proceed to phase n+1.

2.1 Example of Feature-Modelling in the Case

The brief of phase 0 provides a lexicon of design elements that play a role in the building design. In the example we will focus on the concept of space. Spaces such as "hall," "toilet," and "wardrobe" are elements of the lexicon. First, Feature Types have to be defined, after which Feature Instances can be made.

2.1.1 Feature Types

Rather than defining a Feature Type for each kind of space (such as a Feature Type Hall for the hall, Toilet for the toilet, etc.), we will define a Feature Type space of which the various spaces in the brief are instances.

The text in the brief notes the following aspects of a space:

- Function (such as "bedroom" and "bathroom").
- Contained elements (such as "stair" and "toilet").
- Visual relationship (such as "kitchen closed with respect to living").
- Access relationship (such as "doors to garden and bathroom").
- Daylighting (such as "daylighting in kitchen").
- Adjacency (such as "scullery between garage, kitchen and bathroom").
- Rooftype (such as "no glass roof").
- Number of persons (such as "guest room for two persons").

Determining which aspects are to be included in the definition of the Feature Type Space and which aspects are to be defined in other Feature Types is not straightforward. If the aspect concerns only the space itself, and does not refer to other elements, then it can be included in the type. Following this guideline, constraint-like relations such as visual relationship are better defined outside the Feature Type. Function, contained elements, daylighting, rooftype, and number of persons are within the particular space and therefore are included in the type definition. The Feature Type space is defined accordingly, and results in the following:

```
complex BuildingElement.space.Space {
  Has BuildingElement.space.Space contains[0..?];
  Spec User.value.Daylighting daylightIsUsed;
  Spec User.value.Function function;
  Spec BuildingElement.structure.Rooftype kindOfRoof;
  Spec User.value.NumberOfPersons numberOfPersons;
}
```

The first line identifies the Feature Type class, which is 'complex' in this case. The text 'BuildingElement.space.Space' is the Feature identification in the context of a Feature Type library.

The next five lines define the aspects of the Feature Type Space as identified above. They are the contained elements, daylighting, function, rooftype, and number of persons respectively. Each line has a three-part structure: relation, FeatureID, and role. Four of the relations are specifications since they further define the space. The "contains" relation is a decomposition since the contained spaces are part of the space. The FeatureIDs refer to Feature Types that are related to the Feature Type Space. Their definitions follow next. The role describes the role of the Feature in the definition. The numbers in brackets (for example '[0..?]') indicate cardinality of the relation: how many instances of this role are allowed or required in a Feature Instance.

In order to complete the Feature Type definition of space in this phase, the Feature Types User.value.Daylighting, Function, Rooftype, and NumberOfPersons must be defined as well.

```
string BuildingElement.structure.Rooftype {
}
boolean User.value.Daylighting {
  TypeDefault {True}
}
string User.value.Function {
}
integer User.value.NumberOfPersons {
  TypeDefault {1}
  TypeUnit {"person"}
}
```

The Feature Type Space is not only used to make Feature Instances of spaces such as hall and kitchen, but also in all other cases where in the brief of phase 0 a space is mentioned. This is the case for example, in the access relationship between spaces. The corresponding Feature Type is Constraint.access.Space_Space:

```
complex Constraint.access.Space_Space {
  Spec BuildingElement.space.Space access[2..2];
}
```

2.1.2 Feature instances

The Feature Types have to be instantiated into the particular Features Instances used in the design. In the case of Feature Type Space, this means making instances of spaces such as hall, living, kitchen, veranda, etc. We will provide an example of the Feature Instance living.

```
BuildingElement.space.Space Living = {
   contains[1] = Dining
   contains[0] = Sitting
   function = FunctionLiving
}
BuildingElement.space.Space Dining = {
function = FunctionDining
}
BuildingElement.space.Space Sitting = {
   function = FunctionSitting
}
User.value.Function FunctionDining = {
   Value {"Dining"}
}
User.value.Function FunctionLiving = {
   Value {"Living"}
}
User.value.Function FunctionSitting = {
   Value {"Sitting"}
}
```

In subsequent phases the Feature model is established on the basis of the previous phase, which enables to track changes during the design process.

2.2 Feature-Based Description of Phase 0

In this phase, the brief is translated into a Feature model. The brief consists of elements of the design (spaces and objects such as bathtub, toilets, etc.) and of relations between elements. The relations can be viewed as constraints. Some of these constraints concern spatial relations. These can be expressed as: A _adjacent_ B, A _in_ B, A _above_ B, and their logical opposites A _NOTadjacent_ B, A _NOTin_ B, A _NOTabove_ B, with A and B Feature Types.

Other constraints concern access from one element to another. These access constraints can be expressed as: A _access_ B and the opposite A _NOTaccess_ B. Since this relation is reciprocal, only one _access_ constraint has to be defined.

The third type of constraint is expressed in the statement 'kitchen is closed with respect to the living,' meaning that there is no visual relation between the kitchen and living space. The constraints A _visual_ B and the opposite A _NOTvisual_ B are reciprocal, and can be defined in the same manner as the _access_ relation.

The 'fireplace in living' constraint can be expressed in two ways: as a Fireplace_in_Living instance of a Heating_in_Space constraint Feature Type, or by establishing an association-relation in the Feature Instance Living with the Feature instance Fireplace. We have chosen here for the first option, since heating elements are bound to occur in more spaces.

The tables below show the Feature Types and Feature Instances for elements and constraints respectively.

Table 1. Feature Types and Feature Instances of elements in phase 0

Feature Type (super type)	Feature Instance
Space	Hall, Toilet, Wardrobe, Living, Sitting, Dining, Kitchen, Veranda, Scullery, Garage, Bedroom, Bathroom, Shower, GuestRoom
Door (ElementInWall)	DoorBathroom_Bedroom, DoorBathroom_Garden
Floor	FloorLiving
Material	MaterialFloorcovering, MaterialRoofVeranda
Roof	RoofVeranda
Stair	Stair
Garden	Garden
Chair (Furniture)	Chair
Table (Furniture)	Table
Fireplace (Heating)	FireplaceLiving
Bathtub (Sanitary)	BathtubBathroom
ToiletPot (Sanitary)	ToiletPotHall, ToiletPotGuestroom, ToiletPotBathroom
WashBasin (Sanitary)	WashBasin1_Bathroom, WashBasin2_Bathroom, WashBasinGuestroom
Daylighting	DaylightingBedroom, DaylightingKitchen, DaylightingVeranda
Function	FunctionBedroom, FunctionHall, FunctionDining, FunctionKitchen, FunctionSitting, FunctionLiving
Storey	StoreyGroundFloor, StoreyFirstFloor
NumberOfPersons	NumberOfPersonsGuestroom

Table 2. Feature Types and Feature Instances of constraints in phase 0

Feature Type (constraint type)	Feature instance
Space_adjacent_Space (spatial)	Kitchen_adjacent_Living, Veranda_adjacent_Living, Scullery_adjacent_Garage, Scullery_adjacent_Bedroom, Scullery_adjacent_Kitchen
Space_adjacent_Garden (spatial)	Kitchen_adjacent_Garden, Bedroom_adjacent_Garden
Storey_above_Storey (spatial)	Storey1_above_Storey0

Feature Type (constraint type)	Feature instance
Furniture_NOTin_Space (spatial)	Furniture_NOTin_Kitchen
Heating_in_Space (spatial)	Fireplace_in_Living
Stair_NOTin_Space (spatial)	Stair_NOTin_Living
Space_Space (Access)	Bedroom_access_Bathroom
Space_Garden (Access)	Bedroom_access_Garden, Kitchen_access_Garden
Space_NOTvisual_Space (visual)	Kitchen_NOTvisual_Living

2.3 Feature-Based Description of Phase 1

In phase 1, the brief is transformed to a set of spaces in a drawing. Significant changes with respect to phase 0 consist of assigning shapes and their dimensions to spaces, and locating them in the site by means of a grid.

The notion of shape can be implemented as part of the existing Feature Type Space or by defining a new Feature Type for shape, which is associated with the Feature Type Space. When considering phase 1 only, the first option would suffice. However, the notion of shape is very basic in architectural design, and many other kinds of shapes may occur later in the design, each with their own intrinsic properties. Therefore, we have chosen to define a supertype 2DShape of which Rectangle is a subtype. For the definition of a rectangle and its position in a drawing, the dimensions and position need to be defined as Feature Types as well. Therefore, Length, Point, and Coordinate also are new Feature Types:

```
complex Geometry.shape.2DShape {
   TypeDescr {"General shape definition with point of reference"}
   Spec Geometry.topology.Point referencePoint[1..1];
}
complex Geometry.shape.Rectangle(Geometry.shape.2DShape) {
   TypeDescr {"Rectangular shape with dimensions and reference point"}
   Spec Geometry.dimension.Length length[1..1];
   Spec Geometry.dimension.Length width[1..1];
}
real Geometry.dimension.Length {
   TypeDescr {"Linear dimension in m"}
   TypeDefault {1}
   TypeUnit {"m"}
}
complex Geometry.topology.Point {
   TypeDescr {"Point in orthogonal axial system with x, y, z co-ordinates"}
   Spec Geometry.topology.Coordinate xcoordinate[1..1];
   Spec Geometry.topology.Coordinate ycoordinate[1..1];
   Spec Geometry.topology.Coordinate zcoordinate[1..1];
}
real Geometry.topology.Coordinate {
   TypeDescr {"Coordinate along an axis in an axial co-ordination system"}
   TypeDefault {0}
```

```
   TypeUnit {"m"}
}
```

The Feature Type 2DShape is associated with the existing Feature Type Space (bold line shows addition to old Feature Type):

```
complex BuildingElement.space.Space {
TypeDescr {"Space element within which activities can take place"}
   Spec BuildingElement.space.Space contains[0..?];
   Spec User.value.Daylighting daylightIsUsed[1..1];
   Spec User.value.Function function;
   Has BuildingElement.structure.Rooftype kindOfRoof;
   Spec User.value.NumberOfPersons numberOfPersons;
   Assoc Geometry.shape.2DShape shape;
}
```

The shapes are drawn in a grid which is used for co-ordination. The grid can be defined by stating its module and a point of origin relative to which co-ordinates are established (this also accommodates the use of multiple grids). Both the origin of the grid and the position of elements in the grid require a set of co-ordinates. We define therefore, on the instance level, a Feature Instance Origin (instance of Geometry.topology.Point) relative to which measures can be taken and grids positioned. The left-bottom corner of Grid_1 is placed on the Origin.

```
complex Geometry.Topology.Grid {
     Descr {"Origin and module of a grid"}
     Spec Geometry.topology.Point originOfGrid[1..1];
     Spec Geometry.dimension.Length moduleOfGrid[1..1];
}
```

The positive x-axis is oriented horizontally and to the right of the Origin. The positive y-axis is oriented vertically and above the Origin, as is customary in architectural design. For the Feature Type Rectangle, the reference point is defined as the most left-bottom corner of the rectangle, width and length being measured in the orientation of the positive x and y axis.

These Feature Types and Feature Instances suffice to describe the state of phase 1. The Feature Instance Kitchen, for example, is changed because of the additional association to the Feature Type Space of which it is an instance, and the definition of its location and dimensions in the associated Rectangle. In phase 1, the Kitchen has dimensions 3.60 m x 3.60 m, and is located on co-ordinates (6.0, 6.0, 0):

```
BuildingElement.space.Space Kitchen = {
   Descr {"Kitchen"}
   daylightIsUsed = DaylightingKitchen;
   function = FunctionKitchen;
   shape = Rectangle_Kitchen;
}
Geometry.shape.Rectangle Rectangle_Kitchen = {
   Descr {"Rectangular shape for kitchen"}
```

```
    length = Length_Kitchen;
    referencePoint = ReferencePoint_Kitchen;
    width = Width_Kitchen;
}
Geometry.dimension.Length Length_Kitchen = {
    Value {3,6}
}
Geometry.dimension.Length Width_Kitchen = {
    Value {3,6}
}
Geometry.topology.Coordinate Coordinate_X_Kitchen = {
    Value {6}
}
Geometry.topology.Coordinate Coordinate_Y_Kitchen = {
    Value {6}
}
Geometry.topology.Coordinate Coordinate_Z_Kitchen = {
    Value {0}
}
```

In phase 1 not all spaces mentioned in phase 0 are present, and there are four spaces that have not been assigned a function name by the architect. The older instances of spaces that are included in phase 1 change in the same manner as the kitchen example. Space_0 through space_4 are instantiated directly according to the new Feature Type definition.

3. A CLASSIFICATION OF DESIGN CONCEPTS

The description of the design process in the case studies provides a new way to look at design processes. In particular, changes from one phase to the next can be expressed in terms of changes in the Feature model. In this way, design actions can be matched to Feature model alterations.

The changes in the Feature model are based on the case study. The descriptions of the Feature model alterations are very specific for the case. Therefore, it is necessary to classify them into more general descriptions of design actions and associated changes in the Feature model. The classification provides the proper vocabulary for establishing the functional brief for the VR-DIS system. The following table presents the classification and the definition of the terms for changes in the Feature model.

Table 3. Design actions and changes in the Feature model

Design action	Changes in the Feature model
Generalisation	When a group of objects share common properties, define the specific objects as Feature Types, and define a Feature Type (super type) of which they are sub types. The super type functions as generalisation.

Design action	Changes in the Feature model
Concept identification	Terms in the brief that are relations or spatial-, material-, and functional elements, are defined as Feature Types.
Element creation	Terms in the brief that are actual parts in the design ("hall", "floor", "fireplace") can be instantiated directly on the basis of the corresponding Feature Types.
Constraint creation	Terms in the brief that are relations in the design can be instantiated on the basis of constraint Feature Types.
Concept extension	Adding an association relation to a Feature Type in order to include more characteristics.
Shaping	Giving shape to the spatial elements involves element creation of the Feature Types Shape and of Feature Types position and dimension.
Assignment	On the Instance level make an association relation between Feature Instances.
Move	Move means that the co-ordinates that define position have been changed in a Feature Instance.
Substitution	Substitution means that an existing association between Feature Instances is broken and that one of the Feature Instances is replaced.

4. DISCUSSION

In the case, nine design actions have been identified and described in terms of changes in the Feature model. They are either on Feature Type level only (generalisation, concept identification, and concept extension), on the Feature Instance level only (element creation, constraint creation, assignment, move, and substitution), or a mix of both levels (shaping).

In the current approach, the Feature model was constructed on the basis of the case itself without using predefined Features. The 'dynamics' of the design process can be described in a clear way. In a design support environment however, it is not likely that each design task will start with an 'empty' Feature model, and that the whole has to be constructed through the design. Pre-defined Feature Types will probably be used extensively. The Feature Types that are stock and trade in the building and construction industry can be defined in libraries beforehand and used in any design (so-called "generic Feature Types"). Feature Types that are specific for a design project can be easily created or modified (so-called "specific Feature Types").

The classification can be used to define the design functionality of the VR-DIS system in terms of changes in the Feature model. The operations generalisation, concept identification, concept extension, element creation, constraint creation, assignment, move, substitution, and shaping are defined as changes in the Feature model. Since the functionality of the VR-DIS system is going to be based on Features, this description will facilitate development of tools that support the design actions.

In some of the design actions changes are made in the Feature Type. In a design system using Feature models, this raises the problem of consistency in the model. The changes in the Feature Type have to be propagated through all the relevant instances, and it has to be checked whether this creates inconsistencies or not. The theory of Feature-based modelling as developed does not provide this test, but leaves this responsibility to the user of the Feature data.

Two prototypes have been developed (De Vries and Jessurun 1998a; 1998b) that solve the constraints defined as Features in the model (in phase 0). Furthermore, they keep the constraints solved in real-time, so that changes in the VR-environment show immediate feedback in the behaviour of the spaces.

In the case study, geometry is defined in complex Feature Types for points, dimensions, and shapes. The theoretical framework for Feature-Based modelling now includes geometric Feature Types, which can be used for modelling shapes.

The Feature-Based modelling approach used in the research is aimed to support the early stages of design. The stages in design have been defined in Roozenburg and Eekels (1995) with the "basic design cycle". The basic design cycle encompasses all the phases a design process will go through at least once. It consists of a number of activities, each leading to a result. The cycle starts with a function, which is analysed, resulting in criteria; upon which synthesis takes place, resulting in a provisional design; that is simulated, resulting in expected properties; which are evaluated, resulting in an outcome; upon which a decision is taken whether or not to continue (Roozenburg and Eekels 1995, p. 88-92).

It is clear that in each activity, information of the design will be put to different uses. An underlying Feature model therefore, can have various uses. For example, the statement "scullery between garage, kitchen and bathroom" is information resulting from the analysis activity. In a provisional design, it may appear that the scullery is located elsewhere, which is another piece of information that is in contradiction with the previous one. The first piece of information states a constraint, and the second a state of affairs. Future work will have to focus on these different uses and requirements of the Feature model.

5. ACKNOWLEDGEMENTS

The paper reports from work done in a post-doc research project in the Design Systems group, financed by the Eindhoven University of Technology. The authors would like to thank the research team of the Design Systems Group for contributions and critical review.

6. REFERENCES

Achten, H.H. 1997a, "Generic Representations - Typical Design Without the Use of Types", Junge (ed.), *Proceedings of the 7th International Conference on Computer Aided Architectural Design Futures*, 117-133, (Kluwer, Dordrecht).

Achten, H.H. 1997b, *Generic Representations - An Approach for Modelling Procedural and Declarative Knowledge of Building Types in Architectural Design*, Ph.D. Thesis, Eindhoven University of Technology, Bouwstenen 46.

Achten, H.H., and J.P. van Leeuwen, 1998a, "A Feature-based Description Technique for Design Processes: a Case Study", Timmermans (ed.), Proceedings of the 4th conference on Design and Decision Support Systems in Architecture and Urban Planning.

Achten, H.H., and R.M. Oxman, and Th. Bax, 1998b, "Typological Knowledge Acquisition Through a Schema of Generic Representations", Gero and Sudweeks (eds.), Artificial Intelligence in Design '98, 191-208. (Kluwer, Dordrecht).

Achten, H.H., and Jessurun, J., and Leeuwen, J.P. van, and Vries, B. de, 1998c, Virtual Reality Design Information System/Distributed Interactive Simulation, Design Systems Report 1998/1, Eindhoven.

Achten, H.H. and Mallory-Hill, S.M. and Vries, B. de, 1998d, Workshop Case Based Reasoning and Design, Design Systems Report 1998/2, Eindhoven.

Bronsvoort, W.F. and Jansen, F.W. 1993, "Feature Modelling and Conversion, Key Concepts to Concurrent Engineering", *Computers in Industry*, 21(1), 61-86.

Bronsvoort, W.F. and Dohmen, M. and Bidarra, R. and Van Holland, W. and De Kraker, K.J. 1996, "Feature Modelling for Concurrent Engineering", Proceedings of the International Symposium on Tools and Methods for Concurrent Engineering '96, Budapest, 46-55.

Coomans, M.K.D. and Achten, H.H. 1998, "Mixed Task Domain Representation in VR-DIS", Proceedings of the 3rd Asia-Pacific Conference on Computer Human Interaction, APCHI98, Hayama-machi, Kanagawa, 1998.

Coomans, M.K.D. and Timmermans, H.J.P. 1998, "A VR User Interface for Design by Features", Timmermans (ed.) Proceedings of the 4th Conference on Design and Decision Support Systems in Architecture and Urban Planning, Maastricht.

Coomans, M.K.D. 1999, "A Virtual Reality User Interface for a Design Information System", *CCAI : the Journal for the Integrated Study of Artificial Intelligence, Cognitive Science and Applied Epistemology*, Rijks Universiteit Gent.

Cunningham, J.J., and J.R. Dixon 1988, "Designing with Features: the Origin of Features", Proceedings ASME Computers in Engineering Conference, San Francisco.

Dijkstra, J., Roelen, W.A.H. and Timmermans, H.J.P. 1996, "Conjoint Measurement in Virtual Environments: a Framework", Timmermans (ed.), 3rd Design and Decision Support Systems in Architecture and Urban Planning Conference, Vol I: Architecture Proceedings, 59-71.

Dijkstra, J. and Timmermans, H.J.P. 1997a, "Exploring the Possibilities of Conjoint Measurement as a Decision-Making Tool for Virtual Wayfinding Environments", Proceedings of The Second Conference on Computer Aided Architectural Design Research in Asia, 61-71.

Dijkstra, J. and Timmermans, H.J.P. 1997b, "The Application of Conjoint Measurement as a Dynamic Decision Making Tool in a Virtual Reality Environment", *Proceedings of the 7th International Conference on Computer Aided Architectural Design Futures*, 757-770. (Kluwer Dordrecht).

Dijkstra, J., Timmermans, H.J.P. and Roelen, W.A.H. 1998, "Eye Tracking as a User Behavior Registration Tool in Virtual Environments", Proceedings of The Third Conference on Computer Aided Architectural Design Research in Asia, 57-66.

Dijkstra, J. and Timmermans, H.J.P. 1998, "Conjoint Analysis and Virtual Reality – A Review", Timmermans (ed.), 4th Design and Decision Support Systems in Architecture and Urban Planning Conference.

Dijkstra, J., Timmermans, H.J.P. and Roelen, W.A.H. 1999, "Conjoint Analysis and Virtual Reality – Exploring the Possibilities", *Design Systems Reports*, Design System Report 1999/1, TUE.

Groot, E. de and Mallory-Hill, S.M. and Zutphen, R.H.M. and Vries, B. de, 1999, "A Decision Support System for Preliminary Design", Proceedings of the 8th International Conference on Durability of Building Materials and Components, Vancouver (forthcoming).

Leeuwen, J.P. van and Wagter, H. and Oxman, R.M. 1996, "Information Modelling for Design Support". Proceedings of the 3rd Design and Decision Support Systems in Architecture and Urban Planning Conference, Spa.

Leeuwen, J.P. van, and H. Wagter, 1997, "Architectural Design-by-Features". Proceedings of CAAD Futures '97, 97-115. (Kluwer, Dordrecht).

Leeuwen, J.P. van, and H. Wagter, 1998, "A Features Framework for Architectural Information: a Case Study". Gero and Sudweeks (ed.), Artificial Intelligence in Design '98. (Kluwer, Dordrecht).

Mallory-Hill, S.M., 1998, "Building a Case-Based Design Assistant for Workplace Environment Design", Timmermans (ed.), 4th Design and Decision Support Systems in Architecture and Urban Planning Conference.

Markus, Th.A. 1969, "The Role of Building Performance Measurement and Appraisal in Design Method", Broadbent and Ward (eds.), *Design methods in architecture*, (Lund Humphries, London).

Maver, Th.W. 1970, "Appraisal in the Building Design Process", Moore (ed.), *Emerging Methods in Environmental Design and Planning*, MIT Press, Cambridge, Massachusetts.

Roozenburg, N.F.M. and Eekels, J. 1995, *Product Design: Fundamentals and Methods* (John Wiley & Sons, Chichester).

Shah, J.J. and Mäntylä, M. and Nau, D.S. (eds.) 1994, *Advances in Feature Based Manufacturing* (Elsevier Science, Amsterdam).

Shah, J.J 1991, "Assessment of Features Technology", *Computer-Aided Design*, 23(5), 331-343.

Vries, B. de and Leeuwen, J.P. van and Achten, H.H. 1997, "Design Studio of the Future", Proceedings of the CIB W78 Conference Information Technology Support for Construction Reengineering, Cairns, Australia, 139-148.

Vries, B. de and Jessurun, A.J. 1998a, "An Experimental Design System for the Very Early Design Stage", Timmermans (ed.) Proceedings of the 4th Conference on Design and Decision Support Systems in Architecture and Urban Planning, Maastricht.

Vries, B. de and Jessurun, A.J. 1998b, "Features and Constraints in Architectural Design", Proceedings of DETC'98, 1998 ASME Design Engineering Technical Conference, Atlanta, Georgia.

π-Resolution in Design Space Exploration

Andrew Burrow
Department of Computer Science,
The University of Adelaide SA 5005, Australia.

Robert Woodbury
School of Architecture, Landscape Architecture and Urban Design,
The University of Adelaide SA 5005, Australia.

Key words design space explorers, typed feature structures, functional decomposition, mixed initiative

Abstract In studying the phenomenon of design we use models to envision mechanisms by which computers might support design. In one such model we understand design as guided movement through a space of possibilities. *Design space explorers* embody this model as mixed-initiative environments in which designers engage in exploration via human computer interaction.

Constraint resolution provides a formal framework for interaction in design space explorers. Rather than directly providing solutions to design problems, constraint resolution provides a mechanism for organizing construction. Therefore, we are less interested in the set of solutions to a constraint problem than the process by which intermediate steps are generated.

π-resolution is one such mechanism applicable to design space explorers. It describes the solution, by recursive enumeration, of feature structure type constraints. During π-resolution, satisfiers are constructed by the application of type constraints drawn from an inheritance hierarchy. This constructive process provides a strong model for design space exploration. The constraint solver does not do the work of the designer, but rather design efforts are situated in, and organized by, constraint resolution. Therefore, the efficiency of the recursive enumeration in finding solutions is not an issue, since non-determinism in the search is resolved by the human user as design space exploration.

1. INTRODUCTION

Computer aided design environments involve a conflict between human and machine requirements. One approach is to suspend the search for an account of the human designer, and instead to accept metaphors of design that correspond well to formal systems. The intuition is that tools should provide a logic comprehensible to the user, rather than attempt to simulate sophisticated forms of human reasoning, and the hope is that users respond by improvising with the media.

The metaphor of design as exploration within a space of possibilities, suggests mechanisms by which computers might support design. *Design space explorers* are computer media that engage designers in exploration. Appropriate formal systems, on which design space explorers may be implemented, arise in the description of generative processes. One example, is boundary solid grammars in the GENESIS system (Heisserman, 1994).

One difficulty facing shape grammars, and other formal systems focused on the design artifact, is that the extant formalisms give little guidance in structuring the design space exploration. This manifests itself largely in the representation of goals and goal-directed movement in a design space. Lacking a user comprehensible logic in goals and exploration works against achieving clarity in user interaction.

Another approach is to explicitly consider goals in exploration and to structure a formal representation and exploration process using those goals. In such systems, transformations of the goal may become the focus of the interaction between human and machine. Here the user accessible logic is a form of problem decomposition. This paper argues that feature structure π-resolution is a formal mechanism which provides an excellent framework for design space exploration. This fit is improved by extensions which explicitly characterize the constructive resolution process and the intermediate states.

The aim of this paper is to provide an intuition of the formal mechanism of typed feature structures that allow us to both express and generate design representations characterised by complex functional roles. Our experience is that this intuition does not come easily, since most of us are attuned to an object-oriented view derived from object-oriented programming techniques. By contrast the typed feature structure formalism is terse and organised around a global view of the data structure. Therefore, where possible this paper omits the formal aspects of the mechanism in favour of the underlying intuitions. (Woodbury et al., 1999) provides a more detailed introduction.

1.1 Design Space Explorers

Engineering design space explorers involves representing possibilities in the form of symbol structures called *design states*, and *design spaces* of possibilities as a relation over design states. This requires a compact representation of design states with efficient algorithms, and definitions of the space-generating relation which are relevant to the exploration metaphor and which lend themselves to tractable computations.

A pervasive notion within the exploration metaphor is goal-directedness. In essence, goals exist independently of the designs that realize them and the process of exploration is guided by a set of goals that itself may be transformed within the exploration process. Computing with goals requires their representation. Since goals may change throughout an exploration process, design states record both the goals towards which they aim and the artifacts that (partially) realize those goals.

Another notion is that of non-determinism — the goal does not of itself determine the decisions enacting the exploration. An external agent, the user, must resolve the indeterminacy in seeking the goal. In design space explorers, each decision is expressed in the transformation of one design state to another. A design space is implicitly defined by a set of initial design states and a set of state transforming operators. A design space explorer assists a user to unfold this compact and implicational description into an explicit space of recorded possibilities.

Design space explorers provide a mixed initiative environment where user and computer play complementary roles. A user experiences the design space explorer, inter alia, through resolving exploration non-determinism. Support for this role is therefore critical to the external qualities of a design space explorer. Careful presentation of alternatives avoids overwhelming a user, and assists a user to perceive coherent movements in the design space.

The experience from extant spatial grammar based tools is that a necessary modification to their formalisms and thus their interpreters is a mechanism to control the availability of transformations so that construction occurs in coherent phases. The representation of operators should reflect this need to structure and organize alternatives. Alternatives may be mutually exclusive, interdependent or independent. Without a structuring principle for operators, these properties will be evident only in patterns distributed across the design space. With a structuring principle for operators, these properties may be expressed in the presentation and grouping of alternatives.

A design space explorer is then a device that represents goals, operators, and design possibilities. It organizes design states into spaces, and provides means to access states and apply operators to extend the explored space. A designer interacts with an explorer through states, spaces, operators, and the presentation of their interactions. The design space explorer is thus an active participant in a mixed initiative environment based on the exploratory unfolding of design spaces.

1.2 π-Resolution and Design Space Explorers

Feature structures generalize record-like data structures. They are a form of directed graph, where nodes represent domain objects and edges represent associative connections. Object identity is modeled by structure sharing and the structure as a whole models partial information. Since feature structures model partial information there exists an informational ordering with an efficient decision procedure, and an operation to combine information where consistent. In addition, *descriptions* provide a means for calling out collections of feature structures based on *satisfaction*. Satisfaction has the property that, if a feature structure satisfies a description, then so too does every feature structure which extends the information in it.

These properties of feature structures are attractive in design space explorers. As *representatives* for design states, explicit models for partial information lend legitimacy to intermediate exploration states. In particular, we represent functional decompositions of design states by feature structures, and functional roles in these decompositions by features. In exploration terms, extending a feature structure corresponds to answering "How?" questions which transform the problem to include finer levels of subproblem. This is an attractive process as the surface logic of a design space explorer, because it captures a notion of goal directedness in terms of the problem at hand.

π-resolution is a non-deterministic search for solutions to a query description. A solution is a feature structure, which satisfies the query and the *constraint system.* It is the result of a sequence of extension steps corresponding to the satisfaction of constraints, which are organized into an inheritance hierarchy of types. Since π-resolution proceeds by extension, the resultant search space can be order embedded into the informational ordering over feature structures. Since constraints are drawn from an inheritance hierarchy, alternatives may be organized according to notions of abstraction. If feature structures represent functional decompositions, π-resolution non-determinism allows exploration in terms of alternative functional decompositions.

User input accounts for human judgment in the design process, but the intervention associated with this input is also important. Human judgment could be applied without intervention by enumerating the design space and having the user select from the solutions. One issue is computational intractability — in exploration, the user allocates computer resources by guiding exploration. More important is the cognitive load on the user. Design environments should leverage, rather than squander, human skills. Well informed partitioning of solutions, via exploration, depends on exploration choices comprehensible in terms of user-defined distinctions.

This brings us to π-resolution as a model for design space exploration and to the search space of feature structures as a record of exploration. Given that feature structure types and their constraints capture useful distinctions, π-resolution monotonically refines those distinctions, and an element in the search space records the distinctions made locally. Navigation in the space of alternatives then proceeds by asking "which of the decisions important to me was made here?" Since the search space order embeds into the informational ordering the classification and indexing techniques in the knowledge representation literature are available for content based indexing and querying.

1.3 Related Work

Design space explorers have their genesis in the shape grammar literature (Stiny, 1980; Stiny and March, 1981). Later researchers focused on the formal system and human-computer interfaces required to create effective design space explorers (Heisserman, 1994; Carlson, 1993; Coyne, 1991; Carlson and Woodbury, 1994; Woodbury et al., 1992; Harada et al., 1995; Woodbury and Chang, 1995).

Feature structure theory has its provenance in computational linguistics as described by Carpenter (Carpenter, 1992), but owes much to its application to logic programming by Aït-Kaci(Aït-Kaci and Podelski, 1993), who was also the first to realise the potential optimisations for reasoning operations on bounded complete partial orders (Aït-Kaci et al., 1989). The breadth first π-resolution techniques for recursive type constraints in feature structures also grew out of work on feature structures as a term language for logic programming (Aït-Kaci et al., 1993).

The use of order theory for information management and presentation is based on techniques for searching in partial orders (Ellis, 1995) and on the notion of lattices as a media for philosophical discourse as proposed in Formal Concept Analysis (Wille, 1992).

There are numerous knowledge level characterisations of design information stressing the role of function. For example, the SEED knowledge level (Flemming and Woodbury, 1995) and the SHARED object model(Gorti et al., 1998). Some of these are sufficiently explict and parsimonious to admit a symbol level interpretation.

2. FEATURE STRUCTURES

In this section we define a feature structure language(Carpenter, 1992) in sufficient detail to discuss π-resolution. Since the resolution of type constraints may be described independently of the class of feature structures(Carpenter, 1992, p 228), we do not consider type inference, inequations, or extensional types. Furthermore, we do not consider descriptions in detail except to note their use in representing queries and type constraints. Rather than provide formal definitions for subsumption and unification we consider these mechanisms in terms of functional decomposition.

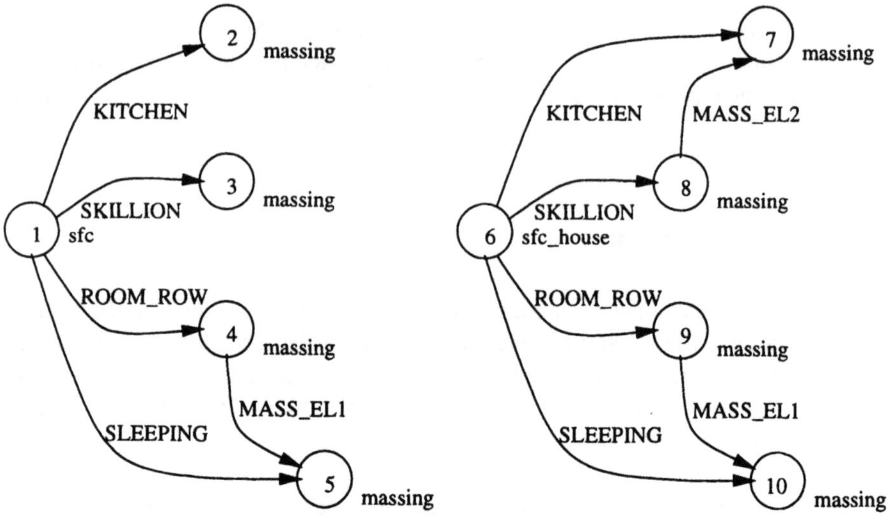

Figure 1: Feature structures in graph notation.

Feature structures generalize record-like data structures and represent partial information. They may be depicted as rooted, directed, finite, node and edge labeled graphs, where nodes represent objects, edges represent functional relationships, and structure sharing is taken to model identity. Therefore, cycles represent reflexive relationships, and informational equivalence does not assert identity. Each node in a feature structure may be interpreted as the root of a feature structure, termed a *substructure*. At nodes, labels are interpreted as types, and on edges, labels name the functional role that the target fills wrt the source.

Types are exploited as a means of efficiently encoding information. They are organized into a multiple inheritance hierarchy, in which information associated with a type is extended in inheriting types, i.e., an informational ordering. It is mandated that this ordering is a bounded complete partial order (BCPO), meaning that for every set of types with a common subtype there is most general common subtype or join. Thus, certain reasoning operations over types are available as the lattice operations $\sqcup$ (join, or most general common specialization) and $\sqcap$ (meet, or most specific common generalization).

Figure 1 depicts a pair of simple feature structures in a subsumption relation. The example represents the decomposition of a building design, the nodes denote building entities and the edges denote functional roles in the design. By characterising feature structures as representations of partial information, it is possible to consider both examples as partial representations of some final design. The second feature structure provides additional information about the final design as recognised by the subsumption relation.

The example demonstrates three sources of ordering in the information. The simplest sources of ordering are the elision of feature paths and the ordering over the type labels. In this case, the first feature structure does not define the feature path SKILLION MASS_EL2, and the type of the first root node is a supertype of the second's. However, since structure sharing is taken to model identity, information on the identity of the represented objects is encoded in the feature paths. Each node is interpreted as an existential proposition that there exists some object conforming to the node's type. The existentially quantified node cannot be interpreted as stating the identity of the represented object, i.e., two distinct nodes may represent the same object or separate objects. Instead object identity is not specified except wrt the feature paths extending to a node. The second feature structure identifies a single object as fulfilling the roles of both KITCHEN and SKILLION MASS_EL2, which is a piece of information not present in the first feature structure.

Since designs are inherently existential representations that are made more specific by proceeding stages of the design process, the notion of partial information captured in typed feature structures is a useful one. This is especially true of the functional decomposition of a design. Monotonic satisfaction between *descriptions* and feature structures allows us to describe a constructive mechanism for functional decomposition. For each description, either an information minimal feature structure exists that satisfies the description, or the description is unsatisfiable. Since every specialisation also satisfies the original description, the union of feature structures satisfying a collection of descriptions must also satisfy every description in the collection and thus satisfy their conjunction. Hence we are able to express information about designs by descriptions, and construct feature structures by *unification* operation as a means of directly processing the descriptions.

2.1 Feature Structures

Given an inheritance hierarchy, a BCPO of types $\langle \mathsf{Type}, \leq_{\mathsf{Type}} \rangle$ and a set of features Feat, a typed feature structure is formally defined as a rooted, directed, finite, node and edge labeled graph.

Definition 1 (Feature Structures(Carpenter, 1992)) *A* feature structure *is a tuple* $F = \langle Q, \bar{q}, \theta, \delta \rangle$, *where*

- Q *is a finite set of* nodes,
- $\bar{q} \in Q$ *is the* root node,
- $\theta : Q \mapsto \mathsf{Type}$ *is the node* typing *function,*
- $\delta : \mathsf{Feat} \times Q \rightarrow Q$ *is the partial* feature value *function, and*
- Q *is the smallest set such that* $\bar{q} \in Q$ *and* $q \in Q \wedge q' = \delta(f, q) \rightarrow q' \in Q$.

A shorthand for composition of the feature value function is useful to denote traversal along a sequence of features. Following Carpenter(Carpenter, 1992), we refer to a sequence of features as a *path*, the collection of paths as $\mathsf{Path} = \mathsf{Feat}^*$, and the *empty path* as ϵ. We then extend the feature value function to paths, such that $\delta(\epsilon, q) = q$ and $\delta(f\pi, q) = \delta(\pi, \delta(f, q))$.

Now, feature structures can be interpreted as structures assigning values to paths. The feature value function assigns nodes to paths. The typing function assigns types to these nodes. A node may also be interpreted as the root of a feature structure. Hence, types and feature structures are also assigned to paths.

Definition 2 (Path Value(Carpenter, 1992)) *Given a feature structure $F = \langle Q, \bar{q}, \theta, \delta\rangle$ and a path π such that $\delta(\pi, \bar{q})$ is defined, the* substructure *at π is the feature structure $F@\pi = \langle Q', \bar{q}', \theta', \delta'\rangle$, where*

$$Q' = \{\delta(\pi', \bar{q}') \mid \pi' \in \mathsf{Path}\} \;, \tag{1}$$

$$\bar{q}' = \delta(\pi, \bar{q}) \;, \tag{2}$$

$$\theta'(q') = \begin{cases} \theta(q') & : q' \in Q' \\ \textit{undefined} & : \textit{otherwise} \end{cases} \;, \textit{and} \tag{3}$$

$$\delta'(f, q') = \begin{cases} \delta(f, q') & : q' \in Q' \\ \textit{undefined} & : \textit{otherwise} \end{cases} \;. \tag{4}$$

2.2 Subsumption

The informational ordering over types is extended to feature structures by the *subsumption* relation. A feature structure's information is represented by the values it assigns to paths. Subsumption is an inclusion ordering over this information. A feature structure F *subsumes* another F', written $F \sqsubseteq F'$, if and only if: for every path π defined in F, π is defined in F' and the type at π in F subsumes the type at π in F'; and for every pair of paths π and π' defined in F, if π and π' identify a single substructure in F then π and π' identify a single substructure in F'. This is expressed formally via a class of maps between feature structure node sets, such that every node in q in F is mapped into a node q' in F', the type at q subsumes the type at q', and q' occurs at a superset of the paths at which q occurs. If $F \sqsubseteq F'$ and $F' \sqsubseteq F$ then F and F' are information equivalent *alphabetic variants*, written $F \sim F'$.

Returning to the example in Fig. 1, the first feature structure subsumes the second because every functional role in the first is present in the second, every object fulfilling multiple functional roles in the first fulfills a superset of these roles in the second, and for every functional role identified in the first the object fulfilling this role in the second is of a matching or more specific type. Each piece of information represented in a feature structure relates to the existence of a feature path or the type label assigned to a substructure at a feature path.

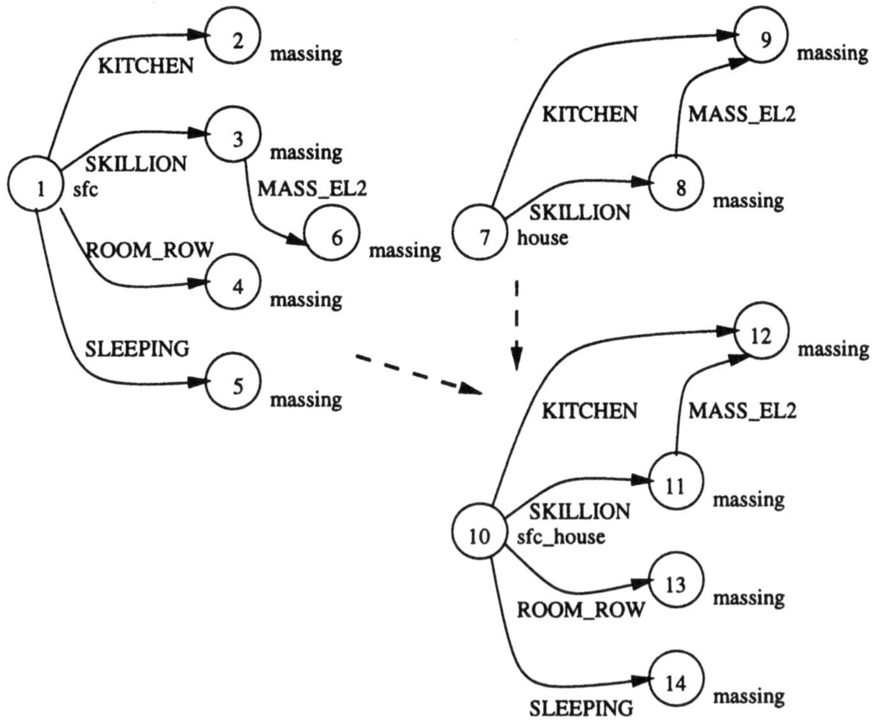

Figure 2: Feature structure unification: the bottom feature structure is the unification of the left and right feature structures.

Subsumption makes it clear that the information associated with a feature structure derives from the collection of feature paths and the type labels associated with substructures. However, subsumption can also be related to abstraction. On the structural side abstraction relates to the elision of functional roles and the uncertainty in their identification, while on the type labelling side it relates to the classification of the represented object. The recursive type constraint system provides a bridge relating these two sources of information.

2.3 Unification

Given the feature structure subsumption order, we seek a reasoning operation to compute the conjunction of two feature structures. Intuitively, this operation seeks the most general feature structure that is more specific than either operand. Unification is such a procedure, that either fails if the two feature structures represent inconsistent information, or constructs the most general specialization otherwise.

The *unification* of feature structures F and F', written $F \sqcup F'$, is a conjunction of the information considered by subsumption. Unification incorporates the collections of feature paths in F and F' as well as their type labelling. It must group feature paths so that sharing is replicated and reassign type labelling. For every pair of paths π and π' in F, if π and π' identify a shared substructure in F, then they also identify a shared structure in $F \sqcup F'$ so that it is subsumed by F. Likewise for F'. This is formalised by a partitioning of the nodes in F and F' into equivalence classes. $F \sqcup F'$ exists only if there exists a common specialisation of the types in each equivalence class.

In terms of functional decomposition, the unification of two design representations simultaneously decides whether some object may exist which is consistent with the two functional decompositions and, if so, provides an information minimal representation true only of such an object. Unification is the single constructive mechanism used to reconcile information from a type constraint system with a query description during design space exploration.

The key intuition to understanding unification is the collection of equivalence classes. During the unification of two feature structures, nodes are collected into equivalence classes according to feature paths. This relates directly to the information derived from feature paths in subsumption. In the simplest case a pair of nodes, one from each operand, are placed into an equivalence class if the same feature path leads to each node. Where there is already structure sharing there are already equivalence classes of paths, and therefore unification leads to more complicated aggregations of equivalence classes.

Figure 2 depicts three simple feature structures which are the operands and result of a unification operation. The example demonstrates a type join at the root node, the inclusion of new feature path value from both operands and the preservation of structure sharing.

2.4 Descriptions

Descriptions serve to call out feature structures through the *satisfaction* relation. Satisfaction is monotonic: if a feature structure satisfies a description, then so too does every feature structure which it subsumes(Carpenter, 1992, p 55). Therefore, a description may be interpreted as a reference to its most general satisfiers. In fact, every feature structure is the most general satisfier of a disjunction free description(Carpenter, 1992, p 56). The relationship between descriptions and feature structures is stated formally in terms of the functions MGSat and MGSats. Given the set of all feature structures $\mathcal{F}$, for the disjunction free descriptions $\mathsf{NonDisjDesc}$ there is a surjection $\text{MGSat} : \mathsf{NonDisjDesc} \rightarrow \mathcal{F}$, and for the set of descriptions Desc, sets of pairwise incomparable feature structures are identified by the function $\text{MGSats} : \mathsf{Desc} \rightarrow 2^{\mathcal{F}}$.

In a design space explorer, descriptions act as compact textual representations of functional decompositions. Since description satisfaction is monotonic wrt feature structure specialisation, descriptions are statements providing a lower bound of specificity on the represented object. By the inclusion of disjunction into descriptions a conjunction of these statements is satisfied by an arbitrarily large collection of functional decompositions. In this paper we are proposing that a form of π-resolution provides a formal framework for the implementation of a design space explorer. The key feature is the organisation of knowledge about functional decompositions into a system of recursive type constraints.

2.5 Recursive Type Constraints

Without additional type information, the type t of a feature structure F asserts only that the represented object is an instance of t or some subtype. By stating restrictions on the feature structures of each type, we assert additional properties on the feature structures and therefore about represented objects. The *constraint system* expresses restrictions on a substructure, according to its type, which determine legal labellings over a finite but arbitrary collection of paths. *Constraint resolution* determines whether a feature structure satisfies a constraint system.

Definition 3 (Constraint System(Carpenter, 1992)) *A* constraint system *is a total function* Cons : $\mathsf{Type} \mapsto \mathsf{Desc}$.

A feature structure is *resolved* if it satisfies the constraint system at every substructure. A substructure satisfies the constraint system if it satisfies the type constraint for its type and all supertypes.

Definition 4 (Resolved Feature Structure(Carpenter, 1992)) *A feature structure* $F = \langle Q, \bar{q}, \theta, \delta \rangle$ *is* resolved *if and only if*

$$\forall \pi \in \mathsf{Path}, \forall t \in \mathsf{Type} \quad : \quad t \leq_{\mathsf{Type}} \theta(\delta(\pi, \bar{q})) \rightarrow F@\pi \textit{ satisfies } \mathrm{Cons}(t) \qquad (5)$$

Satisfaction monotonicity ensures a direct relationship between constraint resolution and unification. A type constraint description is satisfied if a most general satisfier subsumes the constrained substructure. Therefore, constraint resolution is the search for the most general feature structure subsumed by both a constraint satisfier and the current substructure. This is decided by unification. In this paper we make no distinction between a description and its set of most general satisfiers, i.e., we equate constraints with feature structures, and consider satisfaction in terms of subsumption by a most general satisfier, and resolution in terms of unification with a most general satisfier.

2.6 π-Resolution

Given a query description D, *π-resolution* is the search across a sequence of feature structures $P_0 \sqsubseteq P_1 \sqsubseteq P_2 \sqsubseteq \ldots \sqsubseteq P_k$. The initial feature structure in each sequence is a most general satisfier of the query description. The sequence represents the inclusion of type information in the form of constraints — each element extends its predecessor by unification with a type constraint. Since most general satisfiers may occur as collections and unification may fail, the search for resolved feature structures involves a collection of sequences.

Definition 5 (**π-Resolution**) *Given a feature structure $F = \langle Q, \bar{q}, \theta, \delta \rangle$ and a path π such that $\delta(\pi, \bar{q})$ is defined, let the sequence $t_1, t_2, \ldots t_k$ be an enumeration of the types $\{t \mid t \leq_{\text{Type}} \theta(\delta(\pi, \bar{q}))\}$. We take $F \overset{\pi}{\Rightarrow} F'$ if and only if there exists a sequence $F_0 \sqsubseteq F_1 \sqsubseteq \ldots \sqsubseteq F_k$ such that $F \sim F_0$ and $F' \sim F_k$ and*

$$F_{i+1} \sim F_i \sqcup (\pi \mid C_{i+1}) \wedge C_{i+1} \in \text{MGSats}(\text{Cons}(t_{i+1})) \tag{6}$$

The operation in Definition 5 is one step in a π-resolution search. During the search process the current partial satisfier evolves via numerous such resolution steps with the addition of information from the constraint system. Each step acts on a substructure of the current partial satisfier. The notation $\pi \mid F$ denotes the most general feature structure whose value at the path π is F. It is guaranteed to add no information to the partial satisfier P except at the substructure $P@\pi$, so the process can be thought of as acting directly on substructures. However, missing from the definition is a procedure to select π. The π-resolution process recursively enumerates the most general satisfiers of the query description. The process is complete since it can be proven that a feature structure is a most general solution only if it is the result of some sequence of π-resolution steps(Carpenter, 1992).

3. INCREMENTAL π-RESOLUTION

In the proposed mixed initiative environment the user is to provide analytical talents in the comprehension of the design's balance of qualities, within a discourse based on the functional decomposition of the design problem. The machine is to provide feedback on the computable attributes of the current design state, the most fundamental being the current design artifact's satisfaction of the functional decomposition. In this context, design space explorers must transform both the functional decomposition and the corresponding model.

Since model transformation is external to π-resolution, we are interested in identifying possible synchronization points. It is important that these synchronization points define legitimate intermediate states, because they will be parameters to model transformation operations. In particular, their may be further choice points associated with the model transformation. Choice points are another issue in the algorithm. By definition π-resolution generates resolved feature structures, and therefore the solution set is unaffected by the organization of choice points. However, choice points determine the trajectories in feature structure space that resolution may trace. This determines the exploration freedom available to the design space explorers.

As it stands, π-resolution is incremental in the sense that it appears to solve type constraints one substructure at a time. However, it is difficult to formalize this property. A complete search strategy is ensured by a breadth first strategy that selects the unresolved path of minimal length(Carpenter, 1992, p 244). However, a substructure in a cycle can still undergo multiple resolution steps.

Definition 5 suggests a finer level of incremental construction: independently resolving each type constraint introduced at a supertype, rather than resolving the conjunction of constraints on supertypes. By recording the types against which a substructure has been resolved, it is possible to ensure that a substructure is resolved against each type at most once. This is an important synchronization property. It also suggests a useful intermediate state invariant. Note that, irredundant enumeration of types in Definition 5 controls the order in which type constraints are resolved at a substructure. While unification is commutative, this may not be true of the associated model transformations. Therefore, an important class of orderings are the linear extensions of the inheritance hierarchy. In terms of the model transformation, this order is analogous to the order in which supertype constructors are called in C++(Ellis and Stroustrup, 1990) i.e., inherited types are processed before derived types.

Another problem for synchronization is substructure unification. During π-resolution existing substructures may be unified. In terms of functional decomposition, this corresponds to a decision to share an object between functional roles. Clearly, this operation requires model transformations and therefore synchronization, especially since the substructures may be at different levels of resolution.

We propose *incremental π-resolution* as an algorithm with minimal incremental steps and coincident synchronization and choice points. In this paper we characterize the intermediate states in the algorithm. That is, each substructure records a *down-set* of types (a set closed under the addition of supertypes) against which it has been resolved, and may be resolved against any type which extends this down-set and which is also a supertype of the current label on the substructure. The first restriction ensures that each type is resolved once at a substructure, and only after all supertypes have been resolved. The second restriction on type resolution ensures that incremental π-resolution does not perform resolutions outside those in Definition 5. When a substructure is resolved against a type, the type constraint is unified at the substructure.

In incremental π-resolution, synchronization points coincide with choice-points where the user selects a substructure and a type. The down-set of types to which the substructure has been resolved is extended by the addition of the type and a satisfier of the type's constraint is unified at the substructure. This corresponds to one of the intermediate steps in Definition 5. The user experience of this control point is mediated by the model in the design space explorer. For example, having selected a building entity to work on the incremental π-resolution state is consulted to determine the sound functional decompositions available wrt the current state of the building entity. When constraints conflict, the disagreement between the attempted type constraint and the partial satisfier are available to the design space explorer in order to inform the user.

Minimal changes simplify model transformations. The model must be transformed to account for a single substructure which has been incremented by a type, all of whose supertypes have already been handled. New structure entering the partial satisfier from the constraint has minimal impact on the model, since it has not yet been resolved to even the most general type and therefore would not be expected to be represented in the model. Another synchronisation task occurs when existing substructures are unified in the updated substructure. For example, SKILLION MASS_EL2 and KITCHEN from the first feature structure in Fig. 2. The model transformation corresponding to the new structure sharing is informed by the equivalence classes generated in the unification and the previous resolution state of the unified substructures. For example, a certain type resolution may signal a resource allocation in the model so that if two substructures resolved to this resource sensitive type are unified the the design space explorer must relinquish one instance of the resource in the resulting entity.

The trajectories possible in incremental π-resolution capture a notion of goal directedness that is both formal and open. In addition, incremental π-resolution contains a component of shortest unresolved path strategy, since a path does not occur until all substructures on the path have been resolved to include the feature they carry in the path. While the algorithm is sophisticated by the need to track multiple open choice points, this is achieved by the addition of an extra type function that maps nodes to type down-sets identifiers. This information is also useful to the design space explorer in generating model transformations for unified substructures. In this sense incremental π-resolution departs from logic programming in removing the left to right ordering on rule application and throwing the construction of the search space into a completely non-deterministic process.

3.1 Incremental π-Resolution State

An incremental π-resolution state is an annotated partial satisfier. It comprises: the query description, a partial satisfier of the query, and a representation of resolved types. Its purpose is to provide for the computation of available refinements at an incremental π-resolution choice-point.

Definition 6 (Incremental π-Resolution State) *Given a query description D, a feature structure $P = \langle Q_P, \bar{q}_P, \theta_P, \delta_P \rangle$, and a function* $\text{TState} : Q_P \mapsto 2^{\text{Type}}$*, the tuple* $\Pi = \langle D, P, \text{TState} \rangle$ *is an* incremental π-resolution state *if and only if*

$$P \textit{ satisfies } D \tag{7}$$

$$\forall q \in Q_P \quad : \quad \text{TState}(q) \subseteq \{t \mid t \leq_{\text{Type}} \theta_P(q)\} \tag{8}$$

$$\forall q \in Q_P, \forall t \in \text{TState}(q) \quad : \quad P@q \textit{ satisfies } \text{Cons}(t) \tag{9}$$

Because paths do not uniquely define substructures, TState is defined over nodes rather than over paths. This makes (9) awkward so we write $P@q$ for the substructure rooted at q. Otherwise, Definition 6 is straightforward, simply asserting that TState records the state of a partial satisfier in terms of the types resolved at each substructure.

3.2 Incremental π-Resolution

Given a query description D, the algorithm is initialized by having the user select a most general satisfier P_0 from the query description and setting TState to the empty set for all nodes in P_0. Each step in the algorithm requires the user to select a substructure for resolution, a type extending the substructure by resolution, and a most general satisfier of the extending type's constraint.

Definition 7 (Incremental π-Resolution) *Given an incremental π-resolution state $\Pi = \langle D, P, \text{TState} \rangle$, a path π_i such that $q_i = \delta_P(\pi_i, \bar{q}_P)$ defined a type t_i such that*

$$t_i \in \{t \mid t \leq_{\text{Type}} \theta_P(q_i) \wedge t \notin \text{TState}(q_i) \wedge t' \leq_{\text{Type}} t \longrightarrow t' \in \text{TState}(q_i)\}$$

and a most general satisfier $F_i \in \text{MGSats}(\text{Cons}(t_i))$.

We take $\Pi \overset{\pi_i, t_i, F_i}{\Longrightarrow} \Pi'$ *if and only if $P \sqcup (\pi_i \mid F_i)$ exists, and set*

$$\Pi' = \langle D, P', \text{TState}' \rangle \tag{10}$$

$$P' \sim P \sqcup (\pi_i \mid F_i) \tag{11}$$

$$\text{TState}'(q') = \bigcup \{\text{TState}(\delta_P(\pi, \bar{q}_P)) \mid \delta_{P'}(\pi, \bar{q}_{P'}) = q'\} \tag{12}$$

The set of types from which an extending type t_i may be drawn is determined to ensure that the new TState$'$ value is a down-set. Intuitively, a type cannot be selected unless it makes a contribution to the target type on the partial satisfier and all its supertypes have already been resolved, i.e, it subsumes the type on the node and all its supertypes are in $\text{TState}(q_i)$. Equation 12 mirrors the unification operation computation of types for equivalence classes. It is awkward in this definition because the equivalence relation is not available. Intutively, it states that where substructures unify the new resolution state is the union of their resolution states. This effectively prevents type constraints from being repeatedly applied even where this occurs indirectly via unification. This is an important property supporting the model synchronisation process.

Initially there is one possible move for each substructure in the most general satisfier selected for the query description. Since TState is set to $\emptyset$ for all initial substructures, the extending type is restricted to the most general type, which is also guaranteed to subsume any type. It is unlikely that the constraint on the most general type is non-empty. Therefore, each such initial move finishes with the most general type inserted into TState for the selected substructure.

For subsequent moves, the algorithm acts on a user selection at π_i extending to type t_i and with satisfier F_i, by inserting t_i into $\mathrm{TState}(\delta_P(\pi, \bar{q}_P))$ and unifying F_i at $P@\pi$. Each such move introduces new structure or refines existing substructures. The active choice points in an incremental π-resolution state rapidly grow downwards from the root node. This is the intention of the characterisation — minimal incremental changes supporting synchronisation with the model, and a completely open process which may be tailored by the design space explorer with heuristics to suit the complete range of problem decomposition styles.

4. CONCLUSION

Functional decomposition is suggested as a framework for exploration without claiming that functional decompositions equate to designs. Therefore, separate actions are required to ensure a design artifact consistent with the functional decomposition. The proposed algorithm is a modification to π-resolution which makes additional choice points available to the resolution strategy. The advantage of these choice points lies in their relationship to the synchronization needs of an external model transforming process.

In the case of the design space explorer, this external mechanism may involve a hybrid system perhaps including shape grammar fragments. Functional decomposition by incremental π-resolution provides a context for these operations that may be considered a sophisticated labelling strategy. However, additional benefits are realised in exploiting the partial ordering as a system for indexing both whole designs and design fragments. Retrieval is then possible on the basis of design requirements.

One benefit of this approach is that a single object represents the design state's functional decomposition rather than a computational state spread across a database. This can be coupled with a characterisation of feature structures that eliminates alphabetic variance by the generation of persistent identifiers for path equivalences. Since the computational state is characterised by the type hierarchy and path equivalences, persistent storage then extends to computation states. Each such computation state is a table mapping a collection of path equivalences to a target type and a resolution state. Under these conditions design spaces may be stored beyond the execution of the program and their indexing extended to incomplete designs.

This paper introduces the notion of incremental construction by providing a characterization of the intermediate states. The algorithm is not prescriptive in form. Instead it provides a formal mechanism upon which a variety of heuristics can be grafted. This paper does not consider this area, except to note that many very interesting strategies are directly realised as partitionings and traversals of the type hierarchy especially where additional appropriateness constraints are introduced on the combinations of features and types.

The described mechanism has been implemented, as part of a project to engineer a design space explorer based on properties of typed feature structures. A major engineering task has been the management of identifiers for the type down-sets used to represent resolution state. However, this itself has enjoyed benefits arising from the formalism that were unexpected. The experience suggests that even the most formal symbol representations have surprising and useful properties that arise as non trivial consequences of the theory.

5. ACKNOWLEDGEMENTS

The authors would like to acknowledge the support given by the Australian Research Council Large Grants Scheme, The Department of Industry, Science and Tourism Major Grant Scheme, the Australian Postgraduate Award Scheme, The Key Centre for the Social Applications of Geographic Information Systems and The University of Adelaide School of Architecture, Landscape Architecture and Urban Design.

References

Aït-Kaci, H., Boyer, R., Lincoln, P., and Nasr, R. (1989). Efficient implementation of latice operations. *ACM Transactions on Programming Languages and Systems*, 11(1):115–146.

Aït-Kaci, H. and Podelski, A. (1993). Towards a meaning of LIFE. Technical Report 11, Paris Research Laboratory of Digital Equipment Centre Technique Europe, Rueil-Malmaison, France.

Aït-Kaci, H., Podelski, A., and Goldstein, S. C. (1993). Order-sorted feature theory unification. Technical Report 32, Paris Research Laboratory of Digital Equipment Centre Technique Europe, Rueil-Malmaison, France.

Carlson, C. (1993). *An Algebraic Approach to the Description of Design Spaces*. PhD thesis, Department of Architecture,Carnegie-Mellon University.

Carlson, C. and Woodbury, R. (1994). Hands-on exploration of recursive patterns. *Languages of Design*, 2:121–142.

Carpenter, R. (1992). *The Logic of Typed Feature Structures : With Applications to Unification Grammars, Logic Programs and Constraint Resolution*. Number 32 in Cambridge Tracts in Theoretical Computer Science. Cambridge University Press, Cambridge.

Coyne, R. (1991). *ABLOOS: An Evolving Hierarchical Design Framework*. PhD thesis, Department of Architecture, Carnegie Mellon University, Pittsburgh, PA. also available as a TechReport, (EDRC-02-15-91), Engineering Design Research Center, CMU.

Ellis, G. (1995). *Managing Complex Objects*. PhD thesis, Department of Computer Science, University of Queensland.

Ellis, M. A. and Stroustrup, B. (1990). *The annotated C++ reference manual*. Addison-Wesley, Reading, MA, USA.

Flemming, U. and Woodbury, R. (1995). Software environment to support early phases in building design (seed): Overview. *Journal of Architectural Engineering*, 1:147–152.

Gorti, S. R., Kim, G. J., and Siriam, R. D. (1998). An object-oriented representation for product and design processes. *Computer-Aided Design*, 30(7):489–501.

Harada, M., Witkin, A., and Baraff., D. (1995). Interactive physically-based manipulation of discrete/ continuous models. In *SIGGRAPH '95 Conference Proceedings*, volume 29, pages 199–208. ACM Siggraph, ACM.

Heisserman, J. (1994). Generative geometric design. *IEEE Computer Graphics and Applications*, 14(2):37–45.

Stiny, G. (1980). Introduction to shape and shape grammars. *Environment and Planning B: Planning and Design*, 7(3):343–352.

Stiny, G. and March, L. (1981). Design machines. *Environment and Planning B: Planning and Design*, 8(3):241–244.

Wille, R. (1992). Concept lattices and conceptual knowledge systems. *Computers Math. Applic.*, 23(6–9):493–515.

Woodbury, R. F., Burrow, A. L., Datta, S., and Chang, T.-W. (1999). Typed feature structures in design space exploration. *AIEDAM Special Issue on Generative Systems in Design*. manuscript available at http://www.arch.adelaide.edu.au/ rw/publications/progress.html.

Woodbury, R. F. and Chang, T.-W. (1995). Massing and enclosure design with SEED-Config. *ASCE Journal of Architectural Engineering*, 1(4):170–178.

Woodbury, R. F., Radford, A. D., Taplin, P. N., and Coppins, S. A. (1992). Tartan worlds: A generative symbol grammar system. In Noble, D. and Kensek, K., editors, *ACADIA 92*, pages 211–220, Charleston, SC.

Feature-Based Qualitative Modeling of Objects

John S. Gero and José C. Damski
Key Centre of Design Computing and Cognition,
Department of Architectural and Design Science,
University of Sydney, NSW, 2006, Australia
email: {john,jose}@arch.usyd.edu.au

Key words: qualitative modeling, features, object modeling, computer-aided design

Abstract: Objects represent fundamentally important ways with which to perceive and reason about the world. There are very few adequate representations for use at the early stages of designing. Feature-based approaches founded on qualitative representations have been used. This paper extends the qualitative representation developed for two-dimensional shapes to three-dimensional objects. It uses the qualitative representation to detect features. An example of the application of this representation is presented.

1. MOTIVATION

This paper introduces and describes a qualitative approach to the modeling of objects applicable at the early stages of designing. It draws on and extends the work of Gero and Park on the feature-based qualitative modeling of two-dimensional shapes (Gero and Park, 1997).

Objects are the fundamental building block of any building design system. Most modern CAD systems use objects. The problem with current approaches to the representation of objects in CAD systems in the need to be explicit about both object topology and object geometry. This is not a difficulty at the later stages of designing when these have been fixed but it is a problem at the early stages of designing.

Qualitative approaches based on symbolic modeling provide the opportunity to represent objects as classes rather than as instances. As a consequence they can be used at the early stages of designing when objects do not yet have a fixed geometry, only an indicative one based on their topology. Further qualitative representations lend themselves to symbolic reasoning which can be used to convert the representation into one based on features. The use of features adds a dimension to the levels at which a system can reason about objects. Features lend themselves to semantic interpretations with labels which have meanings for human users. Thus, it becomes possible to carry out such reasoning as:

for two objects to mate there must be a match between their protrusions and indentations,

where protrusions and indentations are features. In this paper the qualitative approach to the modeling of objects applicable at the early stages of designing is extended through features to the feature-based qualitative modeling of objects.

We use concepts derived from feature-based modeling to capture design knowledge related to the qualitative character of objects. This will be presented in three steps. The first step is to develop an appropriate representation scheme: this is based on previous work on syntactic pattern and contour representation methods (Gero and Park, 1997). The second step is to turn those qualitative descriptions of objects into a meaningful object semantics. The third step is to discover meaningful object, and then later design, semantics by using syntactic pattern matching of features from the given description of objects. Although the syntactic approach to the representation is not new, our syntactic representation scheme for qualitative characteristics of objects extends current generic methods and specialises them for design-related tasks which have previously either been unavailable or have been difficult.

The term "qualitative" is the complementary word to "quantitative". Qualitative methods normally handle variables within a symbolic system that is not necessarily based on a numeric scale, but based on comparisons and contrasts of relative magnitude and direction. Qualitative representations make only as many distinctions as necessary to distinguish objects and situations. They may be independent of any fixed scale of measurement and a single qualitative value can represent a range over a numeric interval. It represents an object with a set of qualitative symbol codes composed of a finite number of discrete symbols. A qualitative representation represents a class of objects rather than a single instance. In this sense it bears some connection to design prototypes (Gero, 1990).

Setting qualitative values to object attributes follows a strict formalism suggested by researchers from the qualitative reasoning community (deKleer and Brown 1984). Qualitative values are set to variables by mapping their numeric values (where they exist) to a finite and discrete set of symbolic values. Where the numeric information does not exist, a lexigraphic ordering of the concepts can be used. The simplest way of setting qualitative values from the real number range of values is to use a "landmark set" with qualitative values in the range {-↗, 0, ↗}. Then the set of intervals becomes {(-↗, 0), (0, 0), (0, ↗)}, which corresponds to the qualitative set Q with the sign values {+, 0, -} (Wertner 1994). This simple yet effective process of symbolic mapping is useful in modeling design variables and attributes, transforming possible numeric value ranges into small sets of discrete and finite symbol values. This formalism of setting qualitative values can be applied to most design variables since most values can be measured in terms of "polarity" and "granularity".

Whilst a number of object attributes may be used, here we demonstrate results obtained from using only one attribute, namely the angle between the bounding surfaces at their intersections. We take the standard graph theoretic representation of the boundary of solid objects and add angle semantics to the arcs to produce a semantic network representation as the qualitative representation of objects. From this qualitative representation we can reason about the mapping between unique structures in the semantic network and features.

2. SYMBOLIC REPRESENTATION FOR OBJECTS USING Q-CODES

2.1 Basic formalism

Researchers in the qualitative reasoning (de Kleer and Brown, 1984) have defined a set of qualitative values to a shape or object. This set of values is obtained by mapping (existent) geometric features into a set of discrete symbolic values. The commonest way to assign these symbolic values from numeric ones is to use a landmark set. The basic three attributes to represent a landmark of a shape or object are:

- Angle represented by A and is the angle measured between two lines (or surfaces or tangents to surfaces in three-dimensional objects). It is measured from the line that lies in both surfaces. The intervals of values are $\{(0,0), (0,\pi), (\pi,\pi), (\pi,0)\}$ which are mapped on to {Anil, A+, A0, A-};
- Length represented by L. It is the relative length of the line segment (or surface) compared to the previous one in a given order. The intervals of values are $\{(-\infty,0), (0,0), (0,+\infty)\}$ which are mapped on to {L-,L0,L+};
- Curvature represented by K. It represents the relative curvature of the line segment (or surface) with values of "convex", "straight", and "concave" mapped on to {K+,K0,K-}.
-

The qualitative representation of an object is realized by encoding each face of the object following one or more of the above criteria. Figure 1 shows the Q-code encoding of some simple 2D faces.

Those Q-codes are composed in similar way as in any natural language such as English. A sequence of Q-codes forms a word that may represent a shape pattern of significance such as a face of an object. A sentence, as a sequence of words, may describe an entire object. Such an analogy can be extended as we derive a semantic meaning from a sentence in similar way we can derive features from objects. Details of such representation and the reasoning process to obtain them can be found in

Gero and Park (1997), Chuang and Henderson (1990) and Gyankar and Henderson (1990).

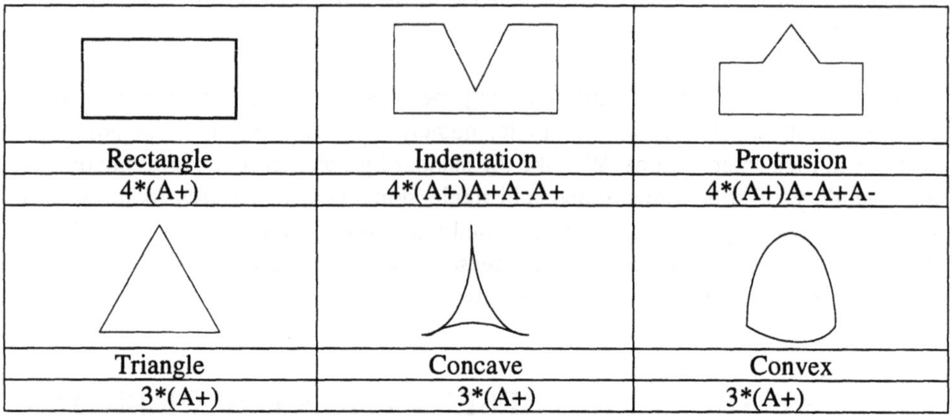

Figure 1. Q-code encodings of some simple faces, adapted from Gero and Park (1997)

The Q-codes representing the shapes in Figure 1 allow us to treat these shapes as exemplars of classes of shapes. Thus, the three shapes in the lower half of the figure are all, in some sense, triangles. All three have the same A-coding even though they appear to be different shapes. In two-dimensional shapes the landmarks occur at the vertices of the shapes. In three dimensional objects, as will be seen, the vertices are not the appropriate locus for landmarks.

3. QUALITATIVE GRAPH REPRESENTATION OBJECT ANALYSIS

3.1 Semantic Net Graph model

The geometry of three-dimensional objects can be represented as a set of vertices (points) and edges (lines) connecting the vertices. In this paper we represent an object as a set of faces and their relative position. This is implemented through a *semantic net* graph where each node represents a face (plan) and the arcs the connection between two faces (Mantyla, 1988). This connection is defined by a common line between planes. For each arc in graph model we assign a label "+" or "-" if the external angle measured between the planes are more (or equal) then 180° or less, respectively.

Figure 2(b) illustrates this graph according for the L-shaped object shown in Figure 2(a). Figure 2(b) shows eight faces “a” to “h” and eighteen edges/arcs. As all angles between plans as more then 180°, all arcs are labelled as "+". We use this semantic net graph to represent the sequence of Q-code encodings of a 3D object and to extract qualitative information, such as features, by parsing this representation.

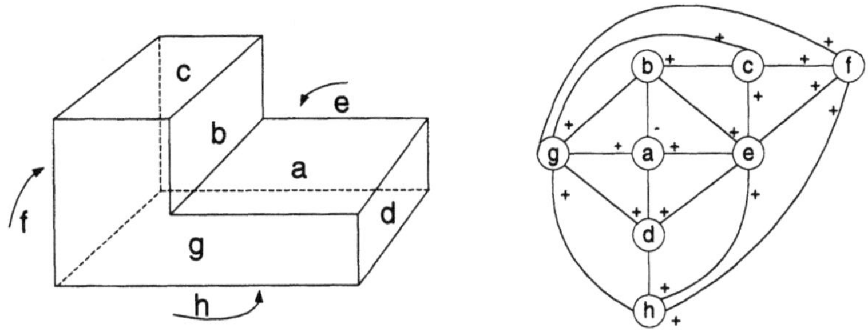

(a) (b)

Figure 2. (a) L-shaped object (b) semantic net representation

Some other examples of this semantic net graph representation of three-dimensional objects are shown in Figure 3.

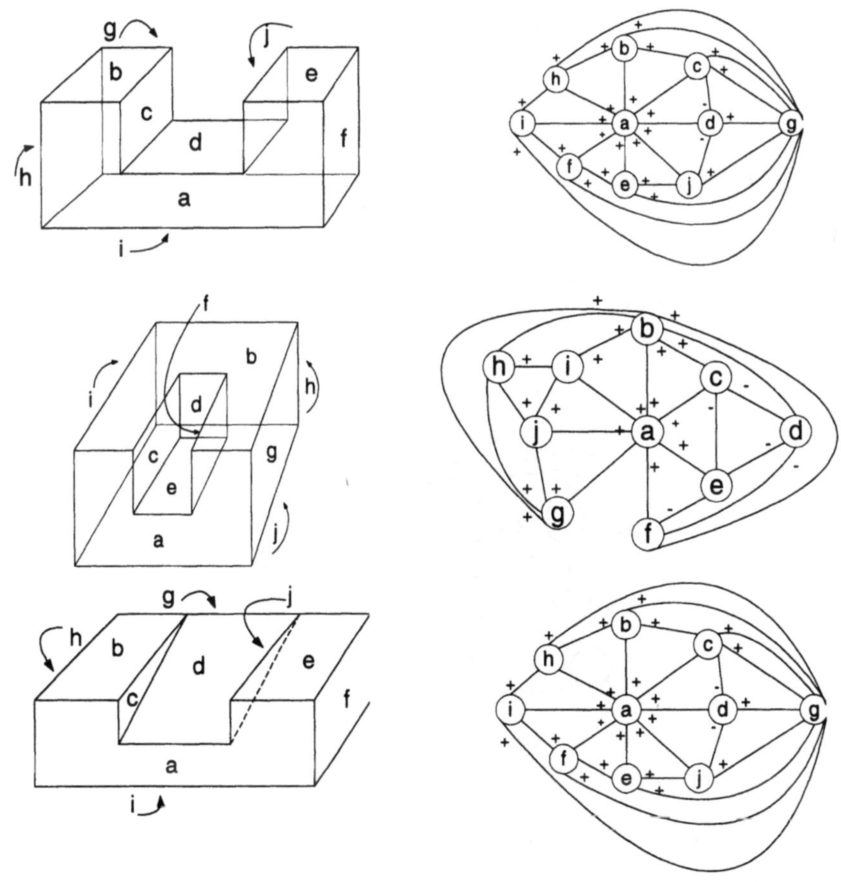

Figure 3. Objects and their semantic net graphs

A vertex of a three-dimensional object is represented by a minimal circuit[1] in the semantic net graph model. As the graph is non-directional, the orientation is not important. In this case a vertex is a sequence of labels "+" and "-" as we traverse a minimal circuit. Let us restrict ourselves to three-dimensional objects where only three planes intersect at each vertex. It is not difficult to generalise this restriction. In this situation there will be 4 types of vertices, labelled *r*, *s*, *t* and *u*:

r	*s*	*t*	*u*
+++	++-	+--	---

Figure 4 shows the four possible types of intersections of three planes. They can be named outer corner (r), outer L shape (s), inner corner (t), and inner L shape (u).

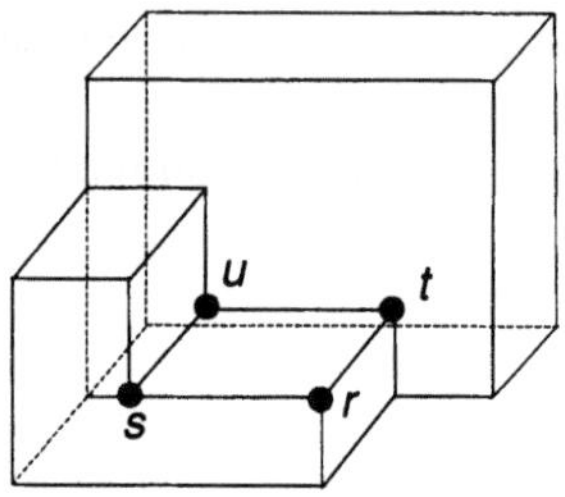

Figure 4. All possible combination of intersection of 3 planes

3.2 Vertex Graph

Given the semantics of vertices *r, s, t* and *u* it is possible to re-represent the semantic net as a vertex graph where nodes are the type of vertex and the arcs the connection between them. Using the example shown in Figure 2, the vertex graph is as shown in Figure 5.

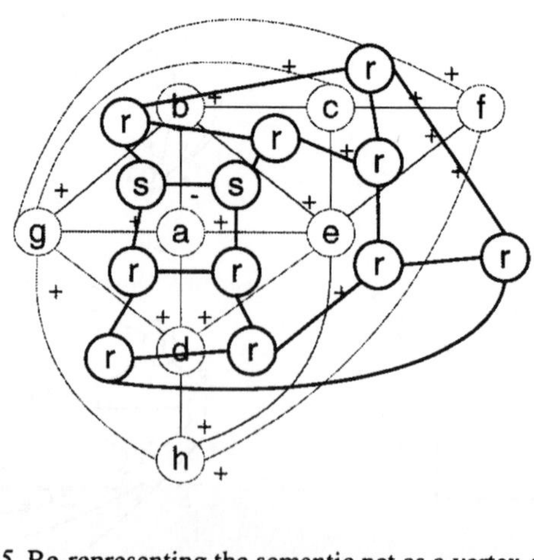

Figure 5. Re-representing the semantic net as a vertex graph

[1] Minimal circuit is a circuit which does not contain any other node.

The graph shown in dark lines in Figure 5 can be re-drawn as shown in Figure 6, which also labels the minimal circuits from a to f as the faces in the original object.

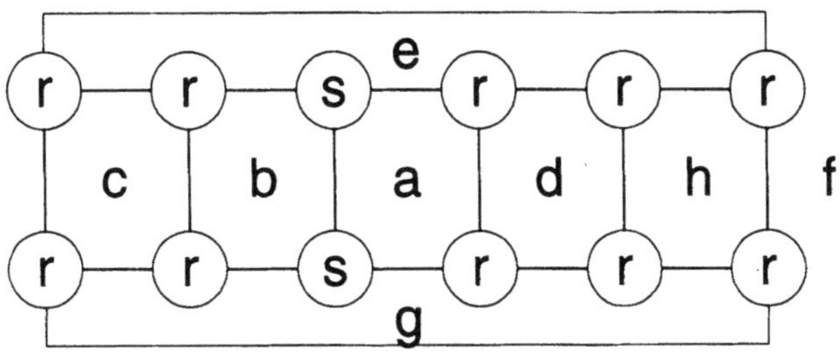

Figure 6. Vertex graph model with faces labelled

In order to analyse the vertex graph model there are some semantic attributions necessary to interpret it. Table 1 shows some of these semantic attributions for objects with vertices composed of three planes intersecting.

Table 1. Some semantic attribution to vertices with 3 planes intersection

Feature	Circuit	Additional constraints
Protrusion	rrrr/tttt	At least one r/t is adjacent to at least one s/u; or All r are connected to t; or All t are connected to r.
Indentation	ssss/uuuu	All s/u are connected to r/t; or All s are connected to u; or All u are connected to s.
Step	rrss/ttuu	
L-Shape	rrss/ttuu	All r/t/s/u are adjacent to r
U-shape	ssss/uuuu	ss/uu are connected to rr/tt the other ss/uu also connected to rr/tt
Hole	ssss	Every s is connected to another s which make another circuit of s.

Using the interpretation shown in Table 1 and the graph in Figure 6 it is possible to detect the following features:

- protrusion for faces c and d
- L-shape for faces b and a.

A more complete example is shown in Figure 7, where a prism sits inside a cavity.

From the semantic net graph shown in Figure 7 it is possible to derive three separated graphs, as shown in Figure 8. The reason for three separated graphs is due to a lack of information about the connection between the vertices of the outside box to the vertices of internal hole. This is a classical problem in computer graphics where the basic solution is to create an artificial link to connect one structure to another. In our representation it is not necessary to create such an artificial link, because the semantic information about the planes are passed over from the representation of semantic net to vertex graph. This is one interesting feature of this re-representation.

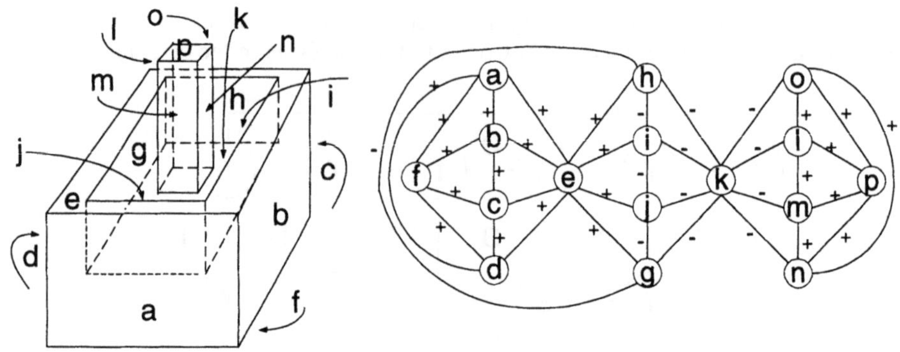

Figure 7. A prism inside a cavity and its semantic net graph

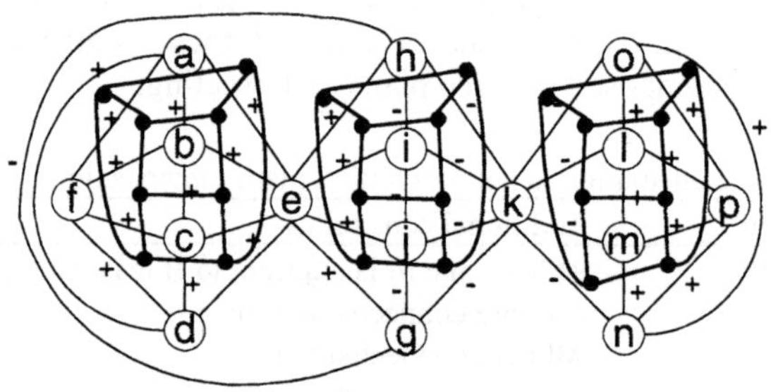

Figure 8. Three separated graphs

Figure 9 show all three graphs and their faces.

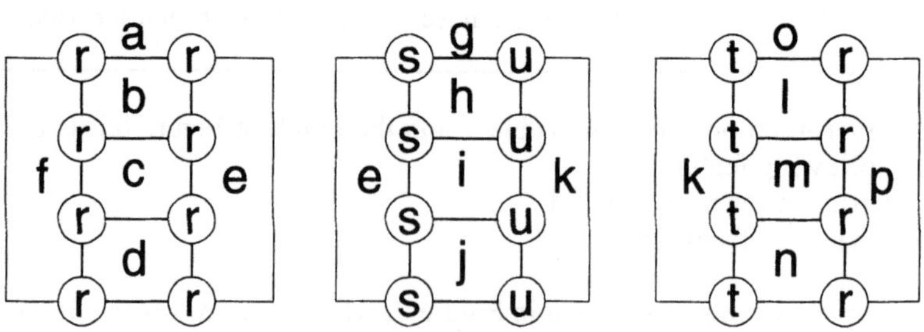

Figure 9. Detailed graphs with planes labelled

Analysing Figure 9 with Table 1 we can infer the following:

- from the left graph: nothing can be inferred with the semantic feature set we currently have;
- from the middle graph: there is an indentation formed between planes e and k; and
- from the right graph: there is a protrusion formed between planes k and p.

It is possible to derive increasingly complex features by concatenating simple features. For example, knowing that an indentation is formed between planes e and k and that a protrusion is formed between planes k and p and that plane k is common to both features, it is possible to create a new feature: an annular indentation.

4. ARCHITECTURAL EXAMPLE

Figure 10 shows The Great Wall of China. A simplified model of the part of the Wall is shown in Figure 11, where the faces are labelled from a to r. A semantic net graph derived from the model in Figure 11 is shown in Figure 12. For simplicity of diagramming the thicker line represents the same label as the one at the end of the line.

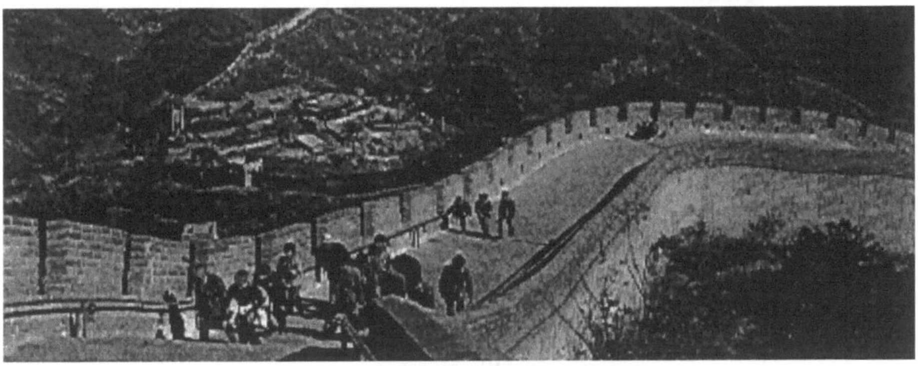

Figure 10. The Great Wall of China

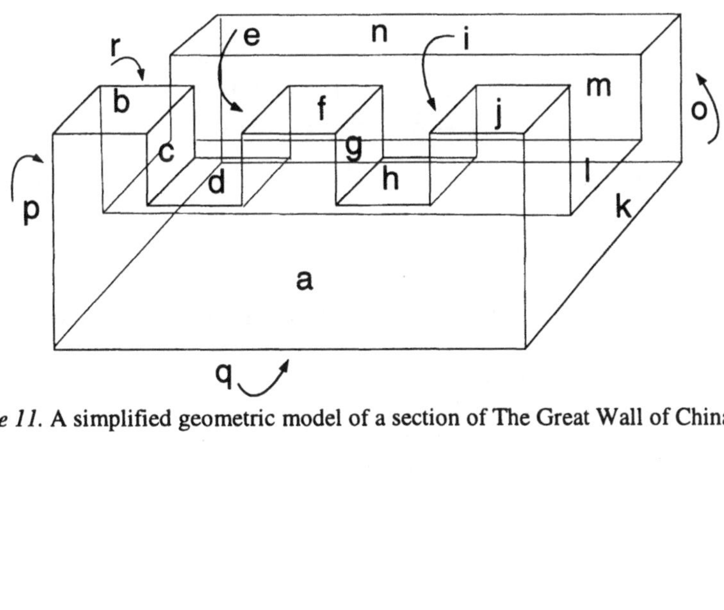

Figure 11. A simplified geometric model of a section of The Great Wall of China

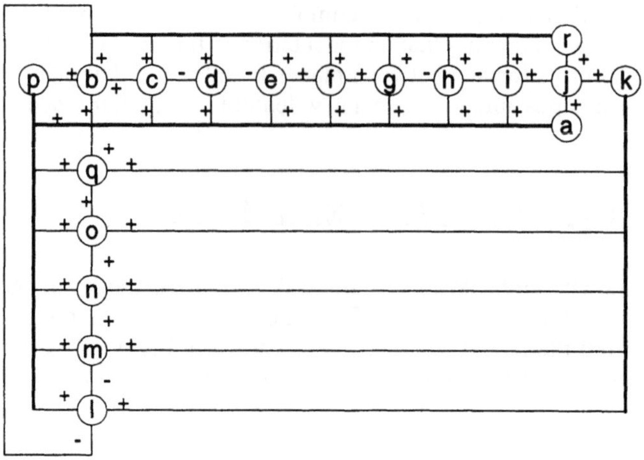

Figure 12. The semantic net from Figure 11

Re-representing the semantic net shown in Figure 12 we obtain the vertex graph shown in Figure 13, with the faces now labelled within the vertices.

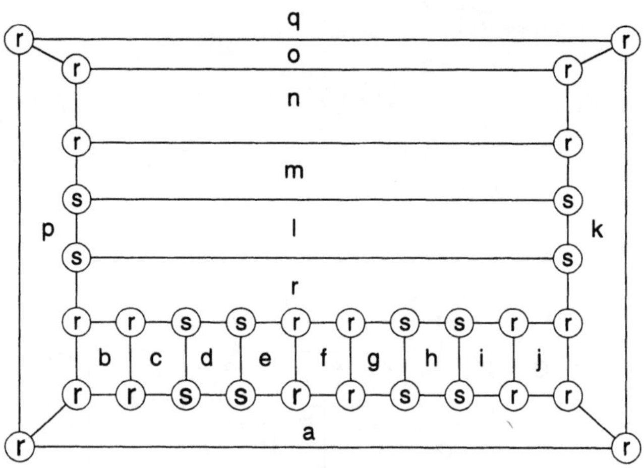

Figure 13. Graph re-representing the semantic net as a vertex graph

Applying the semantic attributes from Table 1 it is possible to detect the following features:

- indentations in faces d, h, l;
- protrusions in faces b, f, j, n
- steps in faces c, e, g, i, m; and
- U-shapes in faces m-l-r, c-d-e, g-h-i.

5. DISCUSSION

Knowledge about shape and objects plays a major role in initial stages of design. Often it is the features in shapes and objects that are as significant as the shapes or objects themselves. Gero and Park (1997) developed a particular Q-Code representation for shapes as an associated reasoning process to detect features in shapes. In this paper we have presented a qualitative extension of that work to three-dimensional objects.

With this representation it is possible to reason about objects through their features unrelated to their specific geometries, since the object knowledge is derivable from the basic description of the object. From this basic description we build a semantic net graph representation. This semantic net is labelled with the relative position of each face of the object compared to its adjacent faces. The graph obtained from re-representing the semantic net shows the property of each vertex. It is interesting to note that the semantic information is carried on from one representation into another.

We showed some basic features that can be obtained from the graph model. More complex features can inferred using a combination of those basic ones. For example, Table 1 shows a feature "hole" which is valid only if the two ends of the hole are directly connected. Figure 14 shows a more complex hole in the form of an L-shape.

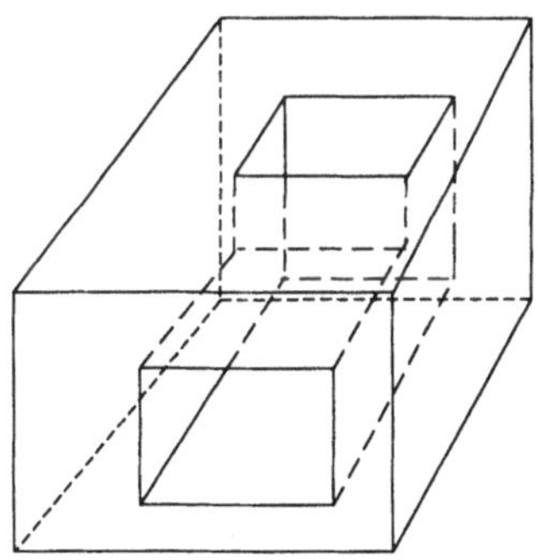

Figure 14. L-shaped hole

Rules to detect such complex features can be built based on the same principles we used to build the basic features. These complex features may be developed targeting particular types of applications or a particular domains.

What we have been able to show is that a qualitative representation of three-dimensional objects provides the foundation for modeling and then detecting features. Feature-based modeling provides a fundamentally different means of describing objects than does geometry. Features may be concatenated to produce more complex features that can then be connected to domain knowledge so that other forms of analysis can be carried out. An obvious next question is whether it is possible to reverse the process and synthesise objects that exhibit desired features. This, however, is beyond the scope of this paper.

ACKNOWLEDGMENTS

This work is supported by a grant from the Australian Research Council, Grant No. A89802104. Computing resources are provided in part by the Key Centre of Design Computing and Cognition.

REFERENCES

Chuang S and Henderson M R, 1990 Three-dimensional shape pattern recognition using vertex classification and the vertex-edge graph, *Computer-Aided Design,* 22(6): 377-387

de Kleer J and Brown J, 1984 A qualitative physics based on confluences, *Artificial Intelligence,* 24: 7-83

Gero J S, 1990 Design prototypes: a knowledge representation schema for design, *AI Magazine,* 11(4): 26-36.

Gero J S and Park S-H, 1997 Computable feature-based qualitative modeling of shape and space, *in* R. Junge (ed.), *CAADFutures 1997,* Kluwer, Dordrecht pp. 821-830.

Gayankar P and Henderson M R, 1990 Graph-based extraction of protrusions and depressions from boundary representations, Computer-Aided Design, 22(7): 442-450

Mantyla M, 1988 An Introduction to Solid Modeling, Computer Science Press, Rockville

Wertner, H. (1994) Qualitative Reasoning: Modeling and the Generation of Behavior, Springer-Verlag, Vienna.

The Front of the Stage Of Vitruvius' Roman Theatre

A new Approach of Computer Aided Design that Transforms Geometric Operators to Semantic Operators

Giovanni De Paoli, Marius Bogdan
University of Montreal, Faculty of Environmental Design, Montreal CANADA

Keywords: architecture, CAD, function, modeling, semantic operator, geometric operator

Abstract: The driving force of all researches where the systems of computation are used, is the utilization of an intelligent method for the representation of building. The use of computer, in design process, is often limited to technical functions (*tekhne*), and what one usually calls computer-aided design is often no more than computer-aided drawing. In this research paper we continue a reflection on the architect's work methods, and suggest an approach to design based on the semantic properties of the object (i.e. semantic operators), rather than by geometric operators. We propose a method of computer aid design using procedural models where the initial state of design is vague and undefined.

We operate from a paradigm that leads to represent a building by means of parametric functions that, expressed algorithmically, give a procedural model to facilitate the design process. This approach opens new avenues that would permit to add the *logos* (semantic properties) and lead to a metaphorical representation. By means of procedural models, we show that, from a generic model we can produce a four dimensional model that encapsulate a volumetric model with semantic characteristics. We use a meta-functional language that allows us to model the actions and encapsulate detailed information about various building elements.

This descriptive mechanism is extremely powerful. It helps to establish relations between the functions, contributes to a better understanding of the project's aim, and encapsulates the building properties by recalling characteristics of common classes which give rise to a new configuration and a completely original design. The scientific result of this experiment is the understanding and confirmation of the hypothesis that it is possible to encapsulate, by means of computing process, the links between design moves during conceptual and figural decisions and transform the geometric operators in semantic operators.

1. INTRODUCTION

The linear approach that permitted the development of the first systems of computation didn't allow a full development of a computer architectural language that takes into account, in phase of design, the architect must develop a complex approach. This method followed the same thought of simplification in the representation used in architecture since the Renaissance. Since that period the drawings are just a part of the information that we want to symbolize and they don't express the design: it is only a geometric representation of building with more or less details.

These considerations allow us to affirm that, in order to obtain a real aid to the design process, the computer tool must not be used like a tool to resolve problems, but like creation tool in design situation. This approach permits all the actors of architectural project to call for the presence of the computer tool in conceptual understanding.

In order to understand this approach we analyzed the distinctive character of the architecture. We studied the results of the researches on the understanding of jamming that limit the utilization of the computer for the design. By way of one procedural model constructed from the description of the Roman theater by Vitruve (First century BC) we asked the computer to act like a tool that allows us to obtain some figurations that can be "governed" by a functional language (De Paoli, Bogdan, 1999).

Subsequently we concentrated our efforts on the assessment and the optimization of various building functions like the acoustics and the visibility in order to see how a functional language could create a system of intelligent representation of building. For it we studied several computer postures: the representation of scenes by the declarative design (Popova, De Paoli, 1998), the functional language for the creation of procedural models (De Paoli, Bogdan, 1999) and finally the complex management of semantic meta-functions presented in this article.

We are going to present our experiences that suggest a change of paradigm for the architectural figuration and the utilization of the data processing during the phase of design. We demonstrate that the computer, by the slant of the utilization of semantic operators, is not only a tool of aid design situation, but also a tool to develop an understanding of the building's properties in a collaborative design situation.

2. LANGUAGE, THOUGHT AND REALITY: AN APPROACH INTELLIGENT TO THE ARCHITECTURE

Every time we study language and we try a principle of its theoretical implementation we have the obstacle of the limitations of linear reasoning:

expressing a model by a language is always an approximation of patterns and an imprecision of the object. For example during the design process we are confronted to some different paradigms: the one is the design of form and matter and the other one the work in the space-time and in the space-world. Presenting these aims and transposing them in terms of conceptual interaction and convenient applications means to change the paradigm of design used until present: «The architecture is like a big sculpture where men penetrate, march, live». (Zevi, 1959)

If we affirm that the architecture is the materialization of combinative ideas of parameters that define the complexity of building, we could suggest to model these parameters, that will be named semantic operators. «Design problems are, as a rule, ill-defined. This implies that the initial state of a design problem may be vague, its goal state not clearly definite, and that no algorithm is available for the "journey" from the initial state to the goal state» (Goldsmith, 1997).

Therefore, like underlines it Goldsmith, we think of the typical design as an indeterministic and complex problem, whose solution necessitates a search. «A solution to a design problem is a design proposal». We want by this approach in the CAO to suggest a way that modifies the present aim of the representation: "manufacture" of utilitarian plans, in truth technique plans for the realization of the architectural work, where these plans are principally a set of resolutions of technical problems. To add semantic problems to technical problems means to choose a figural representation and to analyze, by the way of meta-functional languages, all of the measurements and properties of the space, where men walk and live.

Actually the drawing is transformed, in the workshop, into a simple geometric system that facilitates the realization of the building, while its value as a design tool is disregarded. The same thing happens to computer-based tools where computer-aided design is nothing more than computer-aided drawing, forgetting that the design tool is first and foremost a tool permitting reflection: a non-synthetic and semantic means of developing an idea, to reduce the distance between the representation of the building and its final result. (De Paoli, Bogdan, 1999)

The aim of this article follows some studies during the last ten years by researchers and professionals, on the understanding of the jamming that limits the utilization of the computer during the design process and on the explanation of the links between the visual communication and the data processing. We see a computer as a tool to put together the actors of the building process and to represent the essential operations of the architect during its design. "We want to explain the problems to solve in order to establish a common language between the different actors and suggest some ways of solution, while taking into account that the architect developed a language and some methods of work adapted to the 2D representation, and suggest a modification of language for the figural approach that encourages the full exploitation of this technology." (De Paoli, 1997) Our hypothesis is that it is possible to translate the logic and the history (logos) of the construction process by an algorithmic description, and we examine ways of producing these translations. It is also possible to operate, during the design process, with some semantic operators like acoustics or visibility. Our preoccupation is the study of the methods in order to transmit these translations by a functional language and a figural approach. In fact, researches of this last decade, oriented toward the modeling of the actions and the process of architectural project, reinforce the idea that these operators could be transformed in functions, and give a geometric and representative result, or, like suggest some researchers, a figural representation.

We want, by our gait, to specify and to validate some operators who help in computer design and to build a unique model, where the design passes by the definition of the properties of the object for design like the visibility, the acoustics or the matter, in order to arrive at a complex figural approach.

3. THE SEMANTIC STATE OF THE ARCHITECTURE : THE DESIGN BY DECLARATIVES SCENES AND PROCEDURAL MODELS

A research that we achieved on the methods of work of architects and the computer simulation put in evidence a jamming facing the utilization of the CAO: the impossibility of transmitting to the computer the properties that the architectural model must have in order to extract a figural representation. A figural representation expresses the process of genesis, the constructive logic and the constrained dimensions and the dealings with types of non geometric information but the liberty of the architect's choice is limited by the actual software to the assembly of primitive geometric, that doesn't express the idea. For it, we studied the languages of programming in order to understand some ways of solution. We demonstrated that to design with a computer it is necessary that there is a situation of interaction casual-probabilistic (Popova, De Paoli, 1998). This approach seems very coherent with the characteristic of the declarative methodology, permitting a certain inaccuracy and incompletion of the description. In sharing the means of description by classes it insures the possibility of designing some scenes (models) starting from a partial description of the properties.

With this type of language the rule of the macroinstructions on superior levels provide a method for modeling the design tasks like methodological knowledge. The rule of the macroinstructions of descriptions is a link of causality destined to construct a basis of causal probabilistic knowledge containing the identification of nodes on the causes and effects and the links of causality between these nodes.

In a second research project where the main goal was to model construction techniques for Byzantine vaults, as well as the know-how of the Byzantine themselves, we were able to understand and explain the relationship between the functional and the operational. The functional is the purpose (function) of the vault, and the operational is the expertise (know-how) of the builders (Bogdan, 1997). This allowed us to continue our researches in order to try transferring this knowledge to other moments of history and to transform this meta-function in models that will be named procedural.

By the heritage of this model we have created a procedural model of Roman theater of Vitruve, in order to understand and explain that, by a generic pattern, we could produce a model with all features which belongs to the same family of objects. This pattern allows illustrating the result of processes and proving the relationship with the buildings by their process of construction. In order to express the method of construction, a functional language permits the modeling of the actions. This functional language permits to encapsulate the properties of building. (De Paoli, Bogdan, 1999)

4. MODELING AND DESCRIPTION OF EXPERIENCES

The figuration (figural representation) consists in a series of three-dimensional scenes which metaphor semantic, constructive and temporal reality to validate the hypothesis: it is possible to encapsulate semantic properties and to design, in computer environment, with procedural models. We choose for the end of the experiment the formal description of the Roman theatre by Vitruve, which allows us to interpret the operations of construction and to group them in modules (classes) which we translated into functions by the use of a functional language (SCHEME) and we visualized these modules by a program of solid modeling (SGDLsoft).

Thereafter, we added a semantic operator, the visibility and we proceeded to the declarative design of scenes which interpret these operations. Finally we added our current knowledge in acoustics (for example reverberation) and we created models for design process that could be used in draft and sketches by designers.

The following programs and tables describe our research strategy :

• TABLE A. Program and Representation by geometric primitives;

• TABLE B. Program and Representation by functional language;

• TABLE C. Representation by crossing geometric and semantic operators;

• TABLE D Transformation of functions for semantic properties.

4.1 Program of representation by geometric primitives

In this case we studied the result of a geometric description, it is a pattern which data are "imperatives". The transcription in DXF format is a professional language description. The parameters are frozen, they must be defined to the starting point: the program executes some data which express the geometric relationships only (Figure 1). Some operators described by Vitruve, such relationships between music and astrology could not be interpreted.

TABLE A. Program and Representation by geometric primitives

Operator Analysis

De conformationes theatri facienda (Vitruve)	Operator analysis
The plan of the theatre is constructed as follows: having placed the center, draw a circle whose circumference forms the perimeter of the theatre;	Geometric operator: Circle
Inside this circle, inscribe four equilateral triangles, equally spaced and whose vertices touch the circle and subdivide it as do the astrologers in placing the twelve signs of the zodiac, when they are making computations from the musical harmony of the stars. ...	Semantic operator: Astronomy<->music

Example of Vitruve pattern, by "imperative" and linear computing

0 SECTION 2 HEADER 9 $ACADVER 1 AC1009 9 $INSBASE 10 0.0 20 0.0 30 0.0 9 $EXTMIN............... ... $LTSCALE 40 1.0 9 $ATTMODE 70 1 9 $TEXTSIZE	0.2 9 $TRACEWID 40 0.05...... .... VERTEX 8 LAYER-1 6 CONTINUOUS 62 7 5 39 10 0.0 0 VERTEX 8 .... VERTEX 8 LAYER-1 6 CONTINUOUS 62	7 5 3A 10 7.339467 0 VERTEX 8 .... LINE 8 LAYER-1 6 CONTINUOUS 62 5 52 10 9.181379 .... 11 0.0 0 ENDSEC 0 EOF Etc...	Translation: DXF description

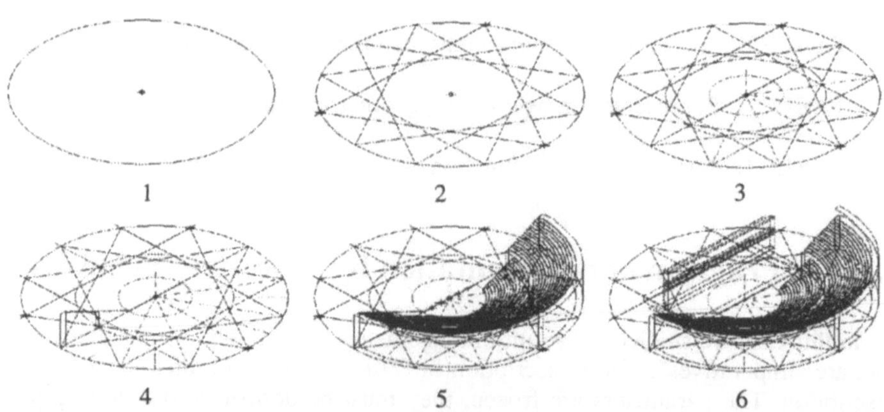

Figure 1. Exemples of Vitruve Description (Geometric Operator)

4.2 Recuperation of knowledge by some functions

The first innovation of our gait was the recycling of functions. This way has two advantages: the first one is a computing fluently and the second one a transfer of know-how. In this case, for example, we took some functions created for the modeling of the Byzantine vaults and, modifying the parameters, we transferred them to the construction of a theater platform.

It is possible to demonstrate that by the slant of a functional language we can model all the vaults constructed according to this functional principle and transferring this knowledge to other types of similar construction. This language is a functional language that permits the description, by the programming, of the logical purpose of design (Figure 2). In this case we transferred a function that shows a know-how in a function for a geometric construction: from the vault construction technique to the geometric construction of the theatre platform.

TABLE B. Program and Representation by functional language

Operator Analysis and functional language (SCHEME and SGDLsoft technologies)

Model of construction's process of the vault:
M-voutebyzantine (n, h, k)
n= width of the arch; h= subbasement; k= number of slices of brick.
The other elements, like the thickness of the arch, the depth or the slant of the slices, could be controlled in the script. (http://esi24.ESI.UMontreal.CA:80/~parisel/Voute/)

The functions (samples):

```
(define Mboules_controle
 (lambda (n h)
  (DLmax
   (DLatt
    (SDcolRGB
  (vector 0 1 0 1))
    (GDsphdis
  (vector 0.1 1)
  (vector 0 0 0 n)))
   (DLatt
    (SDcolRGB
  (vector 0 1 0 1))
    (GDsphdis
  (vector 0.1 1)
  (vector n 0 0 1)))
   (DLatt
    (SDcolRGB
  (vector 0 1 0 1 (GDsphdis
  (vector 0.1 1)
  (vector n 0 n 1)))
   (DLatt
    (SDcolRGB
  (vector 0 1 0 1))
    (GDsphdis
  (vector 0.1 1)
   (vector 0 0 n 1)))
   (DLatt
    (SDcolRGB(vector 1 0 0 1))
    (GDsphdis
     (vector 0.2 1)
  (vector (/ n 2) (- h) (/ n 2) 1)))
)))
etc...
```

```
;;Visualisation of construction of the vault;;

(define Mprovisoire
 (lambda (n h k)
    (DLmax
     (Marc_diagonal n h)
(Mboules_controle n h)
     (M-boulesbleu2 n h k)
     (M-tousmorceaux n h k))))

(pisobs (vector  1.3 0.35 1 0))

(define Voutebiz
 (lambda (n h k)
  (begin
   (pistarget (vector 2 1 2 1))
   (pisbox (vector 12 8))
   (PIswin (vector 900 600))
   (PIview
    (PIgen
  (Mprovisoire n h k )
   "marius")
"marius"))))
```

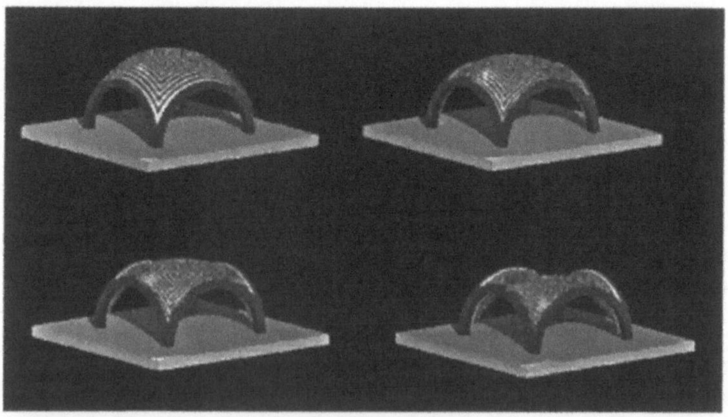

Figure 2. Example of simulation from construction's techniques of Byzantine vault

4.3 Creation of a modeling program by functions

In an article entitled *The Backstage of Vitruvius's Roman Theatre*, we demonstrated that it is possible to cross some geometric parameter with one semantic property and we chose the visibility as the semantic operator. This operator could not be represented by the imperative and linear representation. The description of architect (Vitruve) is vague and it is necessary to create a function that permits the re-interpretation of results according to several parameters such as the height of the spectators (Figure 3 and 4). This would permit, for example, to simulate the model for a children's theater. Therefore these data must be crossed with geometric and imperative data such as the position of the seats. (De Paoli, Bogdan, 1999)

TABLE C. Program and Representation by crossing geometric and semantics operators

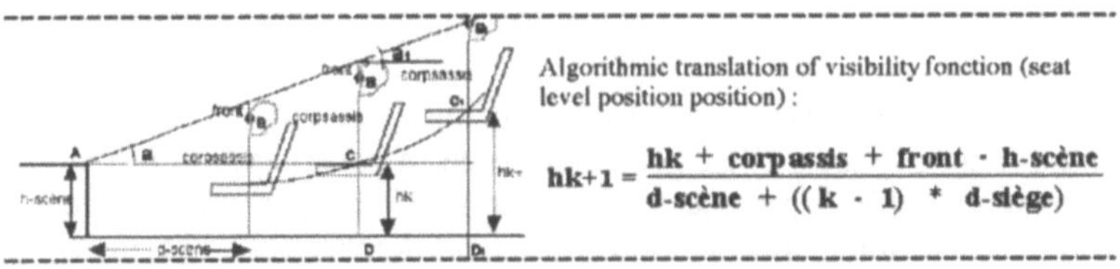

Figure 3. Examples of Vitruve Description (Geometric Operator)

Definition of meta-functions

(define **centrum** (vector 0 0 0 1)) (define **primo** 10) (define **ultimo** (+ primo (* nombre-sieges d-siege))) (define **rapport-vitruve** (/ 2 4)) (define **nombre-sieges** 20) (define **d-siege** 1)	Design parameters by Vitruve.
(define **h-scene** 2) (define **d-scene** 4) (define **front** .4) (define **corp-assis** 1)	Design parameters by new function: Visibility in addition to design parameters of Vitruve (

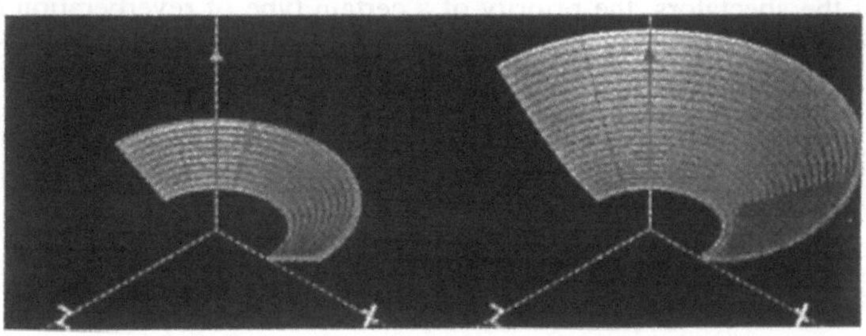

(define M-theatre-vitruve (Dluni (coord_xyz .1)	Theatre with the design profile by Vitruve Theatre with the design profile by Vitruve,

(Dlint (M-gradenes-Vitruve 10) M-forme-vitruve)))	10 seat level and plan of theatre
(define M-theatre-vue (Dluni (coord_xyz .1) (Dlint (M-gradenes-vue20) M-forme-vitruve)))	Theatre with the design profile by Vitruve and visibility function Theatre with the design profile by Vitruve, 20 10 seat level and plan of theatre

Figure 4. Example of two instances of procedural model (Figural and Functional)

4.4 Transformation Algorithmical for Semantic Properties: Visibility and Acoustics

The demonstration algorithmical being made, shows that it is possible to transfer knowledge and cross some geometric and semantic parameters. In this last experience we present the final results of our researche: the creation of procedural models only by semantic properties for a design in declarative scene process. In this approach completely original we added the acoustics to the parameter visibility.

We are aware that, during the definition of the acoustics function, we must take into account a set of factors that goes from time of reverberation to the clarity and type of sound. For this experience we take in consideration only the reverberation factor in relation with the spectators position, and demonstrate that, in encapsulating the data of this factor and crossing it with the visibility already modeled we could simulate several procedural models.

The characteristic of these models is to be a set of scenes for a new approach to the declarative design. Like in the experiences presented on the declarative scenes researches, they permit an analysis casual-probabilistic and a design on a model not clearly defined, where the parameters make abstraction of the geometric data.

These models are based only on the definition of semantic operators. This approach means that the achieved model is not the Vitruvius Theater, but a model that understands some semantic operators. In our case the figural model can be a covered swimming pool, a movie theatre or a stadium for soccer game. We are acting as a designer figuring some forms and modifying some situations, for example: the position of the spectators, the priority of a certain type of reverberation or again of sound transmission time. The end is a figural conceptual sketch. (Figures 5,6 and 7)

TABLE D Transformation of functions for semantic properties

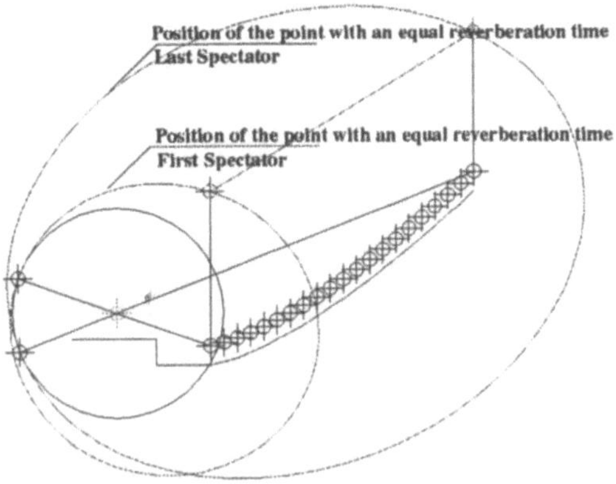

Figure 5. Acoustics and reverberation on multi-position levels.
The two ellipsis are the positions of the points with an equal reverberation time for the first and last spectator

Definition of meta-functions

(define visibility (lambda (view_angle) (m-relation (geometry (space)) (dimensions (spectators)))))	Visibility
(define acoustics (lambda (reverberation-time) (m-relation sound_source (spectators_position (visibility)))))	Acoustics

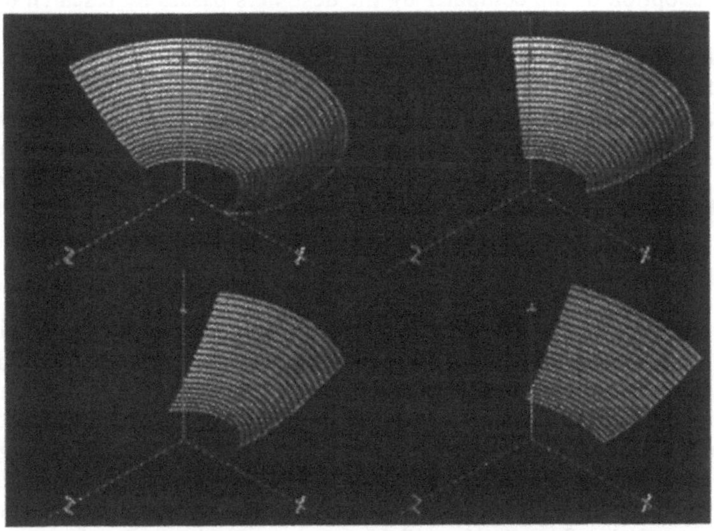

Figure 6. Example of instances of procedural model (Figural and Functional) obtained by interactions with geometrical parameters

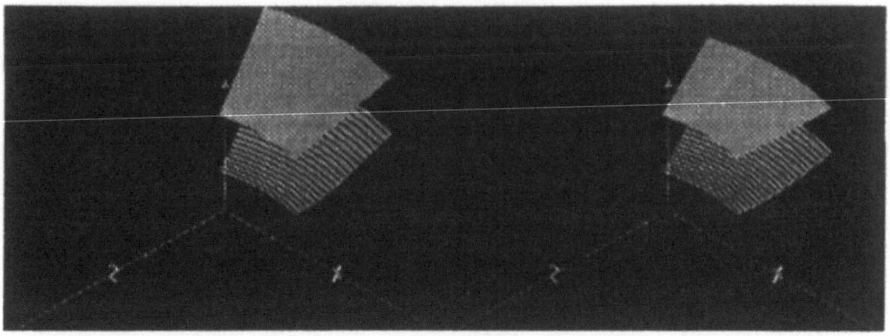

Figure 7. Example of instances of procedural model (Figural and Functional)

Example of semantic interpretations of procedural model: Visibility and Acoustics

(define theatre (lambda reverberation_time visibility_comfort) (Dluni (platform_seats (visibility_comfort) (roof (position_seats (visibility_comfort)) (reverberation_time)))))	By modifying the visibility comfort we obtain an other position for the seats and, consequently, a new position for the roof, that was "linked" to the seats by the reverberation time

5. CONCLUSION

The adoption of the computer by the designers passes by research toward some systems, manipulating at the same time the knowledge, the geometric and semantic properties of the scene and also on the process of design itself. These new systems based on languages of high level distinguish by the type of reasoning based on inferences (deductions, abductions) and on the utilization of an approach by models (scenes) rather than by numeric computing. This renders the process of computer design more explicit and more transparent (Popova, De Paoli, 1998) according to this gait the form is not the resultant of one solution, or a simple list of functions, but it is the result of a cybernetic activity. The functions (activities) are considered like a process of verification of the architectural idea.

Several points of view on this approach demonstrate how we could use the computers in order to change the paradigm. The designer could put back continuously the cognitive interpretations by new interactions, in all moments of the design-realization of model. Once more these interpretations are structuring: they interpreter a model in structuring the form. This feedback is made through an action on the environment and the expectation of the corroboration from the suggested structure. This form could not be exact and precise, because environment evolves; it is finally a whole of predictions, like it was a preliminary sketch composed by the architect on a napkin during a discussion with the client. One more new situation following this approach is the active and simultaneous participation of all the actors in the process of construction.

This mechanism of update recalls the reasoning *abductif* (Bourgine, 1989) which consists in emerging a new plausible hypothesis: as soon as an expected phenomenon is different from one observed phenomenon, we interpret the anomaly and we present a new solution. This is the basis of the design in architecture defined about the concept of representation of the properties, especially semantics.

References

Aprea, M, 1997"L'occhio dell'angolo: note al De Prospectiva Pingendi di Piero della Francesca", *La Natura della visione*, Edizioni Fareneith 451

Bertold, D., 1997, Designing DIGITAL SPACE, John Wiley &Sons, inc., U.S.A.

Bogdan, M., 1997 "La voûte byzantine, modélisation du processus de construction", *Note interne, GRCAO*, Faculté de l'Aménagement, Université de Montréal

Boudon P., Pousin, F.,1988, *Figures de la conception architecturale*, Dunod éd.

Bourgine, P. 1989, "Connaissances abductives et compilations de connaissance", *Actes du 1er congrès européen de systémique*, vol. 2 995-1002, Paris

Brown, F.E., et all.,1995, "An integrated approach to CAD: modelling concepts in building design and construction", *Design Studies*, V.16, Londres: Elsevier Science Ltd.

Dalmas, A., 1986 *Les dix livres d'architecture, VITRUVE*, Éditions Errance, Paris

De Paoli, G., 1997, "La CAO en architecture: modélisation des actions et définition des opérateurs", *Actes de 01DESIGN, les objets en conception*, Sophia Antipolis

De Paoli G., Bogdan M., 1999, "The Backstage of Vitruvius' Roman Theatre : A new method of Computer-Aided Design that Reduces the Gap between the Functional and the Operational", *CAADRIA 99 Proceedings*, Shanghay

De Paoli G., Marty A., 1996, *Architecture et simulation informatique et création industrielle*, rapport final, Ministère de l'Enseignement supérieur, porgramme ETS-1294, Québec

De Paoli G., Pellissier P., 1992 *Dessin d'architecture par ordinateur*, Éditions Berger, Québec

Eastman, C.M., 1992, "Experiments in Architectural Design development using CAD", *CAAD futures '91 Proceedings*, Zürich: Gerhard N. Schmitt

Farel, A., 1991, *Le troisième labyrinthe Architecture et complexité*, Montreuil: Les éditions de la Passion

Forgia, A, 1988 "Les nouveaux systèmes de représentation et le projet architectural", *Les cahiers de la recherche architecturale N.23*, Éditions Parenthèses

Ganascia, J.-G., 1997, *L'intelligenza artificiale*, Il Saggiatore et Flammarion, Milano

Goldshmith, G., 1997, "On Design Reasoning : Contents and Structure", *Rencontres MCX 6*, Poitiers

Mitchell, W. J., 1989, *The Logic of architecture*, MIT Press, Cambridge

Popova V&L, De Paoli G., 1998, "Toward an Object-Oriented Language for the Declarative Design of Scenes", *Digital Design Studios, ACADIA '98*, Québec

SGDLTechnology, 1995, *Manuel de SGDLsoft*, Lyon: S.G.D.L. Technology S.A.

Tidafi, T., 1996, *Moyens pour la communication en architecture — Proposition de la modélisation d'actions pour la figuration architecturale*, Université de Montréal, Faculté de l'Aménagement, Montréal

Tribus, M., 1972 *Décisions rationnelles dans l'incertain*, Éditions c.ie.,Paris

Vitruvius, 1914, *The Ten Books On Architecture*, translated by Morris Hicky Morgan, Harvard University Press, U.S.A.

Zevi, B., *Apprendre à voir l'architecture*, Les Éditions de Minuit, Paris, 1959

Experiments on Animated Visual Forms

Kostas Terzidis
UCLA - Arts and Architecture Los Angeles CA USA

Key words: modelling, representation, and visual systems

Abstract: A series of experiments that investigate and demonstrate the visual logic of three-dimensional representation in animated form through the use of computers is presented..

Perspective systems are designed to construct pictures that, when viewed, produce in the trained viewer the experience of depicted objects that match perceivable objects Panofsky [1] wrote about how our capacities to see are constrained by the perspective system that we use, that is, by our way of depicting what we see. The kind of pictorial spaces are expressed through geometrical models. Each model is expressed as a geometrical transformation applied to Euclidean/Cartesian shapes of the physical environment. These transformations show how shapes are projected in pictorial space [2].

The work of Gauss, Riemann, von Helmholtz, and Clifford [3] showed that geometries could be based on axioms other than Euclidean. This altered the notion of geometry by revealing its abstract postulational character and removing it from any connection with the structure of empirical intuition. Spaces were constructed on the basis of non-Euclidean axioms revealing behaviors closer to our sensations rather than our perception. Luneburg's [4] idea of hyperbolic space was a first attempt to develop a theory of visual space in which geometrical structure is a metric space of constant, probably negative, Gaussian curvature.

On the assumption that visual space is hyperbolic and physical space is Euclidean, it is possible to derive computational numerical models (or mappings) between the two spaces [2]. From a study of these models, we can get a good visual sense of the way our physical environment would manifest itself to hyperbolic vision.

I shall attempt to present below a series of experiments that demonstrate some of these models or mappings for a variety of values for the parameters of the spaces. The experiments address geometric mappings as well as numerical models of projection and their interest lies in the dynamic nature of the continuous computer processing, which is hard to show on a static piece of paper. It is also important to mention that the objects in the physical environment maintain in all experiments their Cartesian coordinates. It is the mappings and models that change.

Projecting Along Curves

The transformation equations for an orthographic parallel projection are straightforward. If the viewing plane is placed at the origin along the *z*-axis, then any point (x,y,z) in viewing coordinates is transformed to projection coordinates (x_p,y_p) as:

$$x_p = x, \qquad y_p = y \tag{1}$$

In parallel projection the z-coordinate value is omitted. In contrast, in perspective projection, the z-coordinate value is involved in the transformation equations of the projection. If the viewing plane is placed along the *z* axis, then any point (x,y,z) in viewing coordinates is transformed to projection coordinates as:

$$x_p = x \cdot t, \qquad y_p = y \cdot t \qquad \text{where} \qquad t = d + \frac{d}{z} \tag{2}$$

and d is the distance of the viewer to the viewing plane. The orthographic and perspective projections can be seen in figures 1 and 2.

In the first experiment, d was assigned a negative value, thus the viewer is assumed to be in the opposite position of the viewing plane and the perspective projection equations become:

$$x_p = x \cdot t, \qquad y_p = y \cdot t \qquad \text{where} \qquad t = -d - \frac{d}{z} \tag{3}$$

From the viewer's point of view, the effect of this alteration was an inverted world where faces of objects appear to be inside out and movements appear to shift surprisingly in the opposite direction. This can be seen in figure 3.

In the second experiment, the equations of perspective projection were scaled by a factor of n. The corresponding equations then became as follows:

$$x_p = n \cdot x \cdot t, \qquad y_p = n \cdot y \cdot t \tag{4}$$

From the viewer's point of view, the projected objects appear to be coming from the vanishing point and then as the objects are rotated they seem to be returning back to the vanishing point. This can be seen in figure 4.

Projection of a point (x,y,z) on a viewing plane does not have to occur along a straight line. In fact, many animals view the world through eyes of different shape and curvature resulting in seeing differently the same objects. The more curved or warped the eye is the more exaggerated the projection becomes. Along this line of thought, the following four experiments employ equations that use exponential and trigonometric functions to map the (x,y,z) points on the viewing surface. The first uses an exponential function. Points are projected along an exponential curve. Distances appear greater as we move away from the center of projection. The first set of equations is:

$$x_p = x^n, \qquad y_p = y^n \tag{5}$$

From the viewer's point of view, objects appear to be elongated along the borders of the viewing plane. This is shown in figure 5.

The second set of equations has the opposite effect. In this case, a point (x,y,z) is transformed through a fractional power

$$x_p = \sqrt[n]{x}\ , \quad y_p = \sqrt[n]{y} \tag{6}$$

From the viewer's point of view, objects appear to be concentrated towards the center of the viewing plane. This is shown in figure 6.

If we rotate the viewing projection lines to become trigonometric curves the projection equations can become as follows:

$$x_p = x \cdot \sin(t) + y \cdot \cos(t), \qquad y_p = y \cdot \sin(t) - x \cdot \cos(t) \tag{7}$$

By involving the z-coordinate in the projection curve the equations can become

$$x_p = x \cdot \sin(y) + y \cdot \cos(t), \qquad y_p = y \cdot \sin(x) - z \cdot \cos(t) \tag{8}$$

These transformations can be seen in figures 7 and 8.

Geometrical Transformations Using Projection Formulas

In the following set of experiments, the same formulas used to project 3D points on 2D surfaces are used to transform 3D points. For example, the perspective formula is used to transform the (x,y,z) coordinates of a point in the following way: (9)

$$x' = x \cdot t, \quad y' = y \cdot t \quad \text{and} \quad z' = z \cdot t \quad \text{where} \quad t = d + \frac{d}{z}$$

In the same spirit, the inverted, exponential, and trigonometric formulas used earlier in equations (1) through (8) to map (x,y,z) point on to (x_p, y_p) projected points are used to produce instead 3D transformations. Some of these results can be seen in figures 9, 10, and 11.

In the next experiment, a spherical projection is used. A spherical projection is based on distributing points on the surface of a sphere of radius r. The mapping and projection equations are: (10)

$$x' = x \cdot r \cdot \cos(\theta) \cdot \cos(\phi) \qquad y' = y \cdot r \cdot \sin(\theta) \cdot \cos(\phi) \quad z' = z \cdot r \cdot \sin(\phi)$$

where $\theta = \dfrac{2 \cdot \pi \cdot x}{npo\operatorname{int} s}$ $\qquad \phi = \dfrac{2 \cdot \pi \cdot y}{npo\operatorname{int} s}$

The result of this mapping is shown in figure 12.

In the next experiment, a hyperbolic projection is used. In hyperbolic space -- one of the non-Euclidean geometries -- the formulas involve the hyperbolic sine and cosine functions (sinh and cosh), that are exponential. Hyperbolic space is infinite in extent, just like Euclidean space. However, it is possible to map the entire infinite space into a finite portion of Euclidean space. It might surprise you that you can map an infinite amount of space with "more room" into a finite piece of a space with "less room", but non-Euclidean geometries have many unexpected consequences for Euclidean intuitions. The equations used are: (11)

$$x_p = x \cdot \cosh(\theta) \cdot \cosh(\phi) \qquad y_p = y \cdot \sinh(\theta) \cdot \cosh(\phi) \qquad z' = z \cdot r \cdot \sinh(\phi)$$

where $\theta = \dfrac{2 \cdot \pi \cdot x_i}{\sum i}$ $\qquad \phi = \dfrac{2 \cdot \pi \cdot y_i}{\sum i}$

and $\sinh(\theta) = \dfrac{e^{\theta} - e^{-\theta}}{2}$ $\qquad \cosh(\theta) = \dfrac{e^{\theta} + e^{-\theta}}{2}$

The result of this mapping is shown in figure 13.

Finally, random disturbance of the position or the rotation of faces in objects is assigned. The equations are simply:

$$x_p = x \cdot random(seed), \qquad y_p = y \cdot random(seed) \tag{12}$$

These produced the effect of "blurring", "wavy", "fuzzy", or "scratchy". The result of this disturbance is shown in figures 14 and 15.

Multiple-view Projections

Projection of a point (x,y,z) on a viewing plane does not have to occur along a single straight or curved line as the previous experiments did. In fact, many creatures view the world by collecting information from different points of view at the same time. The more eyes the viewer and the more offset they are the more collage-like the projection becomes. Along this line of thought, the following two experiments show how a series of objects would appear through multiple views at the same time.

In the first experiment, the viewing plane is divided in areas each one of which projects the scene from a different point of view. As the viewer rotates the scene, the objects appear to be connected and disconnected at the same time. The result of this experiment can be seen in figure 16.

In the second experiment, each face of each object in the scene is projected along a different point of view. Specifically, the equations (2) and (3) are used to project every other face of each object along different values of *d*, that is, either *d* or *-d*. The result of this experiment is shown in figure 17. From the viewer's point of view, faces of objects seem to move in opposite directions at the same time as the scene is rotated.

$$face_{2n+1} \rightarrow view_d \qquad face_{2n} \rightarrow view_{-d} \tag{13}$$

Dynamic Filtering

In the next set of experiments, the geometry of the projection is not affected. Instead, the viewing plane is affected by filtering the screen. The effect is dynamic since the shifts happen as the scene rotates. In the first experiment, the screen is divided into small squares and each one copies its neighbouring pixels. The result can be seen in figure 18.

In the next three experiments, the screen is divided into random size small shapes and each one copies random neighbouring pixels. The result can be seen in figures 19 and 20.

Conclusions

The aim of this paper was to present an experimental tool implemented in a computer system. The main idea is the dynamic transformation of 3-D objects. In brief, the system offers to the user the means to view geometrical models, transform them through a series of different mapping systems and, at the same time, filter them.

The system attempts to introduce the factor of time in the study of visual systems. It takes advantage of the ability the computer has to dynamically display images in real-time, and uses it as a tool where the way objects are viewed plays a much more important role than the objects themselves. The user is placed in the eyes of another creature and visualisation is a process of transformation and filtering of moving images.

Technical Information

The experiments presented in this paper are implemented on a computer program called *Zhapes*. The *Zhapes* routines are written in the both C++ and Java programming languages. The decision to use both languages was based on the wish to ensure graphics processing speed and interaction with the World Wide Web. Since the two programs are tied together, they share common data structures, which are implemented in C++. All routines dealing with network interaction are in Java. Dynamic transformations, such as rotation, translation, and zooming are implemented in C++ in order to take advantage of assembly level code to boost-up the real-time processing.

Zhapes can run both as a 32-bit application and as an applet on any web browser. It resides at the address http://www.cda.ucla.edu/caad/java/x/formProj2/formB.html where it can be downloaded for testing.

[1] Panofsky, E. 1927 *Perspektive als 'symbolische Form'.* In Saxl, (ed.) 258-330

[2] Heelan, P. 1988 *Space-perception and the Philosophy of Science.* (Berkeley: University of California Press)

[3] Clifford, W. K. 1946. First publ. in 1885 *The Common Sense of Exact Sciences.* (Knopf: New York)

[4] Lunenburg, R. 1950 "The metric of visual space." *J. Optic. Soc. Amer. 40* 627-642.

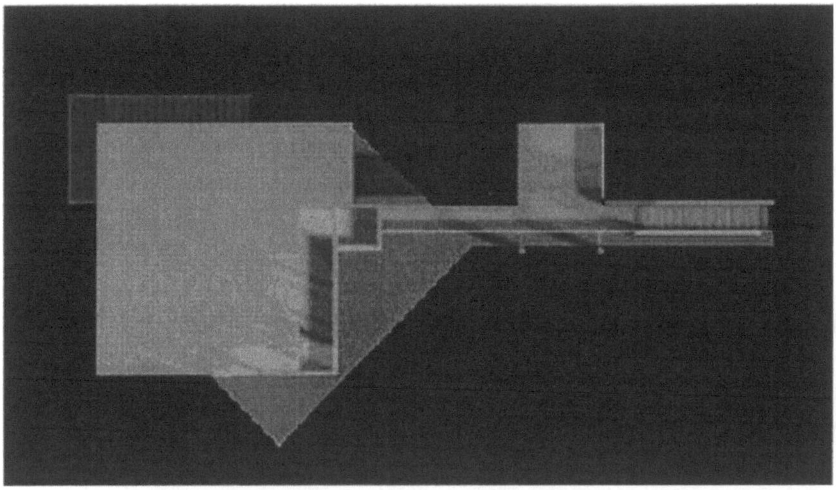

Figure 1. Orthographic View: $x_p = x, \quad y_p = y$

Figure 2 (below). Perspective View: $x_p = x \cdot t, \quad y_p = y \cdot t \quad t = d + \frac{d}{z}$

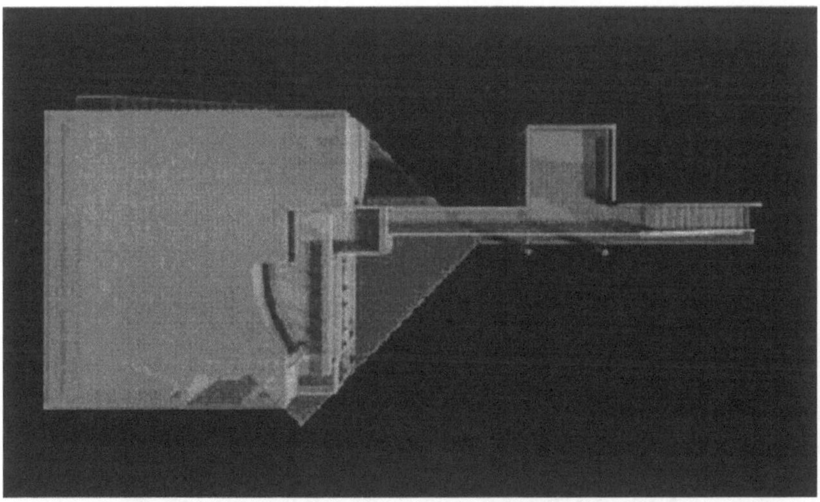

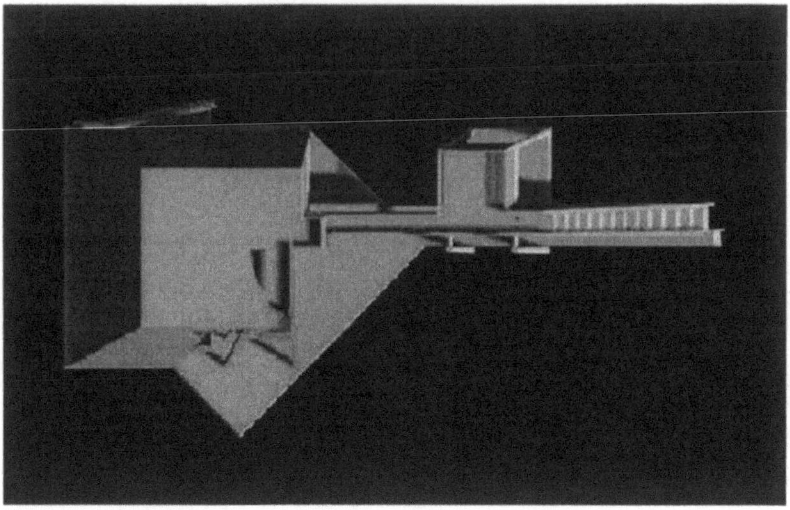

Figure 3. Inverted Perspective: $x_p = x \cdot t, \quad y_p = y \cdot t \quad t = -d - \frac{d}{z}$

Figure 4 (below). Scaled Perspective: $x_p = n \cdot x \cdot t, \quad y_p = n \cdot y \cdot t$

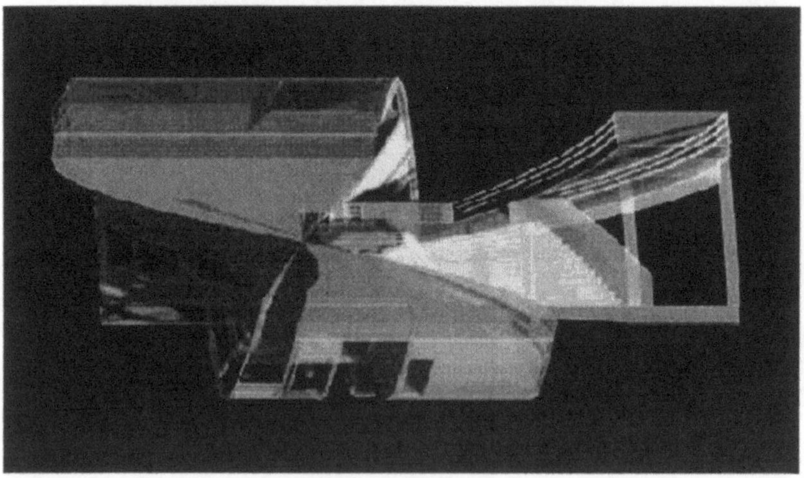

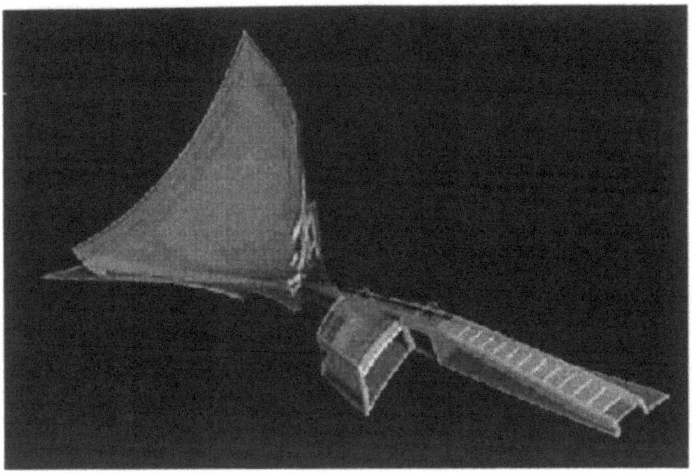

Figure 5. Exponential Projection: $x_p = x^n, \quad y_p = y^n$

Figure 6(below). Exponential Projection: $x_p = \sqrt[n]{x}, \quad y_p = \sqrt[n]{y}$

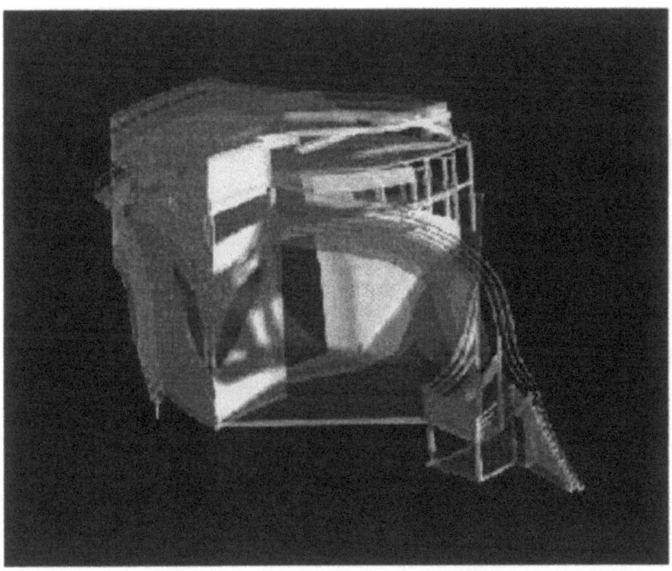

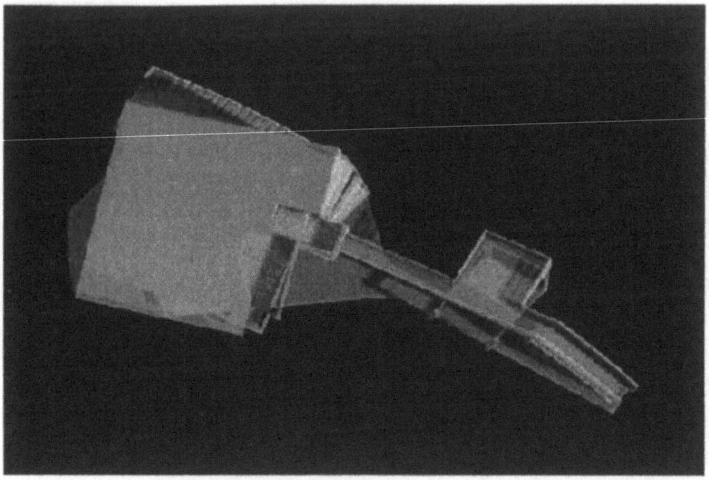

Figure 7. Trigonometric Projection: $x_p = x \cdot \sin(t) + y \cdot \cos(t), \quad y_p = y \cdot \sin(t) - x \cdot \cos(t)$

Figure 8. Trigonometric Projection: $x_p = x \cdot \sin(y) + y \cdot \cos(t), \quad y_p = y \cdot \sin(x) - z \cdot \cos(t)$

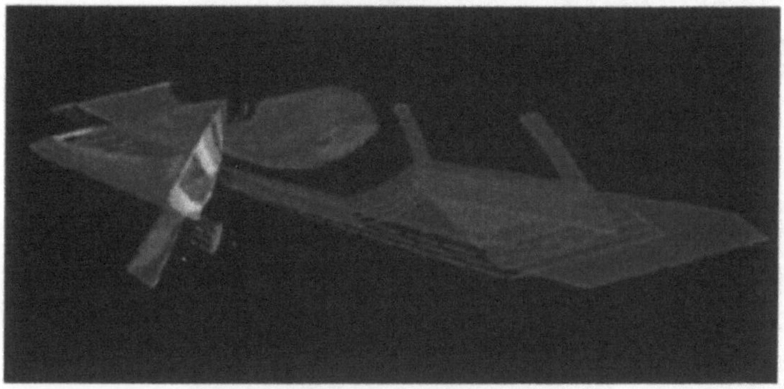

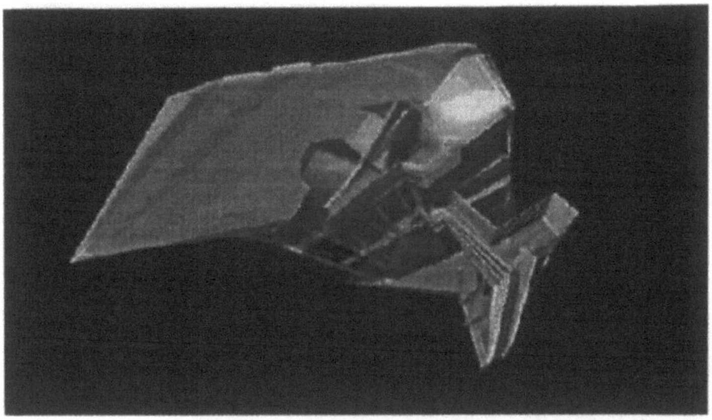

Figure 9. A trigonometric mapping on a trigonometric projection

Figure 10. An exponential mapping on an exponential projection (below)

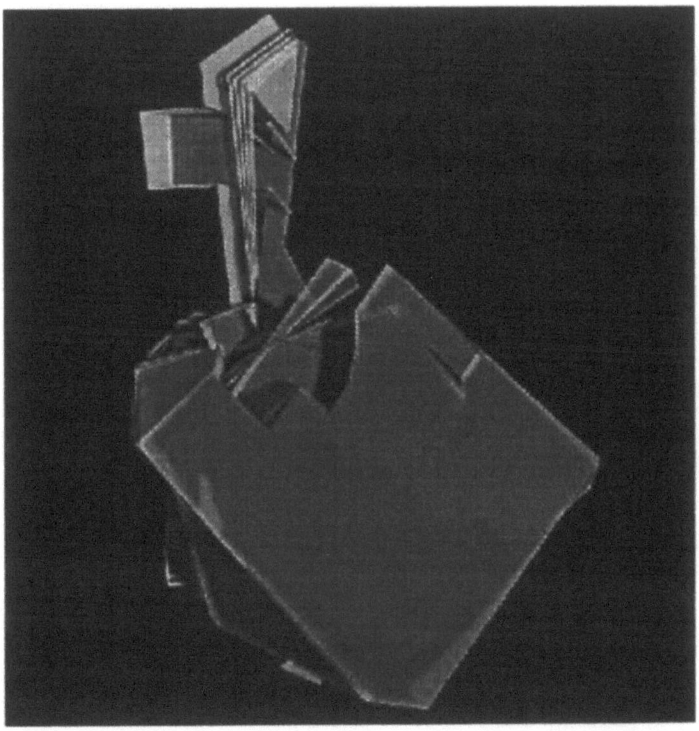

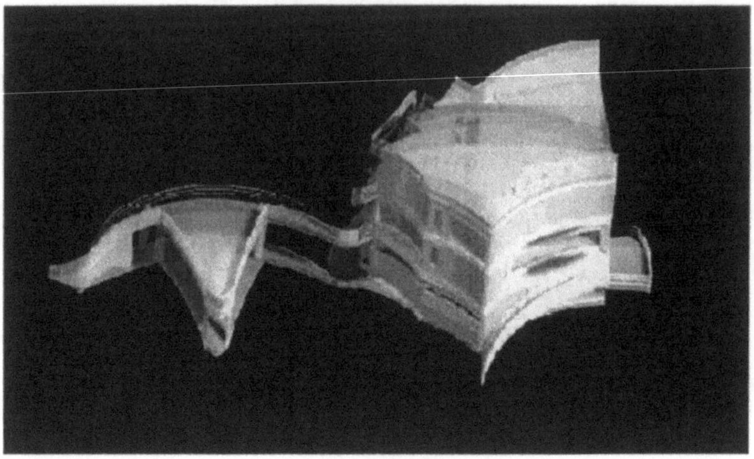

Figure 11. An exponential mapping on an exponential projection

Figure 12. Random Disturbance (below)

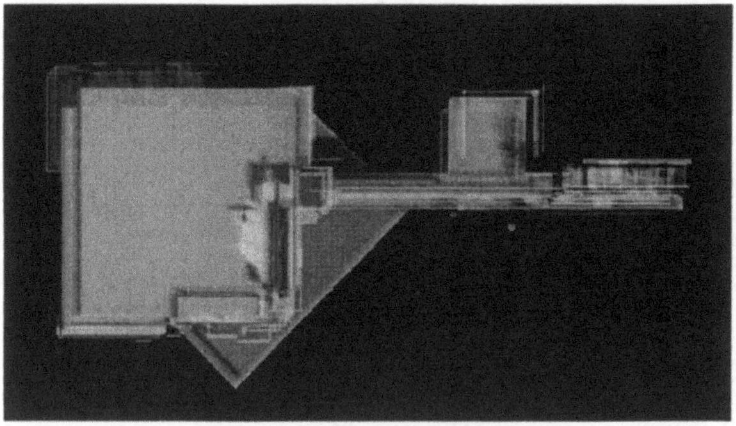

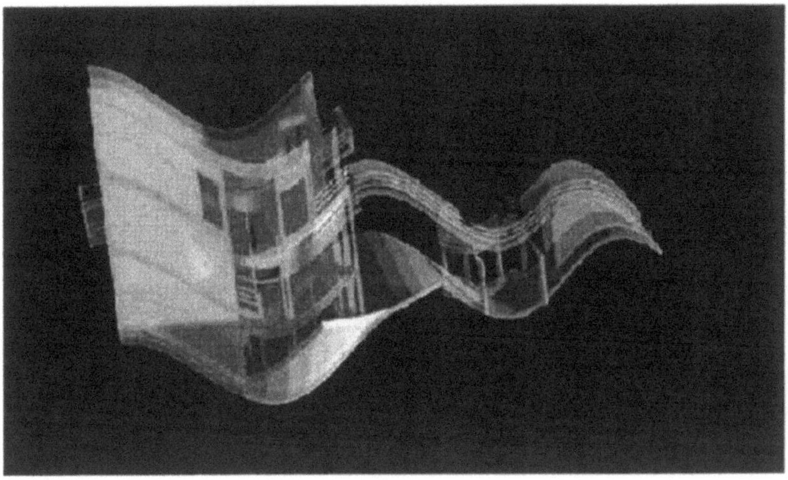

Figure 14. A sine-based disturbance
Figure 15. A spherical mapping on a spherical projection (below)

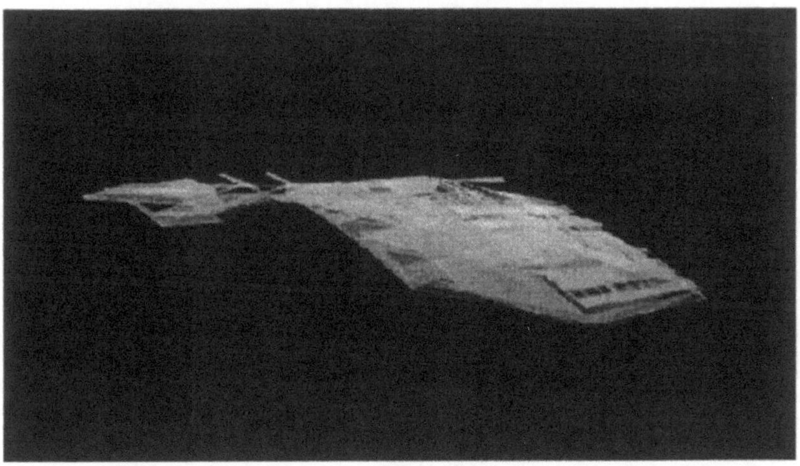

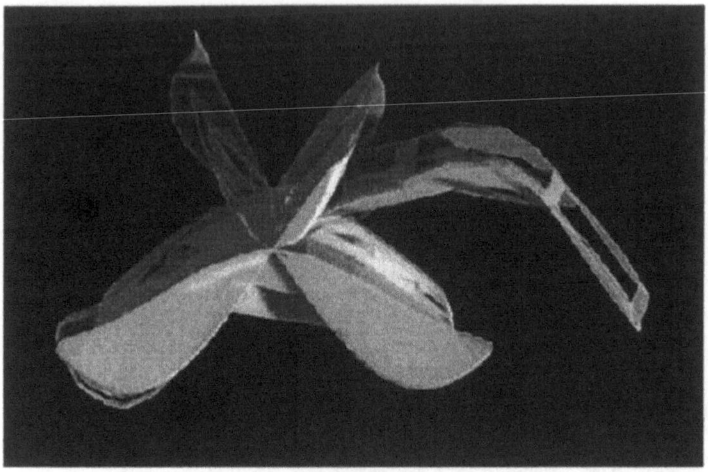

Figure 16. A hyperbolic mapping on a hyperbolic projection
Figure 17. An alteration of perspective view parameters on different portions of the screen (below)

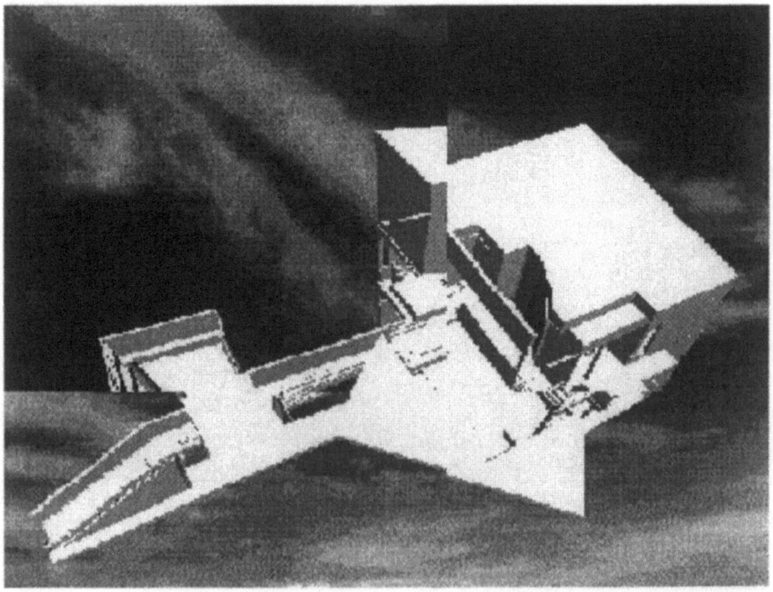

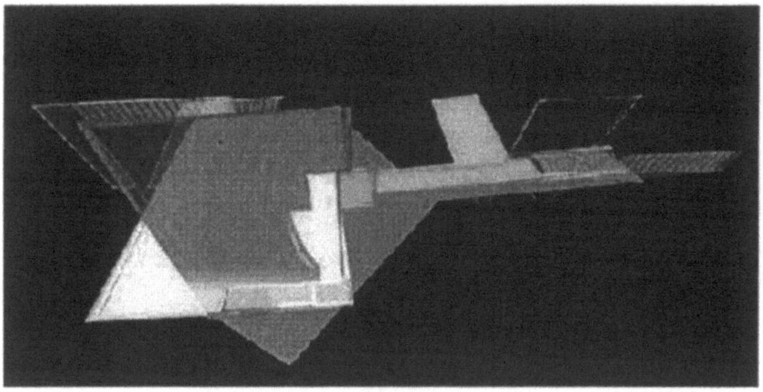

Figure 18. An alteration of perspective view for every face on the object
Figure 19. A grid-like filter (below)

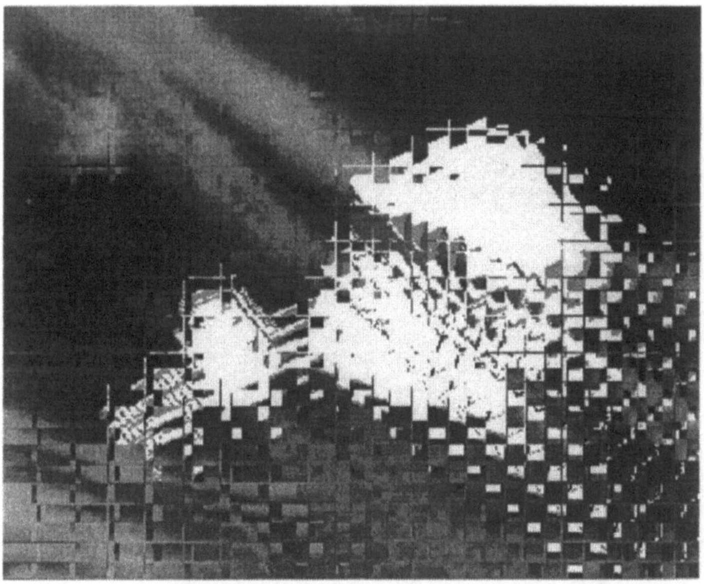

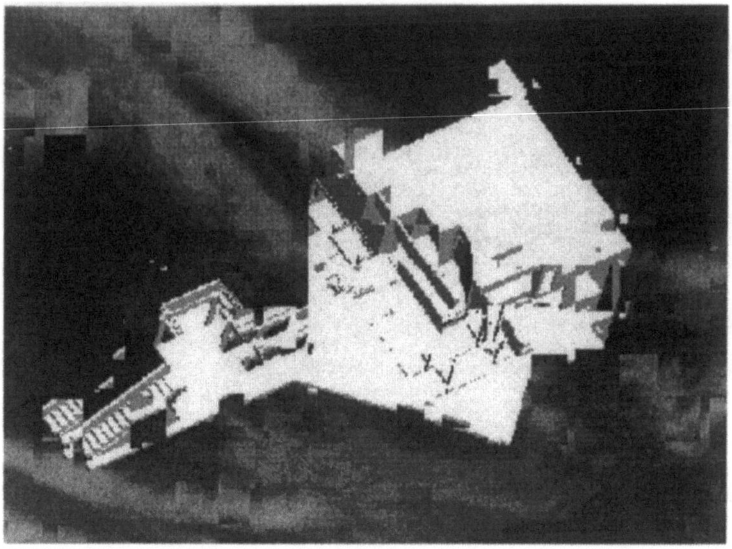

Figure 20. A random size square-based filter
Figure 21. A random size small square-based filter (below)

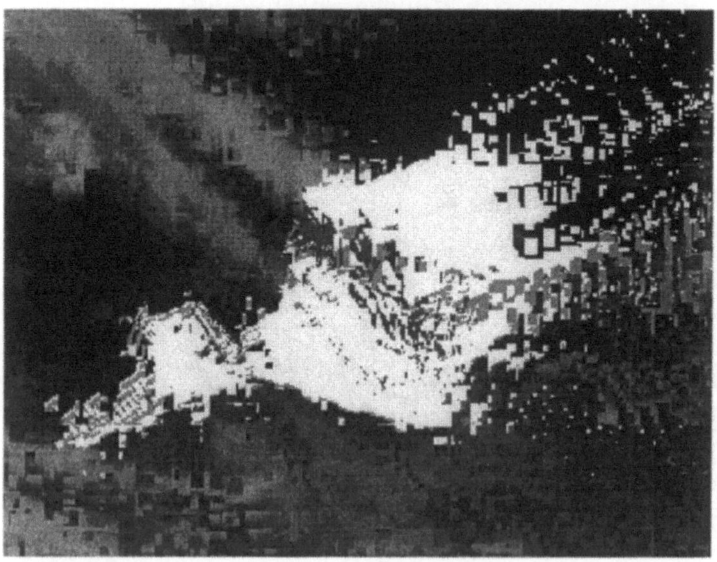

Environmental Life Cycle Assessment in an Integrated CAD Environment: The Ecologue Approach

R. Ries, A. Mahdavi
School of Architecture
Carnegie Mellon University
Pittsburgh, PA 15213 (USA)

Key words: life cycle assessment, integrated computational environmental analysis

Abstract:

Construction and operation of buildings is a major cause of resource depletion and environmental pollution. Computational performance evaluation tools could support the decision making process in environmentally responsive building design and play an important role in environmental impact assessment, especially when a life cycle assessment (LCA) approach is used. The building domain, however, presents notable challenges to the application of LCA methods. For comprehensive environmental impact analysis to be realized in a computational support tool for the building design domain, such tools must *a)* have an analysis method that considers the life cycle of building construction, operation, and decommissioning, *b)* have a representation that is able to accommodate the data and computability requirements of the analysis method and the analysis tool, and *c)* be seamlessly integrated within a multi-aspect design analysis environment that can provide data on environmentally relevant building operation criteria.

This paper reviews the current state of assessment methods and computational support tools for LCA, and their application to building design. Then, the implementation of an application (ECOLOGUE) for comprehensive computational assessment of environmental impact indicators over the building life cycle is presented. The application is a component in a multi-aspect space-based CAD and evaluation environment (SEMPER). The paper describes the use and typical results of ECOLOGUE system via illustrative examples.

1. BACKGROUND

1.1 Introduction

ECOLOGUE was conceived to provide design decision support in environmental impact assessment. Computational tools can play an important role in the environmental impact assessment of a service, product or process especially when a life cycle assessment (LCA) approach is used. LCA requires extensive data collection for each component contained in or each process operating within the boundaries of the assessment. The typical building contains hundreds of products from different manufacturers, which are assembled in ways often unique to an individual building. The designer must be able to acquire, store, organize, and manipulate the data for the components such that it can be used in the required calculations for the situation in question. Computational support tools have the potential to expand the breadth and depth of the information easily available for an assessment. Applications, such as spreadsheets and databases that are stand-alone can be used to store and compute the impact values. To use them in an iterative design task, the designer must, most probably manually, modify and re-enter the building description as the design evolves. If multiple evaluation tools are required, the model must be converted into the respective format. The effort involved in input is a partial reason for the lack of substantive, widespread use of performance simulation applications in the building design industry.

A design environment with integrated evaluation modules would allow the designer to evaluate multiple aspects of a building's performance from the early design stages while describing the building's characteristics to a system only once. This is particularly important for a life cycle environmental impact assessment for the following reasons. A LCA that considers the building during operation can use integrated simulation modules that estimate loads for heating and cooling, lighting, and equipment. The impact assessment module can then determine the impacts of these loads. Alternative solutions, even if similar from a design standpoint, can have significantly different environmental effects. In an integrated environment, alternatives are more straightforward to compare, and evaluations can be accomplished expeditiously. The use of integrated evaluation tools that are able to provide feedback in early design stages can influence siting, orientation, and massing decisions that often effect environmental impact. These types of design changes are more difficult to carry out when the design is more resolved. The integrated analysis and CAD tools of the SEMPER environment allows the user to seamlessly step from design to LCA and back again, while having access to and input from operational phase analysis packages such as energy, airflow, HVAC, and lighting.

Considering the importance of the life cycle of buildings to environmental impact, the data collection and processing requirements, and the multi-disciplinary nature of the evaluations involved in building LCA, the development of integrated computational support tools is one way to provide meaningful design decision support.

1.2 Environmental Analysis Method Overview

The interest in the environmental impacts of products and processes has led to the development of various methods and software tools to assess these effects within the framework of LCA. Early work in this area with regard to buildings focused on the energy used in producing building products. Current assessment methods consider the emissions generated and energy used in the production of a product (from the raw material production to the eventual disposal of the product in a recycling or waste stream). The differences among the methods lie primarily in how the impacts from emissions are evaluated, commonly known as the impact assessment stage, and in the number and types of inventories performed over the product life cycle, commonly known as the inventory stage. The following is a brief description of selected assessment methods. The section is an overview, and is not intended to be an exhaustive review of the field.

The "classic" life cycle assessment (Fava et al. 1991) is defined by the Society for Environmental Toxicology and Chemistry (SETAC) as a three-phase process consisting of an inventory analysis, an impact analysis and an improvement analysis. The three phases are defined as follows:

- Life cycle inventory:
  A documentation of energy and raw material use, air and water emissions, solid waste, and any other releases to the environment throughout the life cycle of a product or process.
- Life cycle impact analysis:
  A quantitative and qualitative process, which characterizes and assesses the effects of the environmental loading of the inventory analysis.
- Life cycle improvement analysis:
  An evaluation of the environmental impacts identified in the previous stages of the analysis in order to determine the appropriate modifications to the product or process.

Impacts are grouped by the media in which they occur (i.e., air, water, and land). No attempt is made to aggregate these into a single indicator value.

Eco-balance (Etterlin et al. 1992, Müller-Wenk 1978) begins by defining the "balance domain", i.e. the parameters relevant for the evaluation, which is similar to many of the methods reviewed here. The data on emissions of pollutants to the air and water, energy and water consumption, and solid waste and residuals generation are grouped into energy consumption (in $MJ \cdot kg^{-1}$ of material) as well as loads to water, air, and land (in $m^3 \cdot kg^{-1}$ of material). The loads to the air and water are converted from units that represent pollutant volume, into a unit that expresses the volume of air or water that would be contaminated to its legal threshold limit by the pollutant. This value is called the "critical volume", which is defined as:

$$\text{Critical volume} = E \cdot T^{-1} \qquad (1)$$

where E is the actual volumetric emission of the pollutant and T represents the legislated legal threshold limit for the pollutant.

Eco-points (Etterlin, et al. 1992) aggregate impacts from emissions by developing "eco-factors" for individual emissions. Eco-factors represent the relationship between the actual emissions of a pollutant, called the *effective load* or

F, and the legal threshold value, called the *critical load* or F_k for a given geographical location:

$$\text{Eco-factor} = C \cdot F_k^{-1} \cdot F \cdot F_k^{-1} \quad (2)$$

where C is a constant and F_k^{-1} represents the "sensitivity" of the environment to a given pollutant and will result in a large eco-factor for a low critical load. The eco-points of a material can be added together to give an estimation of the total environmental impact.

The Eco-indicator (Goedkoop 1995) is the result of a cooperative effort between government, industry, and research groups in the Netherlands, and is an extension to LCA. The goal of the project is to assist in the design of environmentally sound products. The method considers the damage to human and ecosystem health in the European region, and does not consider raw material depletion, the spatial impact of waste disposal, and the impact of land use.

The eco-indicator method involves the following steps: *a)* inventory of impacts (emissions and raw materials), *b)* classification of impacts into 10 effect categories, *c)* characterization of the emissions within an effect category, i.e.: weighting the effects of the different emissions within an effect category relative to each other, *d)* correlation of each effect category with one of three damage categories, *e)* normalization of the effect scores, and *f)* weighting the damage categories relative to one another in order to produce one eco-indicator value.

1.3 LCA Software Overview

There are many stand-alone packages that have a LCA-based assessment capability. The systems are generally built on a database of environmental information, and are used in industry for product analysis and process improvement. A typical example of a stand-alone package is TEMIS (Fritsche and Rausch 1993). The user constructs scenarios that contain one or more processes linked by their input and output types. The analysis results are documented in a text file of emissions, energy and resource use, with an optional graphical display. A comprehensive review of systems is given in (Menke et al. 1996). Examples of systems that are principally targeted toward the building industry are Building for Economic and Environmental Sustainability (BEES) (Lippiatt and Norris 1995) and ATHENA (Trusty et al. 1998). BEES combines economic evaluation and environmental impact with a multi-criteria decision analysis model. The level of analysis is at the product or component level. Plotting the performance results compares alternatives. ATHENA is a system that will calculate environmental impact for a schematic building description. Characteristics of alternative designs can be entered and the results compared.

The LCA tools targeted toward the building industry have begun to address some of the drawbacks of using product- and component-based tools for building design. However, these systems lie outside of a computational design and evaluation environment. The user is required to calculate the quantity of materials and processes used in the building and then enter them into the LCA package for analysis. Operational building loads must be calculated outside of the program, either in other simulation packages or manually (ATHENA is planning links to energy simulation applications).

Integration of design and analysis applications allows the user to improve the scope of the analysis while reducing the effort required obtaining analysis results. This could lead to a wider use of LCA in building design. Another ongoing effort in the area of integrated LCA assessment and CAD is LEGOE (Hermann et al. 1998), a system incorporating quantity take-off, energy analysis, and LCA. A LCA module is also in the future plans for the Building Design Advisor (BDA) (Papamichael et al. 1997).

2. SYSTEM OVERVIEW

2.1 SEMPER

SEMPER is a spaced-based building design and evaluation environment (Mahdavi 1996, Mahdavi et al. 1996). The SEMPER environment consists of a user interface, a building representation that is shared among the application modules, domain specific evaluation modules, and a database for persistent storage.

The SEMPER domain modules currently focus on aspects of building performance; however, the system architecture is designed to be modular, so that additional applications can be added in the future. An object database is used for the persistent storage. This is used for elements of the shared representation, as well as additional attributes of materials and systems relevant for the analyses within individual domains.

The functional requirements for the shared object model (SOM, see *Figure 1*) in SEMPER are to contain the spatial information and the physical description of the project. The spatial information represents the geometry of the spaces in the project, and, in general, the physical description is the location and geometry of the site and building elements, and their type. The type information is selected by the user (such as a user-defined default wall type) and applied when the entity is instantiated. Individual modules provide additional data to their building representations by using type-associated database objects. An evaluation module may refine the type description when it is run and place that information in the database, where it becomes available for re-use or use by the other modules. Type can reference, for example, a representation of a construction method when the entity is a constructed or manufactured element, such as a wall, or it can simply be a reference to an object containing a set of attributes, for elements such as furniture or equipment.

2.2 ECOLOGUE

The ECOLOGUE module's task is to calculate relevant environmental parameters and environmental impact indicators for buildings and provide meaningful feedback to the designer. Minimal intervention from the user is required. The module can "automatically" generate a proprietary building representation for use within its domain, derived from the shared representation, utilizing the inherent "isomorphic" relationship between the two representations. Additional environmental properties and characteristics of materials and components are accessed from the database through the type reference. The user can create or modify process and emission attributes and types, as well as context parameters.

The functional requirement for the domain object model (DOM) in ECOLOGUE is to provide a representation of the building and its context that can provide the information necessary for the calculation of the environmental impact indicators. The primary building representation is principally derived from the SOM model, and contains the basic building information. Additional domain specific requirements include the modeling of the building's context, processes, related emissions, and the implementation of the analysis methods. *Figure 2* shows the descriptive portions of the DOM. The areas of the representation with a correspondence to shared model are shaded.

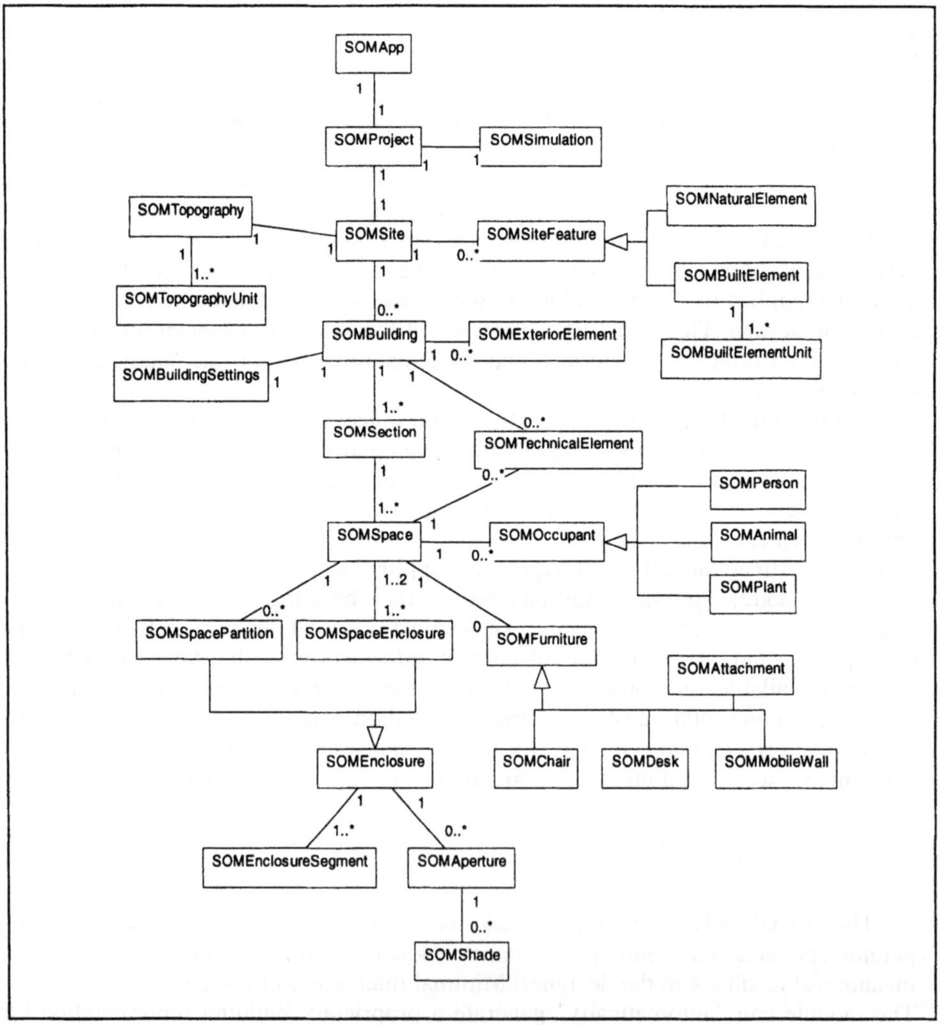

Figure 1. The primary elements of the shared building representation

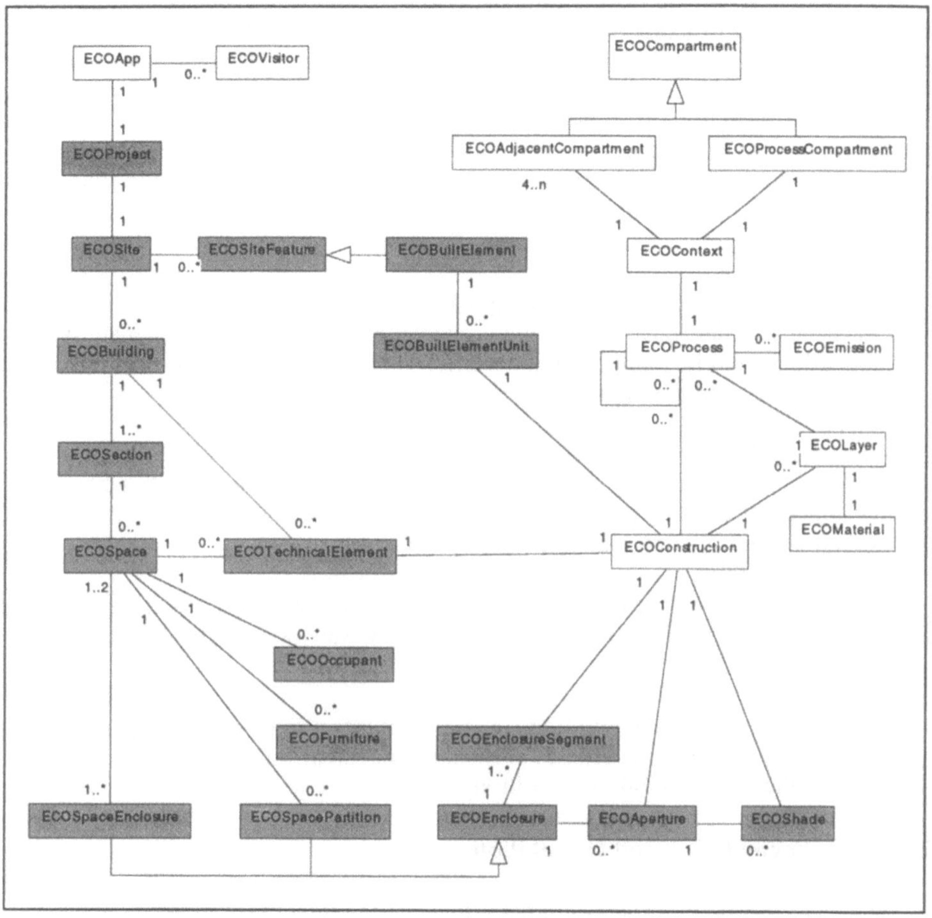

Figure 2. Elements of the ECOLOGUE domain model

2.3 ECOLGUE Analysis

2.3.1 Method Overview

Currently, ECOLOGUE implements three environmental impact methods. These are the Critical Load and Eco-indicator methods described earlier, and a third method developed for ECOLOGUE, called "Affordance". This method is described in more detail in the following section. A code structure was implemented, through the use of Visitor classes (see *Figure 2*) (Gamma et al. 1994) that will allow the addition of future functionality, such as another analysis method, with minimal alteration of the core classes.

Rather than the comparison of alternatives, the goal of affordance is to provide an indicator that will express the performance of a building design in relation to a context. In this way, for the given design and site being considered, the indicator would determine whether the impacts generated by the solution meet or are below the allowable level for the context or contexts effected by the proposed building construction. The advantages of this type of indicator are that when emissions commonly expressed in mass terms are related to a context or eco-system type over a period of time, the ability of the ecosystem to sustain the impacts from the emissions can be included in the indicator. Examples of how such an indicator could be used are:

- Evaluation of the impacts of building materials.
  Using the affordance indicator, environmental evaluation of building material production could be more context-specific. The ecological sensitivity of the region where production processes occur can be considered.
- Land use planning
  When the affordance of a regional area is determined, the spatial allocation and distribution of the load would represent the region's carrying capacity when fully developed. Also, using an affordance indicator for planning would allow for the explicit definition of an area's sensitivity to disturbance.
- Resource planning
  Allocation and use of resources, for example water, could also be expressed in terms of an affordance indicator. Areas with low water availability would have a lower affordance.
- Indoor environmental quality
  Emissions from in-place building materials can effect indoor environmental quality. Given the characteristics of the building, an affordance method could evaluate the potential indoor conditions.

2.3.2 Formulation and Calculation

The characteristics of the preliminary formulation developed to date are an eco-system classification that has a spatial allocation or affordance for substances expressed in $kg \cdot m^{-2} \cdot yr^{-1}$ (Mahdavi 1998). In the following equation, w_i is the weight for emission e_i, and the product is summed from 1 to n emissions.

$$\text{Impact} = \sum w_i \cdot e_i \tag{3}$$

A zeroeth approximation of w_i can be expressed as:

$$w_i = (e_{i,\,max} \cdot n)^{-1} \tag{4}$$

and, as a first approximation:

$$w_i = (e_{i,\,max} \cdot \sum e_{j,\,max})^{-0.5} \cdot n^{-1} \tag{5}$$

In ECOLOGUE, the affordance value of an emission in a context is calculated through the use of an evaluative environment. The fate and transport model is based on fugacity concept (Mackay 1991). The evaluative environment is modeled as compartments, each of which contain four media - air, water, soil, and sediment. The concentration of an emission in the system of compartments is calculated at runtime. With each iteration, the emission rates to the principal and adjacent compartments are incrementally increased until a compartment or compartments reach the threshold concentration level for that emission. This emission rate for the compartment divided by the land area that will potentially be developed, is the affordance for the emission in the compartment. An evaluation of the actual emission rate from the building process can then occur.

3. ILLUSTRATIVE EXAMPLE

3.1 Introductory Remark

To illustrate the operation of SEMPER-ECOLOGUE, the computational life cycle assessment of a building is described below. The example project is a single-family residential house of 2 1/2 stories located in suburban central Pennsylvania (Lee 1997, NAHB 1997). The schematic of the partial representation of the floor construction in the project (*Figure 3*) shows how a portion of the building is described in ECOLOGUE.

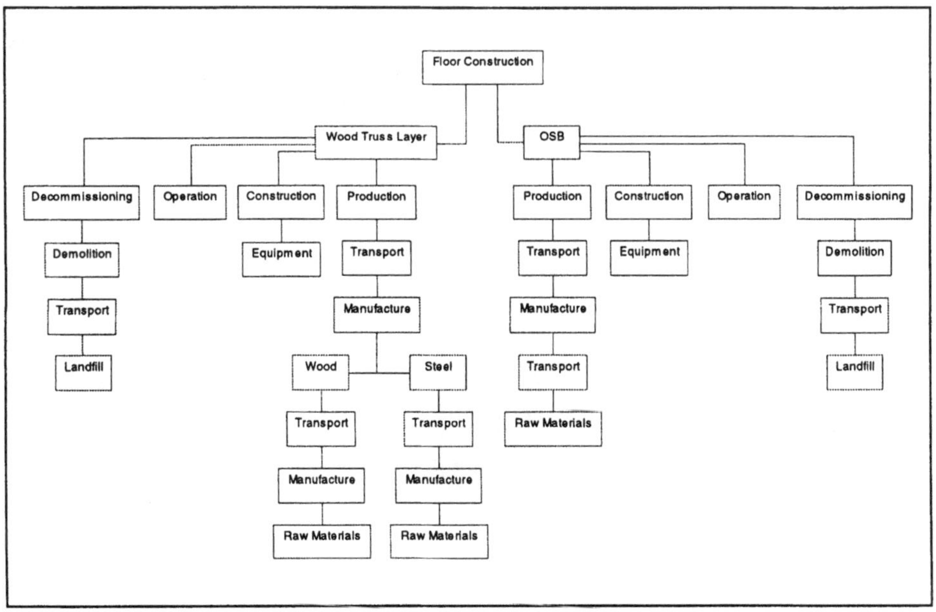

Figure 3. Partial representation of a floor construction in ECOLOGUE

3.2 Results

Figure 4 through *Figure 7* show the summary of the distribution of environmental loads in two media and energy use for the building systems included in the ECOLOGUE analysis in each of the four life cycle phases. The indicator method used is the Critical Load method.

The production and construction life cycle phases include the manufacturing of the materials used in the building, and the assembly of the materials and systems into the finished building. The material production phase includes raw material extraction, transportation of raw materials to the manufacturing facility, and the manufacturing process itself.

The construction phase includes the transportation of the materials or manufactured systems to the building site, the manufacture, transportation, and operation of the equipment used in construction, and the transportation of the required labor force to the site.

The operation and maintenance phase of the building includes all of the activities that occur during the occupancy phase. Included in this area are the estimated loads resulting from the heating and air conditioning of the building, the energy used for lighting, the maintenance (repair or replacement) of building systems, and the energy used for transportation by the building's occupants. Although the inclusion of transportation energy use for the occupants' daily activities is normally not included in environmental analyses of buildings, it is an important aspect, from an environmental perspective, of the loads during the operation of the building. Including this aspect partially addresses the regional issues, such as the location of a development, which contributes to the environmental impacts resulting from building construction.

Decommissioning is defined here as the process of removing the building materials to a local landfill. The two processes included are an estimate of the machinery required to dismantle the structure and transportation to the landfill. It is assumed that all of the material is removed with the exception of the aggregate, and that no material is recycled.

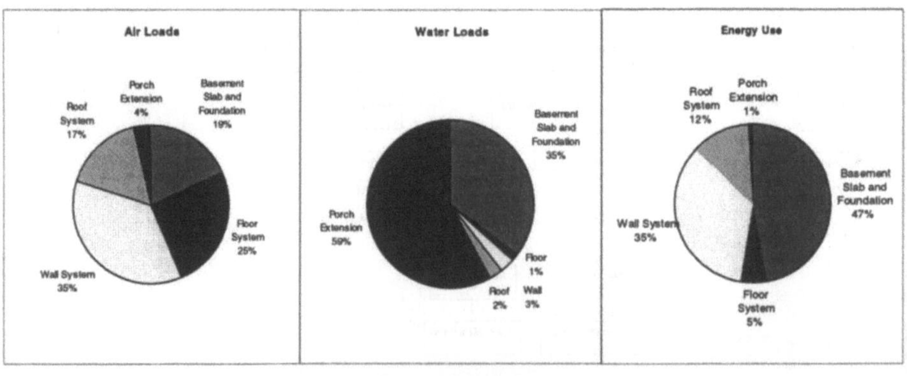

Figure 4. Distribution of environmental loads for material production

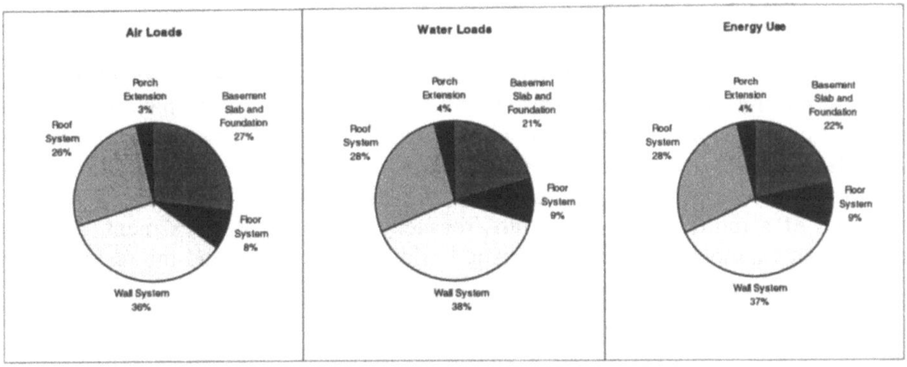

Figure 5. Distribution of loads for the construction phase

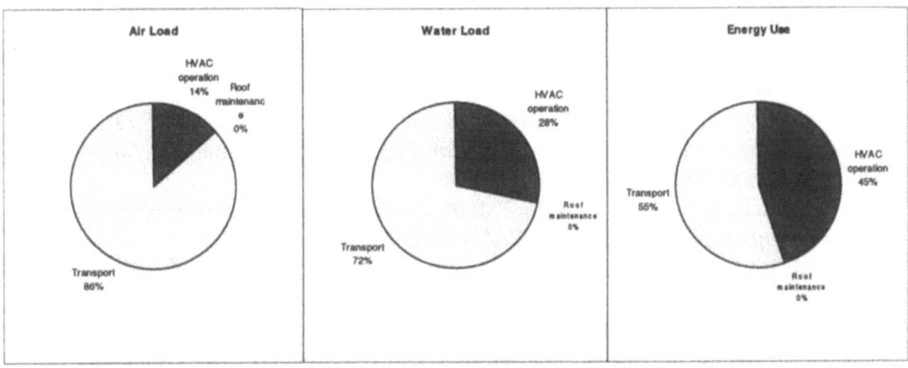

Figure 6. Distribution of loads in the operation and maintenance phase

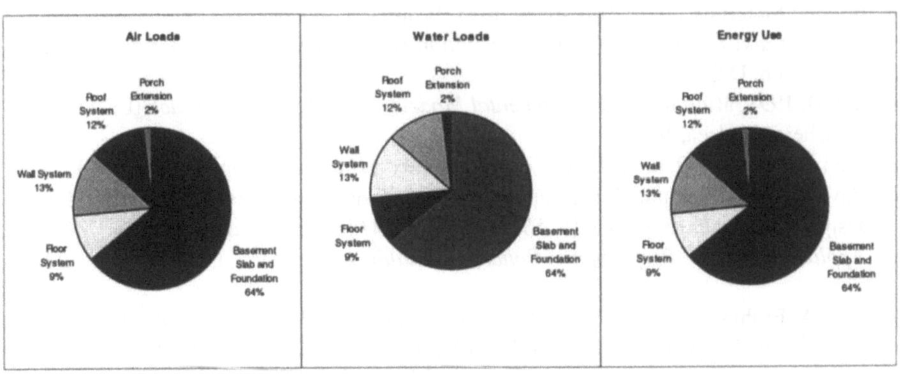

Figure 7. Distribution of loads in the decommissioning phase

4. CONCLUSION

This paper has presented an environmental impact application, ECOLOGUE, within a multi-aspect design environment, SEMPER. The paper has argued that computational design support tools would assist in the effective management and calculation of life cycle environmental impact, and may encourage the use of environmental assessment tools early in the design process. It has shown the relevance of a multi-aspect design environment to life cycle assessment, and has provided an example of the characteristic building model used and the results of an illustrative evaluation.

REFERENCES

Etterlin G, Hursch P, Topf M, 1992, *Ökobilanz, Ein Leitfaden für die Praxis* (B.I. Wissenschaftsverlag, Mannheim)

Fava J A, Denison R, Jones B, Curran M A, Vigon B, Selke S, Barnum J, 1991, *A Technical Framework for Life Cycle Assessment* (Society for Environmental Toxicology and Chemistry, Pensacola, FL)

Fritsche U R, Rausch L, 1993, *Total Emission Model for Integrated Systems (TEMIS), Software for Environmental Analysis of Systems*, (Öko-Institut (Institute for Applied Ecology), Energy Division, Freiburg, Germany)

Gamma E, Helm R, Johnson R, Vlissides J, 1994, *Design Patterns: Elements of Reusable Object-Oriented Software* (Addison Wesley, Reading, MA)

Goedkoop M, 1995, *The Eco-indicator 95 Final Report* (NOVEM, RIVM, and NOH, the Netherlands)

Hermann M, Kohler N, König H, Lützkendorf T, 1998, "CAAD System with Integrated Quantity Surveying, Energy Calculation, and LCA", Proceedings: *Green Building Challenge '98*, Vancouver, Canada. Vol. 2, 68 - 75

Lee S, 1997, "The Susquehanna Project", Proceedings: *Spring Conference of the Architects Research Center*, Atlanta, GA

Lippiatt B, Norris G A, 1995, "Selecting Environmentally and Economically Balanced Building Materials", in: A.H. Fanney, K.M. Whitter, T.B. Cohn, eds., *National Institute of Standards and Technology Special Publication 888* (U. S. Government Printing Office, Washington, D.C.)

Mackay D, 1991, *Multimedia Environmental Models: The Fugacity Approach* (Lewis Publishers, Chelsea, MI)

Mahdavi A, 1998, "A General Theory of Habitability", *Human Ecology Review*, 5 (1), 23-30

Mahdavi A, 1996, "Computational Support for Performance-based Reasoning in Building Design", *Proceedings of the CIB-ASTM-ISO-RILEM International Symposium "Applications of the Performance Concept in Building"*, Tel Aviv, Israel. Vol. 1, 4.23 - 4.32

Mahdavi A, Brahme R, Kumar S, Liu G, Mathew P, Ries R, Wong N H, 1996, "On the Structure and Elements of SEMPER", P. McIntosh and F. Ozel, eds., *Design Computation; Collaboration, Reasoning, Pedagogy: Proceedings of the 1996 ACADIA (The Association for Computer Aided Design in Architecture) Conference*, Tucson, AZ. 71 - 84

Müller-Wenk R, 1978, *Die Ökologische Buchhaltung* (Campus Verlag, Frankfurt)

Menke D M, Davis G A, Vigon B W, 1996, "Evaluation of Life-cycle Assessment Tools" (Hazardous Waste Branch Environment Canada)

NAHB Research Center, 1997, "Construction Monitoring of the Susquehanna House", (NAHB, Upper Marlboro MD)

Papamichael K, La Porta J, Chauvet H, 1997, "Building Design Advisor: automated integration of multiple simulation tools", Automation in Construction, 6, (4)

Trusty W B, Meil J K, Norris G A, 1998, "ATHENA: A LCA Decision Support Tool for the Building Community", Proceedings: *Green Building Challenge '98*, Vancouver, Canada. Vol. 1, 39 - 46

Downstream of Design: Web-based facility operations documents

Mark J. Clayton, Robert E. Johnson, and Yunsik Song
Texas A&M University

Key words: Facility management, Computer-Aided Design, intranet, information delivery, life cycle design, as-built drawings

Abstract: Internet technologies provide opportunities for improving the delivery of facility information to building owners and operators. Discussions with facility operators have led to identification of problems in current practices of delivering facility information using as-built drawings. A Web-based software prototype illustrates how facility information can be automatically structured into documents that support specific facility operations tasks.

1. INTRODUCTION

Beyond the design phase, and even beyond the construction phase, the documentation of a building lives on in support of operations and management. In speaking with building owners, we have uncovered a widespread concern that building documentation as delivered by designers or constructors is inadequate for supporting building operation. There is a need for information to persist throughout the facility life cycle and for project participants to be aware of the downstream uses of information. Facility operators use so called "as-built drawings" in conjunction with test and balance reports, maintenance manuals and many other kinds of information. There is rarely a coherent and integrated strategy to co-ordinate, format and cross-reference the information provided from many sources. To better understand the shortcomings of contemporary practices in delivering building documentation, our research has examined building documentation from the viewpoint of owners and operators. We have focused upon emerging Internet and intranet technologies to develop guidance toward new practices that are likely to reduce the difficulties perceived by building operators.

Our investigations have led us to re-conceive as-built drawings as *just-in-time operations documents*. This new concept for organising facility information exploits Web-based tools to collect, structure and deliver facility information to the point of use. Using automated scripts running on Web servers, generic repositories of electronic data can be reused and linked to support typical work tasks. Although our research has not yet produced a fielded implementation, our working prototype suffices as a proof-of-concept and provides us with confidence in our recommendations for the delivery of facility documentation.

This paper presents the results of several years of research that remains in progress. We will introduce our research questions, the scope of our explorations and our research methods. We will then describe the changing environment of the AEC industry as the background to our research. Next, we will enumerate problems with facility documentation in supporting operations. A short section then introduces the collection of technologies that we have found to be useful. We will then describe the working prototype that we have constructed to illustrate a vision of future information systems to support facility operations. We recommend formats and practices that can be followed by building owners and consultants to facilitate delivery of facility information into Web-based information systems. Finally, we will discuss future research and summarise our conclusions.

1.1 Research questions

The overall question behind our research has been "Can building information be supplied by the AEC team in a more effective way so that it will better meet the needs of facility management work processes?" In turn, this question has suggested three other questions that help to frame this research:

1. What is the structure and content of information that is currently delivered by the AEC team?
2. What is the structure and content of information that is needed for facility management?
3. What is necessary to close the gap between what is delivered and what is needed and can recent advances in information technology create new opportunities to redefine the nature of facility organisations?

At this point, we have conducted several research projects that have focused on these three questions. These studies have produced an improved understanding of the processes that use facility documentation and the emerging information technologies that can contribute to new forms of facility documentation.

1.2 Research methods

The research reported in this paper investigates the information interfaces among building designers, building contractors, and the facilities and services department of a large insurance and financial services company, USAA. In particular, we have focussed upon the needs of the building operators who manage and maintain building services on a routine basis, perform repairs, and make changes to the systems to accommodate remodelling.

We have undertaken several activities to assure that our research can be of broad interest and is influenced by a broad range of input. Site visits with numerous facility management organisations have enabled us to catalogue a range of technology solutions that could potentially be useful in solving these specific problems or, potentially, be useful in proposing changes within the facility management organisation. Discussions with industry groups and several government task groups have acquainted us with related initiatives. Contact with software vendors has enabled us to understand the current state of art.

Concurrently with discussions with our sponsor and other facility management organisations, we investigated Internet technology from the point of view of implementing a corporate intranet to distribute facility operations information. We undertook quick exercises with emerging Web-based tools and constructed software prototypes. Our purpose in these explorations has not been to create a new software product, but instead to create an image of a future facility operations information support system. The prototype enables us to define the requirements for structuring and formatting facility information that can be delivered by design and construction consultants now that can be easily incorporated into the future system.

2. BACKGROUND

This project has been conducted upon a backdrop of an Architecture/Engineering/Construction (AEC) industry that is undergoing tectonic changes. Architects' clients are becoming more sophisticated. They demand different services and a higher degree of understanding of their business practices. Simultaneously, the World Wide Web (the Web) has provided a culture for start-up companies that are providing new services that address current shortcomings in practices in the AEC industry. Existing and new industry groups have recognised that information exchange is a key to improving performance in the AEC industry and are championing different approaches that facilitate this exchange.

2.1 Changes to the professions

Architecture has never been a very homogenous industry. However, a recent industry survey suggests that the profession may be growing even more diverse in response to market pressures. For example, there is continuing evidence from the most recent AIA Survey (AIA 1997) that large architecture firms, particularly those with 50 or more employees, are increasing their share of the market. At the time of this survey, less than 9% of the largest firms accounted for over 60% of total industry net billings—an all-time high. These firms are often multi-disciplinary and generally service clients of large organisations who design, build and operate large, expensive and complex facilities. These clients demand a full range of services to meet their facility needs and are changing the nature of architectural practice from a project focussed process to a client focussed process. As Gutman has stated (1998, 50) the big organisation clients "set the standards according to which architecture is practiced".

These same large clients are turning increasingly toward enterprise resource planning (ERP) systems such as SAP and Oracle to help them manage the large amounts of information associated with operating their corporations. These organisations have learned the value of information and are continually mining data to leverage this information in an effort to become more efficient. For many of these large organisations, information technology "... is integral to processes internal to the firm, to product design, to delivery of services, and to inter-organisational relations" (Allen and Scott Morton, 1994, vi). As a result, information technology is becoming increasingly important as a strategic business tool. Since CEOs and CIOs are seeking to integrate enterprise-wide information, they are also requiring data pertaining to the design, planning and operation of their facilities.

These changes in the services demanded by building owners require architects to have a greater understanding of the role of information in the design, construction and operation of buildings. Such a focus on information is likely to alter substantially the traditional ways of providing architectural services and, perhaps more important, the methods and procedures by which architectural information is supplied. If information is becoming increasingly important to the practice of architecture and if large, corporate clients are "setting the standards" for the practice of architecture, then it follows that architects must better understand information needs throughout a building's life-cycle.

2.2 Opportunities arising from the Web

The World Wide Web is dramatically changing many aspects of professional practice in the AEC industry, particularly in communication and information storage and retrieval. Initial use of the Web by architects included precedent or product research, and firm home pages as a branch of marketing (Padjen 1997). The Web is now being used more actively to improve communications and information management. Some architects are establishing project "extranets" where all project design and construction information can be collected and then viewed by project participants (Mays 1998). At the completion of the project, the repositories can be written to a CD-ROM and delivered to the owner for use in facility management. However, current use of project Web sites does not clearly anticipate facility operations application of the information. The emphasis has been upon supporting the communication among design and construction team members.

Future design support systems may provide more sophisticated functions. Computer-Aided Design (CAD) software is becoming "internet-enabled" (Regli 1997). Drawings can be viewed using a Web browser and links to other documents embedded. Virtual Reality Modeling Language (VRML) 3D models can be distributed across the Web and then explored interactively by participants on a project team. Java applications can perform design evaluation calculations. Design analysis services could be distributed on the Web. Internet-based multimedia communication software, such as NetMeeting, can support design discussions that include graphic sketching and even multi-person interactive CAD sessions (Al-Qawasmi et al. 1999).

Taken together, these new technologies provide extensive opportunities to re-engineer design, construction and facility management information to achieve greater integration and to streamline operations processes.

2.3 Industry initiatives

While Web-based tools have been adopted quickly, other advanced computer methods, although promising and potentially very effective, have taken longer to reach maturity. Web-based tools are relatively generic in that they address distributing, viewing and recording information without placing many demands upon semantics behind the information. Several groups are directly addressing semantics of building information to facilitate information exchange and sharing.

At the simplest level, building information can be shared in common formats, such as building geometry that is represented in IGES or DXF formats. By using a convention for dividing information into layers, such as the AIA CAD Layer Guidelines, a degree of non-graphic information can also be shared (Schley, et al. 1997).

Product modelling under the STEP umbrella aims to define the semantics and syntax of information to enable its unambiguous exchange among computer systems (Watson 1995). Various computer systems may be used by many engineering disciplines throughout the product lifecycle and across geographic and time constraints. In the AEC domain, COMBINE and ICON are two research projects that have developed generic, central models of buildings that may be shared among many users (Augenbroe 1995, Cooper and Brandon 1995).

The Construction Specifications Institute is addressing information integration for the AEC industry that builds from its widely accepted CSI format and Uniformat systems. They have recently published the Uniform Drawing System that provides guidelines for the layout of drawings and the use of symbols and noting on those drawings (CSI 1997). Based upon an industry-wide study, they plan to produce a classification system, a thesaurus of terms used in the U.S. construction industry, and ultimately an "Integrated Information model" that addresses the life cycle of facilities.

In addition, several U.S. government agencies are addressing the creation of standards for facility management information. The Tri-Services CADD/GIS Technology Center is developing standards for the U.S. Armed Forces bases and buildings. A group working out of the National Institute of Health and the National Institute for Building Sciences is examining the use of SGML as a standard for representing building information. A Document Type Definition (DTD) is in development that will define tags and their semantics that can be used to add structure to text files.

The International Alliance for Interoperability (IAI) is a consortium of software vendors, design and construction firms, researchers and government agencies that posits a standard for object definitions for the AEC industry. The concept is to establish a framework by which vendors can produce compliant software that will work effectively with compliant software from other vendors.

At a still more sophisticated level, the concept of individual applications dissolves as software distributed across a network can transparently pass information and obtain processed results. Perhaps by creating component-oriented interfaces or standards using Java, a vision of distributed design or modular CAD can be implemented (Eastman, Chase and Assal 1993, Zamanian 1995).

Among these initiatives there is a predominance of attention to supporting design and construction rather than facility management. The emphasis has also been upon directly addressing the data standards rather than envisioning the software that will make use of those standards. Our research is distinctive in our emphasis upon facility operations and our development of a software prototype of a future information system that uses Web-based technologies.

3. PROBLEMS WITH CURRENT FACILITY DOCUMENTATION

We elected to pursue a problem-driven methodology for this research project. This approach allowed us to ground our research in specific issues that would be of relevance to the AEC-FM community. We conducted our investigations by combining widely targeted surveys, short interviews with a varicty of facility management organisations, and an in-depth series of discussions with facility managers and operators in our sponsoring organisation.

3.1 Identification of Industry-Wide Problems

In our 1997 survey of Fortune 500 facilities organisations (Johnson and Clayton 1997), we found a number of problems that were inhibiting the implementation of information technology solutions and that there appeared to be some differences between service and manufacturing companies. Reported problems included evaluating the costs and benefits of information technology, keeping information up-to-date, customising software, educating staff in new technologies, and acquiring data about a facility for inclusion in information systems. Clearly many facility management organisations are experiencing significant problems as they sought to make better use of information technology to improve their practices.

Other researchers have documented perceived problems with the documents that are commonly produced by designers and constructors when they are used to support facility documentation. A survey of U.S. Army facility managers reached a conclusion that as-built drawings are poorly structured for supporting maintenance and operations and may be missing important information (Liu et al. 1994). This study concluded that more attention to design standards and criteria, warranty information, maintenance records, condition of equipment and facilities, and utility information is warranted.

Clearly, problems with the delivery of facility information to building owners are widespread. However, there are diverse concerns and factors that probably vary among organisations. Our subsequent work with USAA allowed us to focus on specific problems that were of interest to one specific company.

3.2 Identification of Problems at USAA

Our approach with USAA was to identify problems through a variety of mechanisms, including focus-group sessions, interviews, and surveys of operations personnel. From this investigation we identified a number of problems that can be categorised into three groups:

1. Problems with content
   - Operational sequences for running the facility are missing.
   - The original design intent is missing.
   - Documentation of underground services may be missing or inaccurate.
   - Training documents are missing or have proven ineffective.
   - Equipment tags are do not conform to internal standards or are not used consistently. Labels on drawings may not match those on equipment or may use different conventions from labels on other drawings.

2. Problems with format and delivery
   - Information for a specific project is distributed across a wide variety of sources. It is difficult to know what source to consult.
   - Drawings are often too complicated for facility management tasks and contain information that is superfluous.
   - Information is not integrated across functional building systems. For example, mechanical equipment should be co-ordinated with electrical--it is important to know what circuit feeds a particular piece of equipment.
   - Provision of information at point of use is often a problem. For example, cables need to be labeled at both ends. Electrical panels need to be properly labeled and the labels coordinated with plans. Labels on panels may be too general, such as "Lights," or may be too specific, such as "Lennox 3245A."
3. Problems with information collection and up-date
   - Historical information is in paper format or obsolete digital formats that cannot easily be integrated into current information systems.
   - Facility information is difficult and costly to keep it up-to-date.

Further discussions with USAA revealed categories of facility information that are strong candidates for improved documentation (Clayton et al. 1998). Schedules, such as electrical breaker panel schedules, valve tag schedules, door hardware schedules, and lock set schedules were deemed very important yet with low satisfaction. Mechanical design rationale was also frequently mentioned as a kind of facility information that could be crucial to effective operation but was rarely captured and stored effectively.

Design and construction consultants who work with USAA revealed other issues. Each consultant has an independent information delivery approach. While consultants are dedicated to providing satisfactory service, they are also reluctant to change their practices to meet the needs of a single client. A building owner must avoid draconian solutions and requirements that consultants can not or will not meet.

3.3 Problem Summary

Our discussions provide pictures of facility information across processes in design, construction and operation. Designers and constructors tend to create and deliver information to owners in format and with structure that is appropriate to their internal processes. Designers and constructors often have little awareness of the downstream uses of the information. Facility managers and operators put the information to a variety of uses during facility operation and need to locate, retrieve, and filter information rapidly to support a variety of tasks. Maintenance and up-date of the information is a crucial part of an effective facility operations information system.

Three ways of organising facility information appeared important in the overall process. For construction, information is conventionally organised by disciplines and trades such as by the CSI format and the Uniform Drawing System. For training and troubleshooting during facility operation, information needs to be organised by functional system that may span across disciplines and trades. For example, altering a fan coil unit may require information from HVAC drawings, electrical drawings, test and balance reports, control systems and other sources. The CSI UNIFORMAT classification is a precedent for this view. Finally, information also needs to be provided to support specific work tasks, control a sequence of tasks and co-ordinate multiple personnel. Up-dating facility information is a crucial step in many operations work tasks.

Careful attention must be paid to incorporating information into computer systems. A large amount of "data scrubbing" such as conversion of paper drawings to CAD or conversion of CAD drawings to databases could place undue burdens upon consultants or internal resources.

We summarised these problems in prototypical use scenarios. Potential scenarios could be identified by focusing on recurring and routine problems, critical problems, or stand-out headaches. Since critical problems appeared to be handled well by USAA, we decided to focus on recurring and routine problems. We have chosen the following scenarios to guide our implementation of a prototype:

1. Adapting HVAC systems to remodelled offices and expansion projects.
2. Substituting and replacing filters in HVAC system equipment and ducts.
3. Training personnel in equipment shut-down procedures.
4. Trouble-shooting and fixing electrical circuit overloads.

Our next step was to gain an understanding of appropriate information technology in preparation to build a prototype to address the scenarios. We focused upon Internet and intranet technology because it provides hypertext and hypermedia, it is platform independent, and it is relatively low cost.

4. INTRANET TOOLS FOR FACILITY DOCUMENTATION

Our initial effort in attempting to identify future information systems for facility management was to survey Web-based development tools that are entering the marketplace. We then obtained the tools whenever possible and performed experiments with them to develop a feel for their utility and ease of use. This section describes the tools and techniques that most contributed to our development of a prototype:

- **Web browsers and HTML** The fundamental technology in our explorations is the use of Web browsers and HTML to implement an "intranet". These are established, proven and popular technologies for publishing information across a network.
- **Dynamic HTML.** This is not really a technology but is a loose collection of technologies. It includes Java, ECMA Script, the Document Object Model (DOM), Cascading Style Sheets (CSS) and other accepted or proposed formats that enable the appearance of HTML files to be adjusted on the fly. The Document Object Model is a representation of the structural parts of an HTML document or XML document as a "part-of" or container hierarchy. It enables programming additions to HTML to access attributes and formatting.
- **Relational Database Management Systems.** RDBMS systems, such as Microsoft SQL Server and Oracle, are a fundamental repository for information that is structured into fields and records. Open Database Connectivity (ODBC) is a widely supported standard for performing queries on many brands of RDBMSs.
- **Active Server Pages (ASP).** Our explorations relies upon generating HTML pages that are then sent to Web browsers. Microsoft's technology for doing that is Active Server Pages. With ASP, one can insert HTML fragments,

perform database queries, and process the values in user input fields to generate HTML code on-the-fly.

- **Drawing Web Format (DWF)**. Autodesk has continued to refine the DWF format for use on the Web and in intranets. DWF files can have multiple named views, and layers that can be toggled on and off as well as embedded links. The settings can be controlled via Java Script and parameters in the HTML code.
- **XML**. This format is likely to replace HTML for intranet applications. It provides enriched structure that enables machine processing of documents to support multiple uses. Documents can be marked up for content and semantics by using a Document Type Definition (DTD) that defines custom tags. A master document can include sub-documents to promote sharing of boilerplate among several documents. Style sheets may be stored in separate files and shared among multiple documents.
- **Web-based redlining.** A number of tools are becoming available that allow simplified editing of a CAD file to support mark-ups, note errors, and indicate changes. Redlining tools can operate in conjunction with a browser and CAD drawings to enable mark-ups to be managed via the Web.

5. JUST-IN-TIME FACILITY DOCUMENTS

These investigations of Web technologies applied to problems in facility documentation have led us to a concept of just-in-time (JIT) facility operation documents. We have developed a prototype JIT operations document information system to prove the feasibility and potential usefulness of the concept.

5.1 The JIT quality

The acronym of JIT originated in the manufacturing field. The idea is to pull necessary materials for production from a station downstream when needed instead of constantly pushing materials to the next process upstream (Karmarkar 1989, Duncan 1988).

We have applied the JIT concept to facility documentation. Our approach is to store facility information in relatively generic repositories and then collect, format, and filter the information for distribution only when it is needed to support a particular task. Instead of a traditional approach of pushing facility information to the operations personnel in the form of three-ring binders and stacks of drawings, our concept is that operations personnel pull the information from digital repositories using Web tools only as needed.

5.2 Prototype implementation

We have implemented a prototype JIT operations document system. This prototype system generates documents based on requirements of the specific task. We have demonstrated the prototype to field facility operators. Our preliminary conclusion is that it successfully generates facility operations documents, providing information necessary for specific operational tasks.

Our prototype delivers concise and complete information that is focused on specific operations tasks. The following implementation concepts are embodied in the prototype: 1) Generic repositories, 2) Link farm, 3) Task-centric document generation, and 4) Automated workflow management.

5.2.1 Generic repository of facility information:

Existing and emerging standards for electronic documents provide a strong starting point for a JIT operations document system. Our prototype system stores facility information in data repositories that are relatively generic and unstructured. These repositories conform to current practices and proposed standards, such as the Uniform Drawing System, the AIA CAD Layer Guidelines, and the Tri-Services standards. They also apply a principle of using information in "natural" formats. For example, although architects often deliver schedules as drawings, they are more naturally represented as tables that should be delivered as spreadsheets or databases. Table 1 describes our recommendations for delivering facility information and incorporating it into the generic repositories.

Table 1. Generic repository formats and information structure

Types of documents	Required formats and standardisation
Drawings	CAD format and web-enabled format (e.g., DWG or DWF) Sheet organisation standard (e.g., Uniform Drawing System) Label names and equipment tags as provided by owner Layer standard (e.g. AIA CAD Layer Guidelines) Embedded links on all scheduled items (e.g. Mechanical equipment, electrical equipment, control devices) Symbology provided by owner, as suggested by Tri-Services
Product and performance data, such as schedules and test and balance reports	Spreadsheets or databases, with templates and fields provided by owner (Uniform Drawing System provides recommendations)
Equipment operation and maintenance manuals	HTML format and either Web or CD-ROM delivery PDF format if HTML format is unavailable Optical scanning if only paper-based documents are available

Other kinds of information are needed in facility operations but are not currently delivered in any format. Table 2 describes our recommendations for formats and structure for information that is not currently delivered to building owners. These recommendations will require further research and refinement to establish effective standards that can be widely adopted.

Table 2. Recommendations for facility information for which there are no clear standards

Types of documents	Required formats and standardisation
Design intents	XML document type definitions and authoring tools
Work processes	Documentation of processes through IDEF0, pseudo code or Web scripts

5.2.2 Link farm

A "link farm" is a database that stores hyperlinks, providing for improved management of hyperlinks. We have derived our link farm concept from a link management technique described in an XML reference (Light 1997). The link farm has several purposes:

- Defer assignment of physical location of files that are targets in links that are embedded in other documents;
- Facilitate global adjustment of targets;
- Implement bi-directional links and one-to-many links;
- Classify links by information content; and
- Provide for auditing of traversals of links.

Our implementation uses Microsoft Access databases and Active Server Pages (ASP) to insert a level of indirection upon the hyperlinks in our JIT facility operations system. Images of the tables are illustrated in Figure 1. The database contains records for each hyperlink that store the physical location of files and markers for each end of the link and classify the link by information content. Links embedded into drawings and HTML documents point to ASP documents that perform queries on the link farm to resolve links. The queries can initiate relatively simple traversals or complex searches of information. They may return compound documents that combine information from multiple sources.

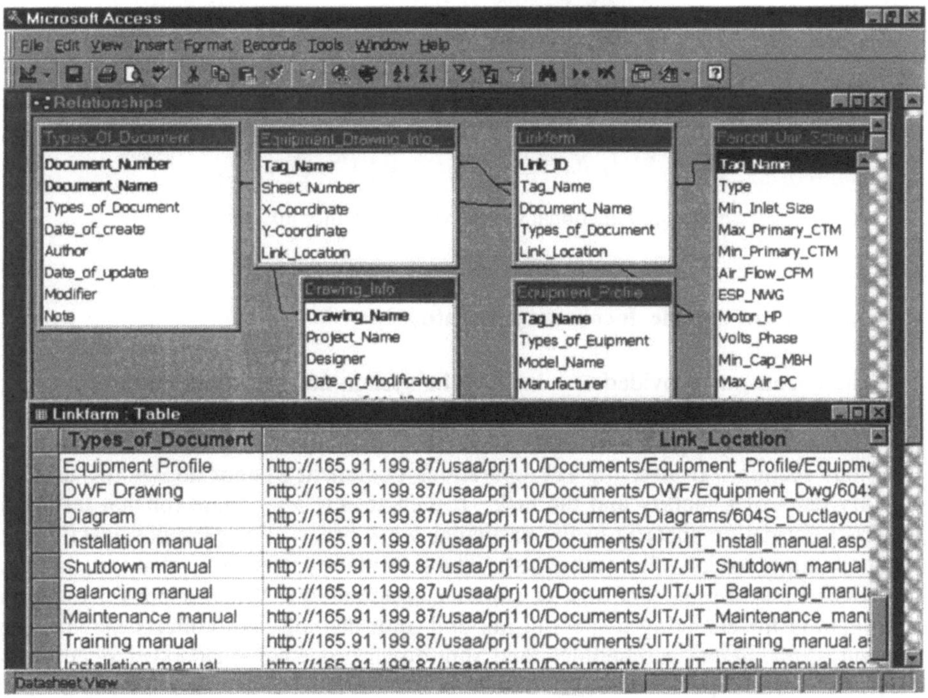

Figure 1. Link farm database tables used in the prototype

Fortuitously, the link farm also provides a way to audit the use of the information system. When invoked, the ASP scripts create new records in an audit database that records the person requesting information, the time, and the source link for the query. From these audit records, the personnel in charge of maintaining the facility information system can detect patterns of use and anticipate important information to include.

The link farm also easily accommodates discipline-oriented, hierarchical models of the facility. Figure 2 illustrates such a view in our prototype system.

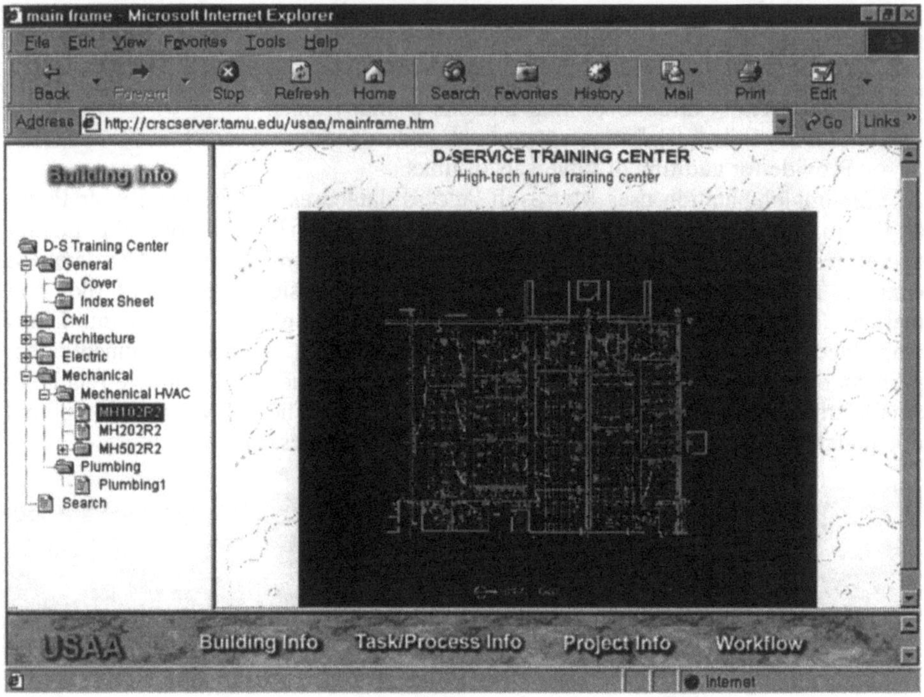

Figure 2. Functional system view of facility information

5.2.3 Task-centric document generator

The documents provided to a user of the system are generated on the fly using ASP scripts. The scripts record knowledge of what information is needed to perform a facility operations task, such as one of the scenarios described in a previous section. In use, the JIT operations document system allows the user to specify parts of a facility and the task. It then selects a template for that task and inserts information specific to the particular work order. For example, installing a fan coil unit is a relatively routine task. The information needed for that task and the process of performing that task is stored in a template. When a person requests information for changing a specific fan coil unit, the software collects the specific drawings, maintenance schedules, and shut-down procedures and other documents necessary for the specific task.

In our prototype, the task-centric templates have been created with Active Server Pages in conjunction with Java Script and VB Script. They generate HTML documents on the fly by querying the link farm for the needed link targets. A script retrieves records from a database, drawing files, diagrams, images, text and many other kinds of formats of data and generates a concise document in a predefined format. Figure 3 illustrates a JIT operations document generated to provide training information for a particular air handler.

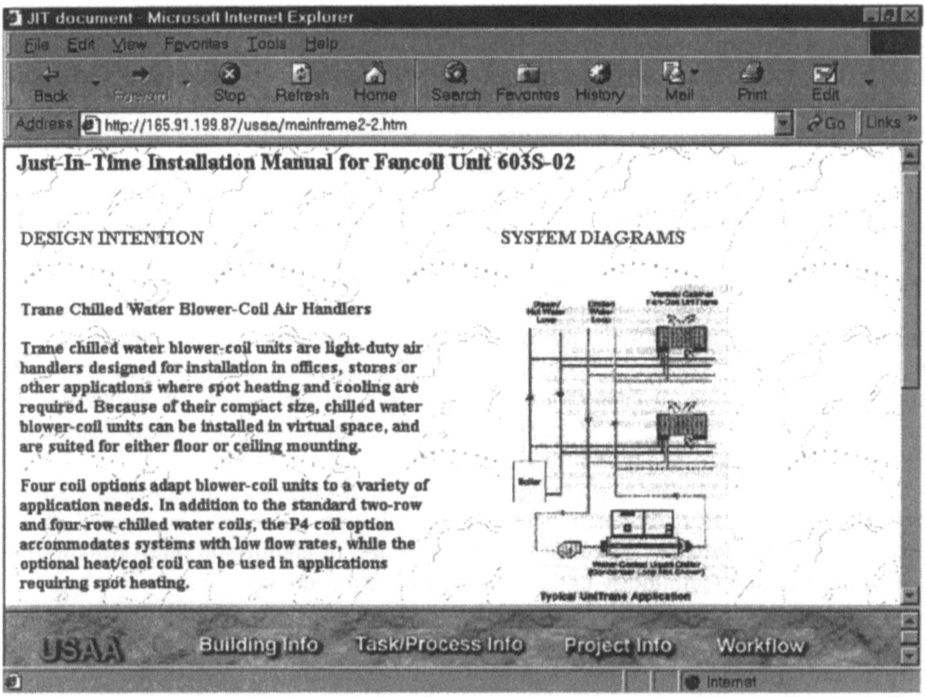

Figure 3. JIT operations document for training information for an air handler

5.2.4 Automated workflow management

The workflow management component of our prototype delivers work reports, notices, and feedback to the responsible people. Reports are usually sent to the facility managers for their management purpose or strategic planning. Notices are sent to the people affected by shutdowns, remodelling or emergency repair. The feedback is notification that information must be up-dated, due to either changes in the facilities that must be recorded in documentation or errors that are detected in the documentation.

The workflow management has been implemented using scripts embedded in the Active Server Pages. The scripts include automatic generation of e-mail messages. They also include execution of redlining software so that a user can annotate drawings with update information. We have used CADViewer by ArNoNa Internet Software Inc. redline DWF files (ArNoNa 1998).

6. CONCLUSIONS

By discussing with operations personnel, exploring related initiatives, observing innovative practices at several facility operations organisations, and constructing software at an experimental level, we have clarified the issues and solutions by which designers and constructors can deliver facility information to support operations. Our research also raises additional questions that can be addressed in further research. However, our efforts have already produced contributions upon which other researchers and facility managers can act.

6.1 Further research

More research could further sharpen our image of future facility operations support systems and address problems that currently lack clear solutions. In particular, further research should address the development of industry-wide standards, exploration of representations of design rationale, and quantification of costs and benefits through trial implementations. Opportunities exist to team with other organisations to leverage research and development investments.

Our continuing studies involve contacting a broad range of companies that either deliver or use facility information that are employing what may be termed "best practices" in the industry. Because we are interested in identifying innovators in the use of information technology in the AEC/FM industry, the people and organisations we plan to contact cannot be determined prior to our survey. We plan to begin by contacting those people and organisations who are involved in thinking about or implementing innovative solutions in facility management and real estate. We then plan to ask those individuals for their recommendations as to whom we might talk with next. If the opportunity and need arises, we may conduct site visits to learn more about an organisation or particular solution or problem.

Our concepts of a just-in-time operations document system could be explored more thoroughly by constructing a more advanced prototype. By focusing upon just a few work processes or tasks in operations, additional knowledge could be gained regarding cost of implementation and benefits.

Knowledge of intranet development tools should be distributed among key staff in facility management organisations. A practical understanding of the technologies underlying those systems will enable better decisions. It can also equip personnel with the ability to maintain or even customise vendor-supplied systems to better conform to internal practices. We are examining the viability of facility management graduate programs and continuing education courses that could focus upon applying information technology.

This study has begun to give us an informal understanding of the costs and benefits of new facility operations information systems. A next step in moving toward practical systems may be to develop a qualitative model of costs and benefits. The qualitative model could then be tested by quantifying it for several small test cases.

Additional research is necessary to establish useful standard structure and content for design rationale. We are exploring the use of XML as a medium for recording design rationale for use in activities downstream of design.

6.2 Contributions

Web technology can be used to produce on demand facility documents that are tailored to particular tasks and needs. A just-in-time facility document system could meet a variety of needs, such as training, trouble-shooting, and work-flow management.

The facility information for such a system may be recorded in digital formats that are not dramatically different from current industry practices. By holding the information in relatively generic repositories, the costs of delivering information from numerous external consultants can be reduced. It is reasonable for a building owner to establish standards for delivery of facility information from consultants. Facility information should be acquired as digital drawing files, spreadsheets, and Web documents. Scripts and Web server programming can then adapt the information into documents that are tailored to particular operations uses.

AKNOWLEDGEMENTS

This research has been conducted as a partnership among CRS Center, USAA and Texas Engineering Experiment Station.

REFERENCES

AIA. 1997. *AIA firm survey report.* Washington, DC: The American Institute of Architects Press.

Allen, T. and M. Scott Morton . 1994. *Information technology and the corporation of the 1990's.* New York: Oxford Press.

J. Al-Qawasmi, M. J. Clayton, L. Tassinary, and R. Johnson. 1999. Observations on Collaborative Design and Multimedia Usage in Virtual Design Studio. In *Learn Virtually ... Anywhere*, Proceedings of the Sixth Annual Distance Education Conference. Ed. J. Woosley and T. Adair. College Station, TX: The Center for Distance Learning Research, Texas A&M University.

ArNoNa. 1998. http://www.cadviewer.com/, ArNoNa Internet Software.

Augenbroe, G. 1995. Design systems in a computer integrated manufacturing (CIM) context. In *Integrated construction information*, eds. P. Brandon and M. Betts, 194 – 212. London: Chapman & Hall.

Clayton, M. J., R. E. Johnson, Y. S. Song, and J. A. Al-Qawasmi. 1998. *A study of information content of as-built drawings for USAA.* College Station, TX: CRS Center, Texas A&M University.

Cooper, G. S., and P. Brandon. 1995. Integrating points of view through information modeling. In *Integrated construction information*, eds. P. Brandon and M. Betts, 251 - 261. London: Chapman & Hall.

CSI. 1997. *The Uniform Drawing System.* Alexandria, VA: The Construction Specifications Institute.

Duncan, W.L. 1988. *Just-In-Time in American Manufacturing*. Dearborn, MI: The Society of Manufacturing Engineers Publications Development Department.

Eastman, C. M., S. C. Chase and H. Assal. 1993. System architecture for computer integration of design and construction knowledge, *Automation in Construction,* 2(2): 95-107. Amsterdam: Elsevier Science Publishers B. V.

Gutman, R. 1998. *Architectural practice: a critical view.* Princeton, NJ: Princeton Architectural Press.

Johnson , R. E. and M. J. Clayton. 1997. *Impact of information technology on facility management practice.* College Station, TX: CRS Center, Texas A&M University.

Karmarkar, U. 1989. Getting control of just-in-time, *Harvard Business Review, Sep.-Oct.*

Light, R. 1997. *Presenting XML.* Indianapolis, IN: Sams Net Publishing.

Liu, L. Y., A. L. Stumpf, S. S. Kim, and F. M. Zbinden. 1994. Capturing as-built project information for facility management, *Computing in Civil Engineering*, vol. 1, 614 - 621. New York: American Society of Civil Engineers.

Mays, P. 1998. Longer life for drawings, *Architecture*, 87(3): 152- 154. New York: BPI Communications, Inc. 1998.

Padjen, E. 1997. Spinning your Web page, *Architecture,* 86(6): 168-172. New York: BPI Communications.

Regli, W. C. 1997. Internet-enabled computer-aided design, *IEEE Internet Computing* 1:39-51. Los Alamitos, CA: IEEE Computing Society.

Schley, M., R. Buday, K. Sanders and D. Smith, eds. 1997. *AIA CAD layer guidelines.* Washington, DC: The American Institute of Architects Press.

Watson, A. 1995. Product models and beyond, *Integrated construction information*, eds. P. Brandon and M. Betts. London: Chapman & Hall. 159 –172.

Zamanian, M. K. 1995. Beyond Object-Oriented Programming: Applications of Component Software Technology in Computer-Aided Engineering Systems. In *Computing in Civil and Building Engineering, Vol. 2* 1259 - 1266. New York: American Society of Civil Engineers.

The Piranesi system for interactive rendering

Paul Richens

The Martin Centre for Architectural and Urban Studies, University of Cambridge Department of Architecture, UK

Key words: computer graphics, non-photorealism, painting, interaction, rendering

Abstract: Photorealistic rendering requires special skills to achieve, but the result is not necessarily ideal for architectural communication. The Piranesi software provides an alternative way of producing a much wider range of images from the same geometrical model. Piranesi appears to be a 2d paint program for editing raster images, but its input includes a z-buffer. This allows painting and and pasting to reflect the perspective of the image, and many other depth-related effects. The result is a new, and enjoyable, way of producing architectural images.

1. INTRODUCTION

1.1 Photorealism

Computer rendering has come a long way in the last twenty years. But is it going in the right direction? Is the glossy photo-realistic image the only goal worth pursuing? And does the process of making it contribute enough to the design, or the ongoing dialogue with the client?

There certainly are alternative modes of image-making. Frank Lloyd Wright, according to legend, could conceive a whole building in his head, and set it down rapidly, in plan and section. He would leave these drawings overnight with his assistant, who would set up a perspective. In the morning, FLW would spend an hour or two completing the rendering, ready for a lunch-time meeting with his clients. Today, many architects use their computers in the same way as FLW used his night staff, to set-up an outline perspective, over which a rendering is produced by hand. Students, in particular, can often be observed attempting to complete a basic rendering by using a paint program such as Photoshop to apply textures and entourage in a kind of electronic collage.

Photorealism has been espoused by computer graphics researchers because of its easy determinacy. If you specify all the light sources, the optical characteristics of all the materials, and simulate the physics of light with the utmost fidelity, then the resulting image is fully determined; there is nothing left to the imagination, and it is the ideal subject for a computer program of the old-fashioned kind. But this hardly meets the needs of a visually oriented profession, which is accustomed to using images day in and day out, and sweating the maximum benefit from each one. Imagination has to be let back in.

The photorealistic image is problematical in practice. It tends to be read as a complete, definitive statement. It is non-negotiable. Sometimes this is appropriate, but there are many times when something fuzzier, more tentative is called for, either because ideas are only partially worked out, or to encourage discussion. In our work images can contain any amount of "noise" which people will tend to interpret according to their own predilections. As Gombrich(1960) tells us, such images stimulate the imagination, rather than suppress it; and as an added benefit, require less geometrical detail in the original CAD model, and so are both economical and especially appropriate at the early stages.

Photorealistic practice ignores the *selectivity* which is fundamental to all kinds of freehand drawing. Everything is rendered in the same way, with the same amount of detail. Piranesi offers infinite ways of foregrounding the subject, by modifying its weight, colour or textural detail, and reducing background by fog, blur or vignetting; and this hardly requires conscious effort. The way we judge these renderings has shifted – we are no longer talking about technical virtues such as the smoothness of shading, the sharpness of textures, or the size of highlights, but are concerned with their pictorial qualities, of how they focus an issue and stimulate the imagination or support an argument.

1.2 Rendering in practice

Computer rendering of architectural proposals is becoming increasingly common, and increasingly specialised, usually being carried out by an expert group within the office, or very often by an external firm. The software employed for the most advanced photo-realistic images is often complex, slow, expensive and demanding.

If we look at the workflow seen in a commercial rendering house, it moves through recognisable stages. Site photography is commissioned and scanned to provide a background. The proposal is modelled in 3d, specifically for visualisation, and only to the extent needed to achieve the required views. It is then rendered, often in several layers such as background, elevation, and as much interior as is visible through windows. Then these layers are combined with the site photography in

Photoshop, entourage added, and the whole thing finished by manual editing. Each stage may take several days of skilled operator time to perform.

We are particularly interested in the last two stages, that of rendering and photo-finishing. Rendering is far from interactive - it includes placing lights and textures, adjusting many parameters, and carrying out many trials. By contrast the photo-finishing stage is done by direct manipulation; this is a much better environment for the artist, and it is here that the visual balance of the final image is achieved.

2. THE PIRANESI PROJECT

2.1 Interactive rendering

The Piranesi project at the Martin Centre was set up to address the two issues of allowing more appropriate style of rendering than the photorealistic, and finding a mode of interaction that is a better fit to the cognitive style that visually creative people employ. The result is a new kind of program; it produces images of three dimensional objects comparable with traditional renderers, but with an interactive interface akin to a painting program such as Photoshop or Fractal Design Painter. It could be termed, with equal justification, a 3D paint system, or an interactive renderer.

Piranesi is a software system that succeeds in shifting the boundary between the rendering and the finishing stage, making much more of the process direct and interactive. To use it a basic rendering is made, without textures, and exported to Piranesi with its z-buffer, which gives the distance to each pixel, and a material buffer which encodes the layer or material in the model from which each is derived.

Piranesi itself is structured like a paint system, with the usual sort of brushing and bucket-fill operations. But the presence of the z-buffer gives it many new capabilities. By using the z coordinate to modulate painting effects, it becomes sensitive to depth. By differentiating the z-buffer, it can reconstruct the surface normal, and so become sensitive to orientation. By differentiating twice it can recover curvature, and detect physical edges. Finally, by combining pixel position with depth, and passing it through an inverse perspective transform, Piranesi can reconstruct the original 3d coordinate at any point. Some of the more interesting effects enabled by the z-buffer are:

1. Restriction of a stroke to a particular plane, or to planes with a particular orientation. Combined with restriction based on material, this removes the need for selection masks. As the typical Photoshop user spends more time making masks than doing anything else, this is by itself a substantial advance.

2. By making the paint effect proportional to depth, we get the effect of fog or aerial perspective. By using various functions of depth we can achieve haloes around a point, or the effect of mist rising from the ground.

3. Textures can be brushed on in a conventional 2d sense. But by using the z-buffer we can map textures by projection, either frontally or by normal projection onto the surface being painted. This changes texture generation from a hard to control batch-mode process in the renderer, to an easy piece of direct manipulation. Textures can be applied experimentally and built up by layers.

4. By relating paint effect to the surface normal, we can achieve lighting effects outside the renderer.

5. Scanned images can be pasted in at a definite depth in the image. So they scale according to the perspective, and can obscure and be obscured properly. They can also be fixed onto surfaces, giving the effect of a poster or a creeper growing up the wall.

6. Edge detection in the z-buffer leads to a peculiarly sensitive kind of line-drawing. Silhouette edges are naturally emphasised over internal edges, while shadow edges do not register at all.

Piranesi is constructed as a re-rendering system rather than (as Photoshop) an editor of a single image. The original image is retained through the process, and is sampled in various ways as the new image is painted. So a texture may provide its own colours, or a constant colour may be used, or the original image may be sampled to provide the colour. Sometimes the output image is initalised to the input image, but on other occasions we may prefer to start with a monochrome screen.

Piranesi can achieve most of the effects of a conventional photorealistic renderer, but is even more succesful with non-photorealistic rendering. It is not governed by the laws of optical physics, but by the eye and judgement of its user. This vastly increases its creative potential.

Non-photorealistic rendering will typically proceed in two passes. In the first, colour and shading are added to the original image, plus entourage and background, defining the composition and overall tonal balance. This becomes input to a second pass which starts with a blank sheet. Edges are pulled through, like a pencil outline. Then textures which sample the input colours are applied. These may be 2d or 3d procedural textures, or vey often hand-drawn and scanned samples, which are mapped into perspective by Piranesi. A technique akin to half-toning can produce effects like pen and ink drawing, or wood-engraving. The amount of detail, and hence the density of such a texture, is controlled by the tonality of the input image.

2.2 Related work

Some of the fundamental ideas of Interactive Rendering (in particular the separate treatment of geometry and appearance, and the enriched pixel) were suggested by Perlin's Pixel Stream Editor of fourteen years ago (Perlin, 1985). However, it was far from interactive, and required that each pixel should be processed independently, which prohibits many desirable operations. The ideas of fractal noise, turbulence and solid texture included in the same paper have been immensely influential.

Perlin was rather vague about what data should be associated with each pixel; Saito and Takahashi (1990) by contrast define a comprehensive G-buffer that contains identifier, parametric coordinates, world coordinates, depth and direction cosines. This is used to support perspective space hatching and edge finding, with the idea of making technical illustrations "comprehensible". We differ in having some interest in ambiguity as well as clarity, and preferring an interactive process to one that batch treats a complete image. We have found it possible to reduce the G-buffer to a pair of numbers, without serious loss of effect.

The idea of image space rendering using synthetic paint marks was important in the genesis of our project. Similar ideas have been published by Haeberli (1990), and have found their way into commercial products such as Fractal Design Painter. It was about this time that the term non-photorealism came into use. Three-dimensional painting by means of reverse texture mapping was first described by Haeberli & Hanrahan (1990). It is beginning to appear in commercial products (such as Painter 3d), but requires high performance hardware to be interactive. It works purely in perspective space.

A group at the University of Magdeburg has been looking at computer-generated 'sketches' and, more recently, at rendering techniques for medical illustration, another discipline where photorealism is less than ideal(Strothotte & Strothotte 1997). The work on pen and ink drawing simulated by stroke textures by Winkenbach & Salesin (1994, 1996) and Salisbury, Wong, Hughes & Salesin (1997) at the University of Washington is closer in spirit to ours, and has been applied specifically to architectural rendering. Originally a batch mode renderer, recent papers indicate a growing interest in making the process interactive. Some of the latter work has been commercialised by Inklination Inc, while the former has resemblances to the LiveArt 98 software produced by Viewpoint Data Labs International Inc.

The modern world of digital image-making has been comprehensively surveyed by Mitchell (1993). Early work on this project has been described by Richens (1994, 1995). A comprehensive review of non-photorealism can be found in Schofield and Lansdown (1995), and additional discussion in Strothotte & Strothotte(1997) . There will be a workshop on the subject at SIGGRAPH 1999.

3. EXAMPLE

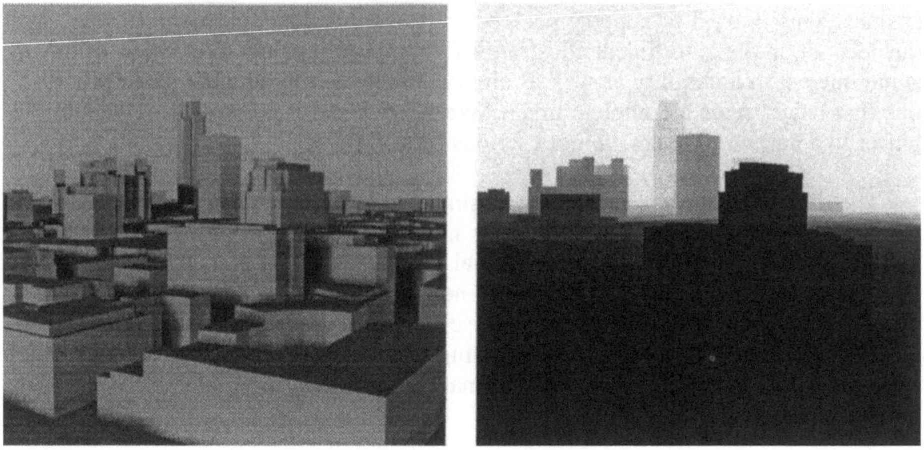

Figure 1,2

The input to Piranesi is a kind of extended TIFF file, containing an initial rendering (Fig 1) and the z-buffer (Fig 2). The latter encodes, for each pixel, how far away the object rendered is. So near objects are dark, distant ones light. It also contains a material index for each pixel.

The rendering we are using here was made by 3D Studio Max, using a freeware plug-in renderer to export the interface file. The scene was generated by the London architects Miller+Hare, from their urban model of the city. Most of the buildings are simply extruded from their footprints on a map, though a few of the more prominent have been modelled in more detail. The model is simply lit and untextured, which is of no consequence, as both will be reconstructed in Piranesi.

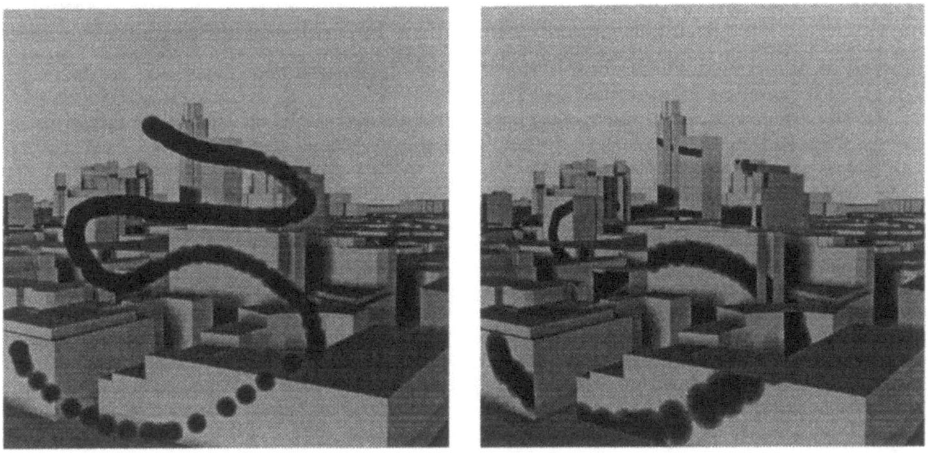

Figure 3,4

The z-buffer is what makes Piranesi unique among painting systems. The first consequence, and perhaps most startling to a Photoshop user, is that is does not need selection tools and masks to limit painting to particular areas. Instead it uses *Locks* to constrain painting either by material, or by geometry. The most commonly used is *Plane Lock*, which constrains paint to pixels which lie in the same plane as that at the start of the stroke. (The relevant plane equation is determined by calculating the gradient of the Z-buffer). Here (Figure 3) we show an unconstrained stroke, compared (Figure 4) with one made with *Orientation Lock*. This selects the planes facing in the same direction as the one where the stroke started. In Fig 3 we use a standard 2d brush, but in Fig 4 the stroke is repeated in 3d. So the brush outline becomes elliptical, and its size diminishes as it gets further away.

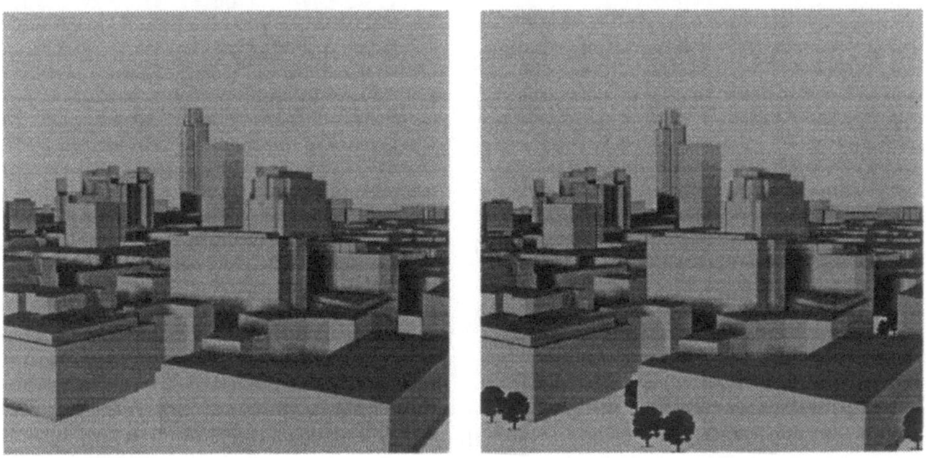

Figure 5,6

The first step in a Piranesi rendering is usually to correct small modelling errors using a tool which operates on the z-buffer by extending existing planes. Here

(Figure 2) it is used to modify the foreground building so that the street behind is more visible. We are not concerned with realism, but with making a picture.We need to be able to see into the street, so we demolish some building.

In Figure 6, we paste in a few cut-out rees, to establish the scale. Pasting in Piranesi is in perspective, true to scale (by default), and deals with obscuration. So trees know what they obscure, and what obscures them, and the distant ones are smaller than those nearby, according to the perspective of the scene.

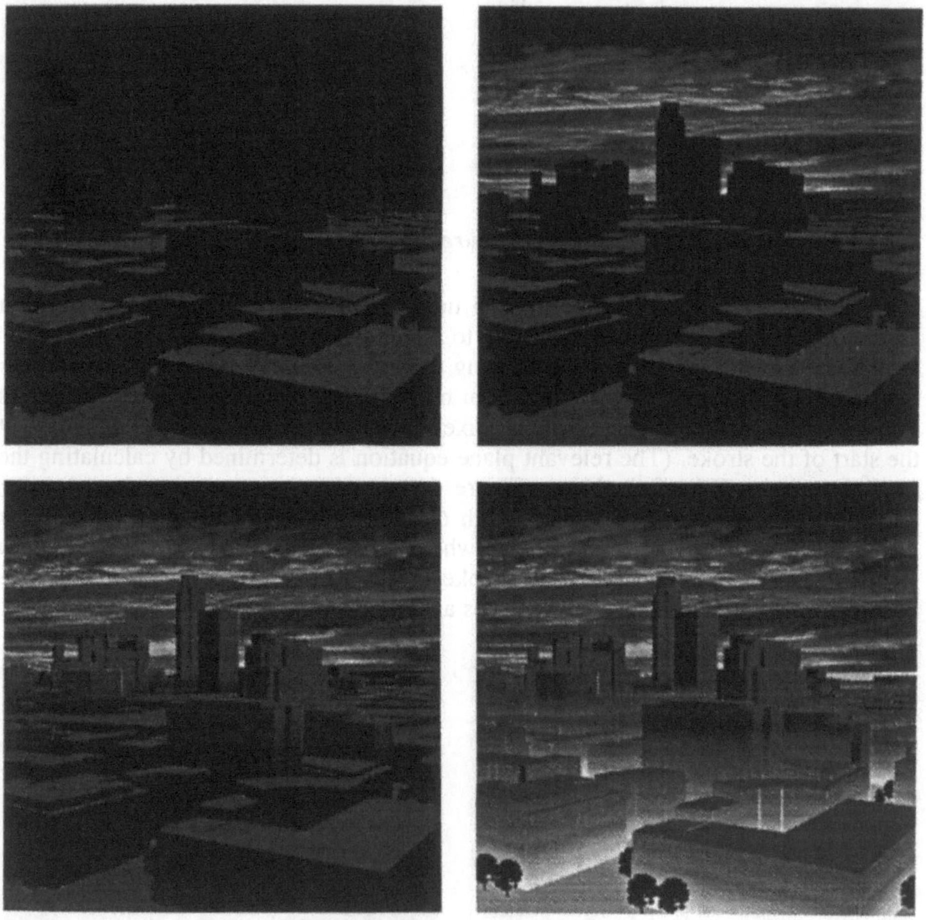

Figures 7,8,9,10

We now start to paint the lighting, with the intention of capturing the general effect of a city just after sunset, when the electric lights come on. Starting with a black screen (Fig 7), we allow a little light onto the rooftops, by painting with *Orientation Lock.* Then, selecting a sky shot as a texture, we paint in the background (Fig 8). The high buildings will catch the last of the sunset glow (Figure 9), so we paint them with two grading effects. One keeps the light stronger as it is higher, and the other takes the plane orientation into account, so giving the effect of directional lighting from the back right. Finally we add an overall approximation to the artificial light in the scene, which is generated at street level, and fades out with height. This light, in reality the result of thousands of indivdual sources, is approximated by another vertically graded fill.

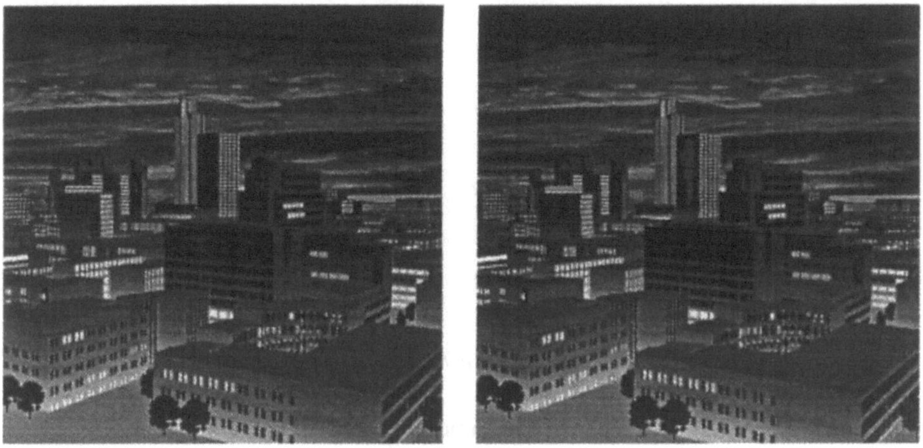

Figure 11,12

In Fig 11 we start to paint windows into the scene, by using very simple black and white texture tiles representing various fenestration patterns. The actual colour used is decided when the paint is applied; here it is sometimes light, sometimes dark. As the textures are brushed on, this sort of detail variation is perfectly natural. Piranesi automatically puts the texture into corrrect perspective according to the depth and recession of the plane on which it is being painted.

In Piranesi it is very easy to build up layered effects. For example the figure of a model in Fig 12 was pasted onto the face of the tall building, but transparently. This operation changed the material buffer only. Next she was textured with a pattern of lights, using the material lock, so picking out the silhouette.

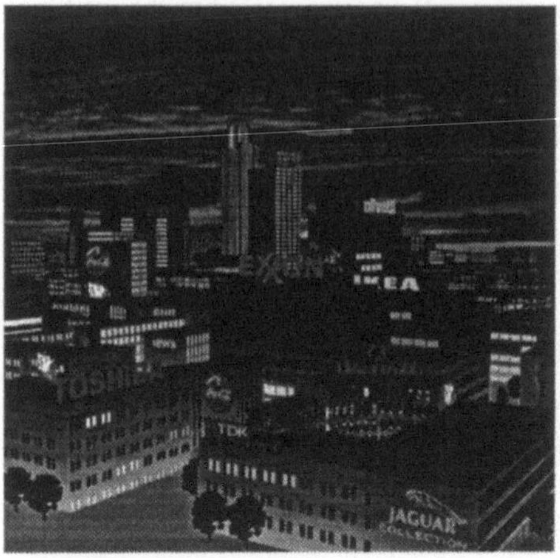

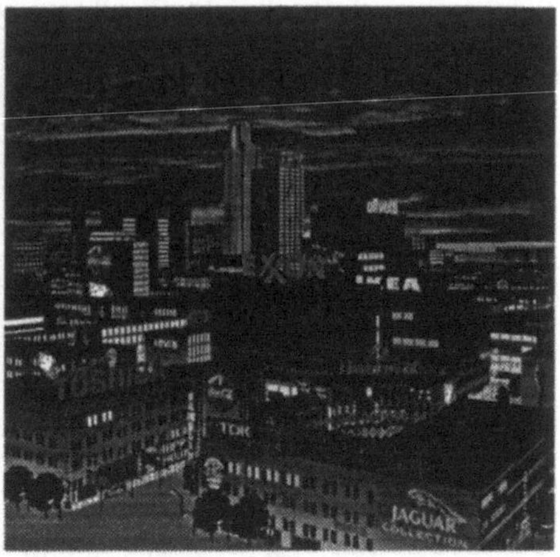

Figure 13,14

Neon signs are added in much the same way as trees were at the beginning. In figure 13 rooftop signs are generally aligned with a wall, and then slid into place on the roof. These cut-out elements may be fully coloured, or simply greyscale, in which case the colour is provided at the time of insertion. When one is inserted, the z-buffer values are inspected. Only when the insert's z value is less than the existing, is it painted in. Where it is painted, the z and material buffers are also updated. This ensures that future cut-outs will correctly obscure or be obscured by the first one, and that it may be repainted or textured subsequently to its insertion.

The use of cut-outs saves a huge amount of modelling effort in dealing with details such as trees,people, railings and fences, or, as in this case, neon roof signs. In Figure 14 street lamps have been added in the same way, and a lot more signage at street level. This is usually applied flat onto the wall, but for little more effort can be rotated to project out from it.

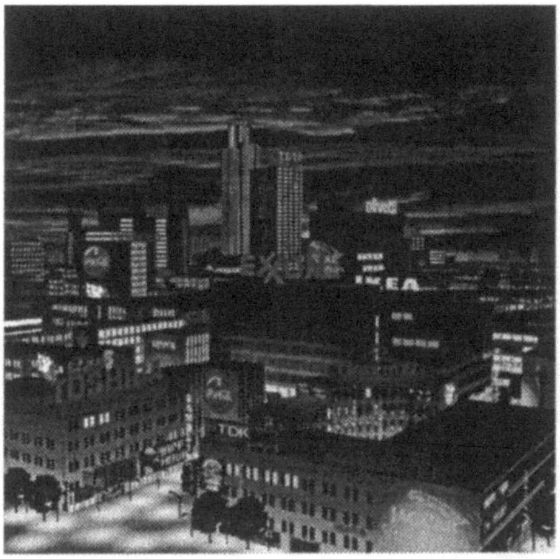
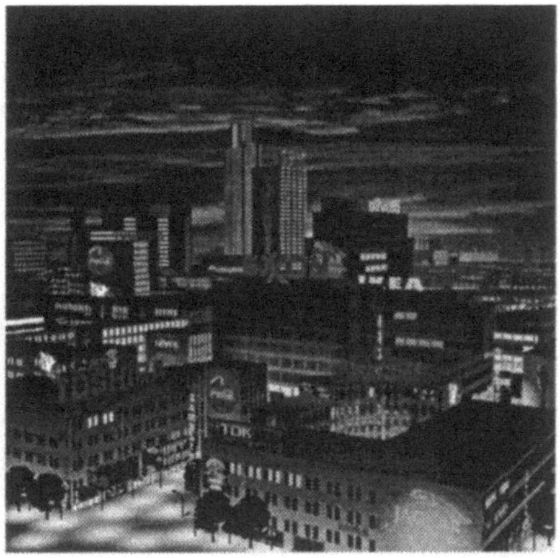

Figure 15,16

A first approximation to lighting was painted in earlier, before it was motivated by any apparent sources. Now it is refined by painting a glow around individual sources. The pools of light beneath street lamps are done using a soft-edge circular brush. Because Piranesi works in perspective, the brush marks are rendered by perspective into appropriate ellipses. Glows can also be arranged by defining a source position, and then using a painting effect that is graduated by distance from that source. Piranesi does actually include algorithms equivalent to those for conventional computer graphics point (omni) and directional light sources, but everything here is mediated by the brush.

Another way of providing glows is to use a blur filter. In Fig 16 the horizon glow as been arrived at by using some distance-dependent mist, plus some distance-dependent blurring.

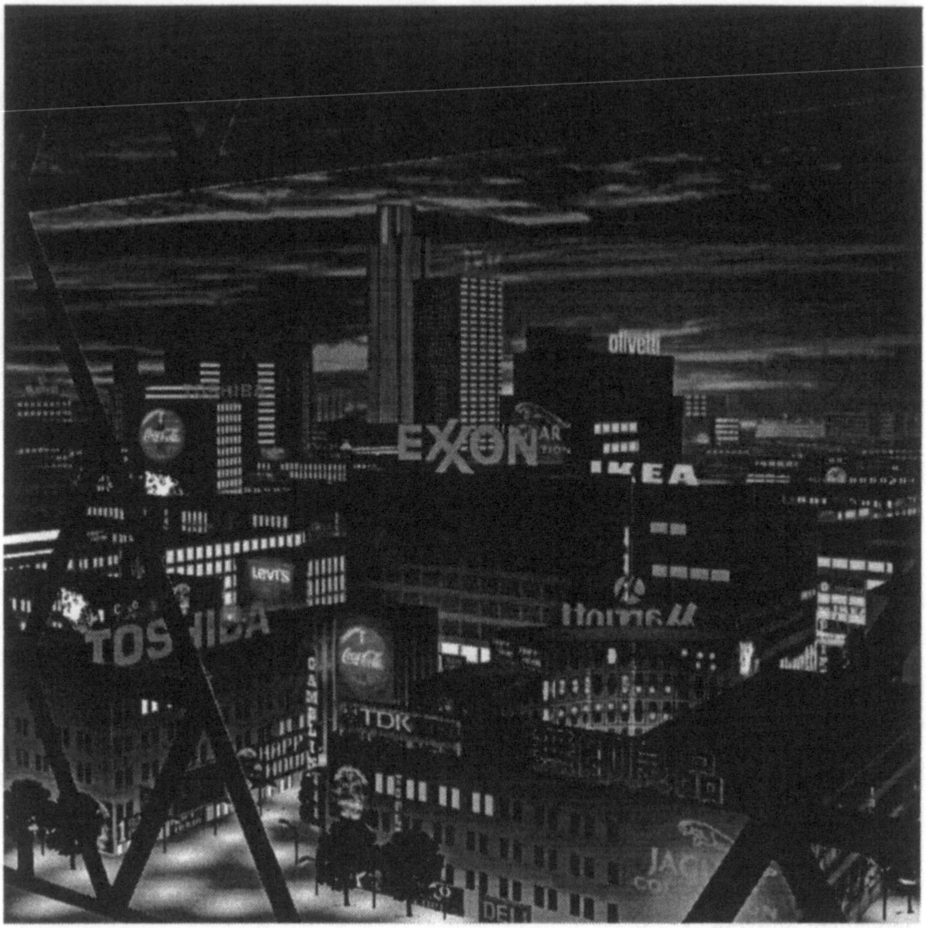

Figure17

This concludes the development of the lighting, but pictorially there is something wrong – there is no foreground. This is quite readily rectified by constructing a girder pattern from a greyscale texture. Again this is located in perspective, and so is really there in the z-buffer. The edge tool can find it, and as a finishing touch the edges are allowed to pick up a little reflected light, which instantly gives the flat texture a three-dimensional feel (Fig 17).

3.1 Re-rendering

So far, everything has been done using the original Studio Max render as input. But it is possible to go further, using the new rendering of Figure 17 as input to further stages. As the work so far has been done in colour, and this paper will probably be published in black and white, some of the techniques in Piranesi for producing non-photorealistic black and white graphics will be explored.

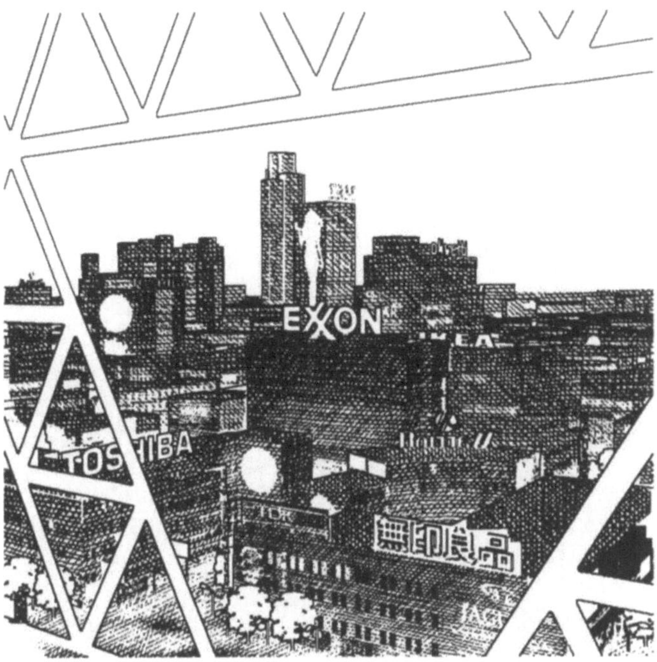

Figure 18

The first example (Fig 18) is a binary bit map, highly suitable for output on a laser printer. It is generated using a hand-draw pen and ink cross hatch texture, which is capable of responding to the darkness in the input image by varying the thickness and density of the hatch lines it produces. This technique is called thresholding, and can produce a number of duotone image effect, including etching and engraving. Here the image retains the characteristics of pen and ink sketch. The texture has been applied in a 2d sense in the background of the image, but in the foreground it has been texture-mapped onto individual planes of the buildings, so that the hatch lines respond to the perspective. This ability to work in both perspective space and on the surface of the paper, in any combination, is a unique feature of Piranesi.

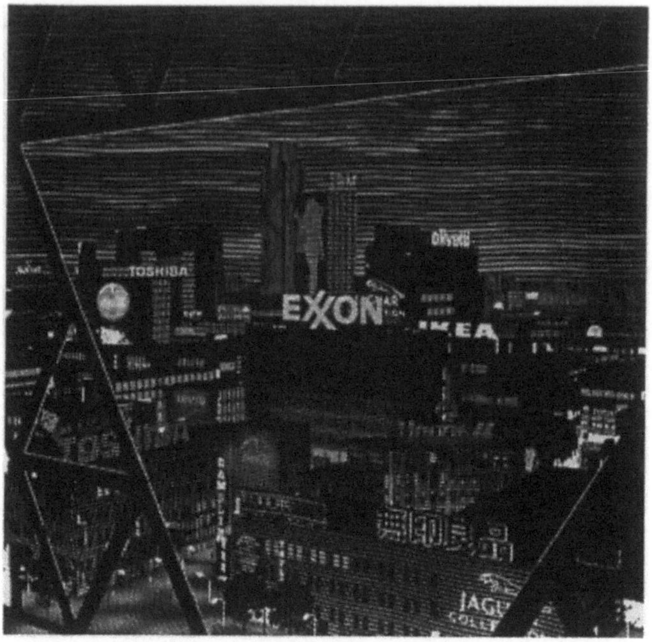

Figure 19

Figure 19 is another example of using a texture, in this case a computer generated screen texture of slightly wavy horizontal lines. This time it is applied with a softer sigmoid thresholding function rather than a sharp step. It gives a result quite similar in its chiaroscuro to a nineteenth century wood engraving. The sky was rendered first, using the texture in 2d, at a fairly large scale. Then the foreground was painted plane by plane, using the texture in perspective. The background cityscape was finished using the texture rotated ninety degrees, and at a smaller scale than was used for the sky, but similarly in 2d.

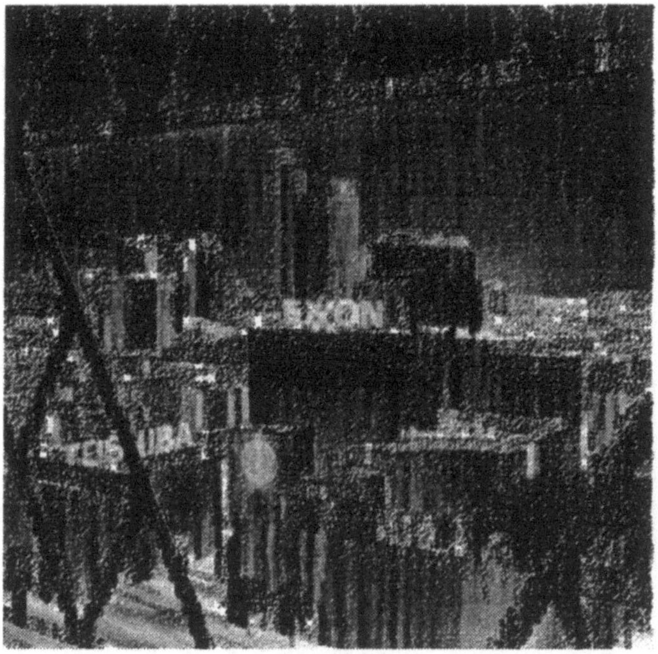

Figure 20

Figure 20 is generated by quite a different technique. First a grain texture was selected, with a surface similar to water-colour paper. This texture is used in 2d, over the whole surface of the picture, and basically affects how readily the output image becomes coloured. The rendering was done using automatically generated marks, vertical lines with a slight random wobble, whose colour is generated by sampling the input image at the middle of each stroke. The marks themselves are clipped to plane boundaries. The resulting image has low definition, and is full of noise. There is a definite resemblance to charcoal or graphite stick drawing. As one would do when using these media, the highlights (such as the neon signs) are picked out afterwards using an eraser, constrained using the *material lock*.

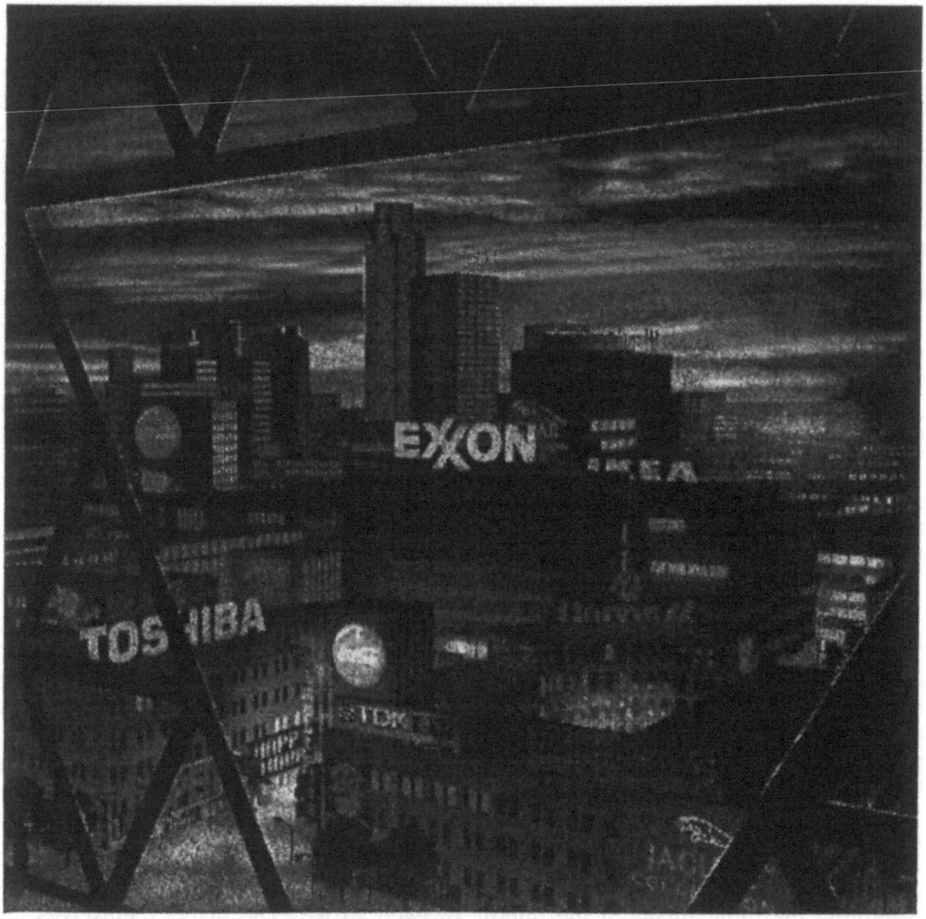

Figure 21

The final impression of this scene (Fig 21) is rendered in a manner analogous to mezzotint, and has a similar effect. In mezzotint, the copperplate is roughened all over until it prints a perfect black (a very lengthy process), and then the image is formed by burnishing away the roughness to draw out the light tones. In Piranesi you start by filling the screen with black, and then select an appropriate grain – in this case a fractal noise function was used. A first pass is done automatically, at very low intensity, so that you can see the ghost of the image. Then, keeping the intensity quite low, the lights are brushed over repeatedly, gradually building them up to strength. This is best done using a tablet and pressure-sensitive stylus. Because of the low-intensity setting, and the use of the grain noise, it takes many passes to bring a light up to the strength it has in the original image. This gives great control over the development of the image. The lock functions can be used, but are not so necessary in this sort of slow image making. Here we used material lock at first, to develop the sky independently of the buildings. Towards the end the edge tool was used a little, to pick out silhouettes in white.

4. CONCLUSION

Using Piranesi has a profound effect on how you go about making images. The difficult stage of setting material parameters and texture positioning in a batch-mode renderer is eliminated. Instead the image is directly constructed, in a surprisingly short time (typically 30 minutes). For the first time, it is possible for the rendering artist to sit down with his client architect or developer, and try out 3 or 4 presentational styles before his eyes. This brings computer presentation back in line with the traditional architectural approach of talking while drawing, in a very satisfactory way. The software itself is no more complex than Photoshop, and needs similar hardware resources and operator skills.

Piranesi uses the computer as a medium - not a simulation of a traditional medium, but something unique and new, a three-dimensional paint-box. Behaving as a medium is quite different from behaving as a heavy duty analytical engine, which is what the normal photorealistic renderer is. The fundamental difference is that there is an immediacy to the result, you see the effect not a few minutes or a few seconds after you make a move, but continuously, within milliseconds. This feeling, of a complete connection between what you are doing and what you are seeing, is even stronger if you use a tablet with a pressure sensitive stylus. You now focus entirely on the image, rather than on the process of making it, and the work "flows". This is the cognitive shift to "right brain mode" described by Betty Edwards(1982). With traditional paint programs this flow keeps breaking down, because you have to make selection masks for nearly every operation. Piranesi uses *Locks*, and does not even have a selection tool. It also avoids any kind of numeric input (sliders or direct manipulation instead), and is very close to being keyboard-free. The consequence is that Piranesi is a fast, expressive, absorbing, and uniquely enjoyable way of rendering images on a computer.

5. ACKNOWLEDGEMENTS

We acknowledge with gratitude the initial inspiration for this work, which was the talk by van Bakergem and Obata (1992) at the CAAD Futures conference in Zurich, on "wiggly pen" plotting, which demonstrated conclusively that computers, just like draughtsmen, need to "loosen up" before they can produce interesting drawings.

Piranesi was developed in prototype form at the Martin Centre CADLAB by Simon Schofield (an artist), with support from Informatix Inc. A commercial version was released by Informatix Software International in January 1998. We are particularly indebted to Masanori Nagashima, President of Informatix, for his long-term support, and to Brian Woodward of Informatix Software International, who led the implementation team. A number of the techniques described in this paper will not be available until the release of the second version, expected in the fall of 1999. Part of the software is protected by US Patent Application 4,558,302, and equivalents is other countries. For a downloadable demo and details of the commercial version please see the Informatix web site at www.informatix.co.uk.

6. BIBLIOGRAPHY

Bakergem, W.D. van and Obata, G. (1992). *Free-hand plotting - Is it live or is it digital?* In (ed) G.N. Schmitt, CAAD Futures. Vieweg, Wiesbaden.

Edwards B. (1982) *Drawing on the Right Side of the Brain.* Fontana, London, p57.

Gombrich, E.H. (1960). *Art and Illusion: A study in the psychology of pictorial representation.* Phaidon, London.

Haeberli, P. (1990). Painting by numbers: abstract image representation, Computer Graphics, vol. 24, no. 4, ACM SIGGRAPH, pp. 207-214.

Haeberli, P. & Hanrahan, P. (1990). *Direct WYSIWYG painting and texturing of 3-D shapes*, Computer Graphics, vol. 24, no. 4, ACM SIGGRAPH, pp. 215-224.

Mitchell, W.T. (1993). *The Reconfigured Eye.* MIT Press, Cambridge MA.

Perlin, K. (1985). *An image synthesiser*, Computer Graphics, vol. 19, no. 3, ACM SIGGRAPH, pp 287-296.

Richens, P. (1994). *Does knowledge really help?* In (eds) G. Carrara and Y.E. Kalay, Knowledge-Based Computer-Aided Design, pp. 305-325. Elsevier, New York.

Richens, P. (1995). *Interactive computer rendering*, Architectural research quarterly 1:1 Autumn 1995. London, Emap press, pp82-95

Salisbury, M., Wong, M., Hughes, J., & Salesin D. (1997) *Orientable Textures for Image-Based Pen-and-Ink Illustration.* In Computer Graphics Proceedings, Annual Conference Series, 1997. ACM SIGGRAPH, pp. 401-406.

Saito, T. and Takahashi, T. (1990). *Comprehensible rendering of 3-D shapes*, Computer Graphics, vol. 24, no. 4, ACM SIGGRAPH, pp. 197-206.

Schofield, S. and Lansdown, J. (1995). *Expressive rendering: an assessment and review of non-photorealistic techniques*, Computer Graphics and Applications (May), IEEE.

Schumann, J., Strothotte T., Raab A., Laser S. (1996). *Assessing the effect of non-photorealistic rendered images in computer-aided design*, in Proc. CHI '96, Human Factors in Computer Systems, Vancouver BC, pp 35-41.

Strothotte C. & Strothotte T. (1997) *Seeing between the pixels.* Berlin, Springer-Verlag.

Watts, A. (1962) *The Way of Zen.* Penguin Books pp214-5

Winkenbach, G. & Salesin, D.H. (1994). *Computer generated pen-and-ink illustration.* In Computer Graphics, Proceedings, Annual Conference Series, 1994. ACM SIGGRAPH, pp. 91-100.

Winkenbach, G. & Salesin, D.H. (1996). *Rendering free-form surfaces in pen and ink.* In Computer Graphics Proceedings, Annual Conference Series, 1996. ACM SIGGRAPH, pp. 469-476.

MIX
Papier aus verantwortungsvollen Quellen
Paper from responsible sources
FSC® C105338

If you have any concerns about our products,
you can contact us on
ProductSafety@springernature.com

In case Publisher is established outside the EU,
the EU authorized representative is:
Springer Nature Customer Service Center GmbH
Europaplatz 3, 69115 Heidelberg, Germany

Printed by Libri Plureos GmbH
in Hamburg, Germany